MONOGRAPHS ON THE PHYSICS AND CHEMISTRY OF MATERIALS

General Editors

C. E. H. BAWN, H. FRÖHLICH,
P. B. HIRSCH, N. F. MOTT

EXPERIMENTAL HIGH-RESOLUTION ELECTRON MICROSCOPY

[Second Edition]

JOHN C. H. SPENCE

New York Oxford
OXFORD UNIVERSITY PRESS
1988

Oxford University Press

Oxford   New York   Toronto
Delhi   Bombay   Calcutta   Madras   Karachi
Petaling Jaya   Singapore   Hong Kong   Tokyo
Nairobi   Dar es Salaam   Cape Town
Melbourne   Auckland
and associated companies in
Beirut   Berlin   Ibadan   Nicosia

Published by Oxford University Press, Inc.,
200 Madison Avenue, New York, New York 10016

Oxford is a registered trademark of Oxford University Press

Library of Congress Cataloging-in-Publication Data
Spence, John C. H.
Experimental high-resolution electron microscopy/John C. H. Spence.—2nd ed.
p. cm—(Monographs on the physics and chemistry of materials)
Includes index.
ISBN 0-19-505405-9
1. Electron microscope, Transmission.   2. Electron microscopy.
I. Title.   II. Series.

QH212.T7S68   1988   535′.3325—dc19   87-31300

9 7 5 3 1 2 4 6 8

Typeset in Northern Ireland by The Universities Press (Belfast) Ltd;
and Printed in the United States of America
on acid-free paper.

PREFACE

In the ten years which have passed since the first draft of the first edition of this book was completed, the point-resolution of the highest-performance high-resolution electron microscopes has improved from about 0.38 nm to 0.16 nm, an improvement of about 2 Ångstroms in a decade. This spectacular progress has transformed the field of high-resolution electron microscopy and has enormously increased the range of materials amenable to study by this technique. In the first edition I wrote that I was not aware of the existence of any true 'structure images' (see Appendix 4 for a definition) in the literature. Contemporary instruments now allow genuine structure imaging in a wide variety of materials, and the futuristic design parameters which I analysed in the previous editon (Appendix 3) have now been comfortably surpassed. I have tried to summarize some of these applications of high-resolution electron microscopy in a new chapter (Chapter 12). The sheer volume of literature which now exists on the uses of high-resolution electron microscopy has meant, however, that this new chapter can at best serve a selective bibliographic function.

In addition to improvements in resolution, the last decade has seen substantial progress in the interfacing of computers to electron microscopes, both for quantitative data acquisition and for the automated control of electron-optical parameters (focusing, alignment, astigmatism correction, etc.). The increasing use of the convergent-beam mode in conjunction with high-resolution imaging is a particularly promising development. In addition, we have seen the widespread adoption of video-based image-recording systems (allowing both the artificial enhancement of image contrast and the reduction of radiation damage) and a general improvement in microscope vacuum systems (ultra-high-vacuum transmission electron microscopes are now commercially available). Parallel detection systems capable of detecting the arrival of every electron are now a reality, and it is now possible to undertake reproducible experiments in surface physics in the electron microscope.

These impressive developments have also served to highlight the fundamental limitations of the high-resolution technique, the limits of which we may now be approaching rapidly. In particular, elastic relaxation effects in the very thin samples used (which render them unrepresentative of bulk material), the absence of chemical and three-dimensional structural information, difficulties in accurately aligning samples for work at the highest resolution, and radiation-damage problems at the higher voltages now in use, all appear to pose perhaps intractable problems.

I have therefore added a second new chapter (Chapter 11) on some of the spectroscopic and other new techniques (such as microdiffraction, energy-loss spectroscopy, and X-ray microanalysis) which address these difficulties, and which are compatible with high-resolution imaging. Each of these new detectors fitted to our machines gives rise to a new sub-discipline, and the excited states responsible for these spectroscopic signals may be difficult to relate to the ground state observed in high-resolution images. I have also greatly expanded Chapter 5, with new sections on Bloch-wave methods, symmetry reduction of the dispersion matrix, sign conventions, absorption effects, dynamically forbidden reflections, and computational methods.

In preparing this second edition, I have been greatly assisted by many correspondents and colleagues. In particular, I wish to thank Drs D van Dyck, V. N. Rozhanskii, A. Crewe, J. L. Hutchison and K. J. Hanssen for pointing out errors in the first edition and suggesting improvements. J. M. Zuo has also kindly assisted with some of the new work, and both the Russian and Chinese translators of the book have made many helpful suggestions. Diane Stiles has produced a faultless manuscript with remarkable efficiency and speed from my disorganized notes.

Tempe, Ariz.   J. C. H. S.
November 1987

ACKNOWLEDGEMENTS

The bulk of the first edition of this book was written during the year following my term as a post-doctoral fellow at the Department of Metallurgy and Materials Science at the University of Oxford. I therefore owe a great deal to all members of that department for their support; in particular Professor P. B. Hirsch, Dr M. J. Whelan, and Dr C. Humphreys for their encouragement and optimistic conviction that the methods of high-resolution electron microscopy can be useful in materials science. I am also grateful to Dr J. Hutchinson for his advice and encouragement.

I also wish to thank members of the Melbourne University Physics Department (Dr A. E. Spargo and Dr L. Bursill in particular) for sharing their experience of practical and theoretical high-resolution work with me during 1976. My debt to the Australian groups generally will be clear from the text.

The quality of technical support is particularly important for high-resolution work. I have been fortunate both at Oxford (with Graham Dixon-Brown) and at Arizona State University (with John Wheatley) to work skilled and enthusiastic laboratory managers who have each taught me a great deal.

I have also learnt a great deal since coming to Arizona State University from Dr Sumio Iijima, to whom I express my thanks.

I owe a particular debt of graditude to Dr David Cockayne for his interest in the book and painstaking review of some of its chapters.

My debt to Professor John Cowley will be obvious to all those familiar with his work. His major contributions to high-resolution imaging have laid the important theoretical foundations in this subject and I have used many of his results. His secretary, Yvette Auger, has dealt with a disorderly manuscript in her characteristically efficient, accurate, and pleasant manner. Dr Peter Buseck and Dr Leroy Eyring have freely discussed their experiences with high-resolution work. Harry Kolar and Dr Michael O'Keefe have each also been kind enough to assist in the book's production.

Finally, I am grateful to Drs A. Glauert and D. Misell for their detailed comments on an early draft of the manuscript.

CONTENTS

LIST OF SYMBOLS

Δ	Chromatic damping parameter (see eqn (4.9))
Δf	Defocus. Negative for weakened lens
	Special cases—Scherzer focus (eqns (6.16) and (4.12))
	Stationary phase focus (eqn (5.76)), Gaussian focus
θ_c	Beam divergence
C_s	Spherical aberration constant. Positive
C_e	Chromatic aberration constant. Positive
$\chi(u)$	Phase shift due to lens aberrations
Φ	Complex wave amplitude
ϕ	Magnitude of Φ
v_g	Fourier coefficients of crystal potential (in volts)
U_g	Similar (in Å^{-2}). $U_g = \sigma V_g/\pi\lambda = 1/\lambda\xi_g = 2m|e|V_g/h^2$
S_g	Excitation error. $S_g = -\lambda g^2/2$ in axial orientation
ξ_g	Extinction distance
F_g	Structure factor
$f(\theta)$	Atomic scattering factor
θ	Scattering angle. $\theta = 2\theta_B$
θ_B	Bragg angle
λ	Relativistic electron wavelength (see eqn (2.5))
m	Relativistically corrected electron mass
m_0	Rest mass of electron
V_0	Accelerating voltage of electron microscope
$\phi(\mathbf{r})$	Crystal potential (in volts)
W	Total relativistic energy of electron ($W = mc^2 = \gamma m_0 c^2$)
V_r	Relativistically corrected accelerating voltage
γ	Relativistic factor; $\gamma = m/m_0$
$\gamma^{(j)}$	Eigenvalue of dispersion matrix. May be complex
$\mathbf{k}_i, \mathbf{k}_e$	Three-dimensional vacuum electron wavevectors
$\mathbf{k}_0$	Three-dimensional electron wavevector inside crystal corrected for mean potential
$\mathbf{k}^{(j)}$	Labelling electron wavevector for Bloch wave j inside crystal
$\mathbf{K}$	Two-dimensional scattering vector
	$\mathbf{K} = u\hat{\mathbf{i}} + v\hat{\mathbf{j}}$, with $(u^2 + v^2)^{1/2} = |\mathbf{K}| = \theta/\lambda$
$\mu^{(j)}$	Absorption coefficient for Bloch wave j.

The 'crystallographic' sign convention, which takes a plane wave to have the form $\exp(-i\mathbf{k} \cdot \mathbf{r})$ is used throughout this book, except in Section 5.7. This convention, and its relationship to the 'quantum-mechanical' convention, is discussed in Section 5.12.

Atomic resolution transmission electron micrograph of the $YBa_2Cu_3O_7$ superconductor recorded at 400 kV. The black dots are columns of atoms about 10 nm thick viewed along the [100] projection. Inset are shown a computer-simulated image (left) and the optical diffractogram (right). The point resolution is 0.17 nm. (From Ourmazd and Spence, (1987) *Nature*, **329**, 425.)

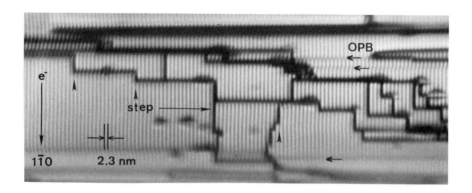

Reflection electron microscope lattice image of the 7×7 reconstruction of the silicon (111) surface obtained at 600°C in ultra-high vacuum. The fringe spacing is 2.3 nm, the beam azimuth is [110], accelerating voltage is 200 kV. Single-atom height steps and out-of-phase boundaries (OPB) are indicated. Note fringe shift at OPB, and coherence of fringes across steps. (From Takayanagi, Tanishiro, Kobayashi, Akiyama and Yagi, (1987) *Jap. J. Appl. Phys.* **26**, L957.)

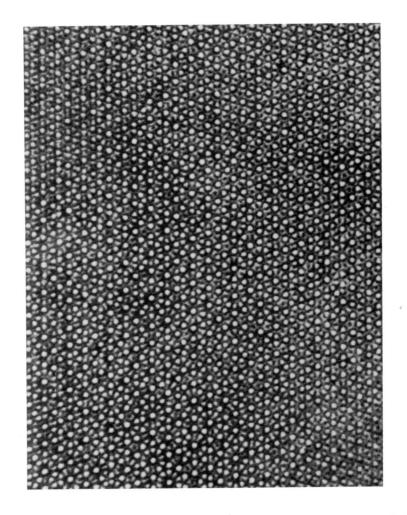

High-resolution image of rapidly quenched $Al_{74}Mn_{20}Si_6$ 'Quasicrystal' alloy, viewed along the fivefold axis. The diffraction pattern shows tenfold symmetry. (From Hiraga, K. Hirabayashi, M., Inoue, A. and Masumoto, T. (1985). *J. Phys. Soc. Japan.* **54,** 4077.)

EXPERIMENTAL HIGH-RESOLUTION ELECTRON MICROSCOPY

INTRODUCTION

This book is intended for electron microscopists interested in using and developing the techniques of high-resolution transmission electron microscopy and in the interpretation of high-resolution images. A section on high-resolution imaging of perfect crystals using scanning transmission electron microscopy (STEM) has also been included. STEM imaging of small molecules has been adequately covered in recent reviews (see Section 5.10), however, much of the material in this book also applies to this field. The book contains both theoretical and practical information. 'High resolution' is taken to mean specimen detail on a scale less than about one nanometre. An undergraduate training in science has been assumed and some knowledge of elementary mathematics.

The success of the electron as a useful probe for the investigation of the structure of matter has depended on many factors. Unlike X-rays, lenses for electrons are easily made, allowing the local structure of isolated defects to be studied. Electrons are strongly scattered by solids, allowing scattering and imaging experiments to be performed in which useful signals and excellent statistics (by the standards of X-ray diffraction and high-energy physics) can be obtained in a conveniently short time. The existence of efficient electron detectors (phosphors, scintillators, and film) has been another major factor, as has the development of electron sources far brighter than those which are possible in other fields of particle physics. Finally, the variety of electronic transitions which can be stimulated by a focused electron probe no more than a few nanometres in diameter has opened up entirely new possibilities for chemical analysis over the last two decades by allowing the decay products from these transitions (or the energy-loss electrons themselves) to be analysed.

Despite these impressive advantages, electron scattering and imaging experiments suffer two main limitations. Firstly, as a result of their strong interaction with matter and consequent multiple scattering, the interpretation of electron diffraction patterns and images usually requires the use of a sophisticated theory which normally leaves no simple relationship between the image or pattern recorded on film and the specimen structure. A second important limitation arises from the two-dimensional nature of high-energy electron diffraction, which usually means that electron images and diffraction patterns are insensitive to the movement of specimen atoms in the direction of the electron beam. The recent development of three-dimensional convergent beam techniques, together with the use of the closely related 'forbidden reflection' method may go some way to overcome this limitation.

Because of these characteristics, the electron microscope has provided the ideal—indeed the only—tool for the investigation of the atomic and electronic structure of isolated defects. These defects, of largely unknown structure, are generally thought to control most of the mechanical, electrical, chemical, and thermal properties of solids. Beginning in the mid-1950s, instruments with the necessary resolving power were first developed for the observation of linear and planar defects in crystals. The theoretical groundwork needed for the interpretation of these images, based on the powerful 'column approximation' was extensively developed and the highly successful technique of medium-resolution diffraction contrast microscopy has been applied to an increasingly wide range of problems in materials science ever since.

The same period has also seen the development of many new uses for electron microscopes, some of which include the microanalytical techniques of X-ray analysis and electron energy loss spectroscopy and the techniques of cathodoluminescence and electron-beam-induced conductivity (EBIC) imaging of interest to geologists and semiconductor physicists. The use of high-voltage microscopes, both for *in situ* studies of chemical reactions and for the stereoscopic observation of 'thick' biological sections has also rapidly increased in popularity. The parallel development of scanning transmission electron microscopy over the last twenty years, and the spectacular images of individual atoms produced by these machines also hold great promise for the future. Recent results also offer the hope that, for the first time, really useful images of surfaces will be obtainable by the reflection high-energy electron diffraction (RHEED) technique (Osakabe, Tanishiro, Yagi, and Honjo, 1980).

Most of these varied techniques do not require particularly high resolution capabilities, however the demand for electron microscopes arising from these applications has led to a steady improvement in their quality and, particularly, their resolution. The current generation of 100 kV instruments has a point resolution of between 0.3 and 0.4 nm. Perhaps the greatest impact of high-resolution work at this resolution has been in solid-state chemistry and mineralogy. Here unit-cell resolution is often sufficient to elucidate reaction mechanisms or to distinguish polytypes and phases which account for departures from stoichiometry at the atomic level. An extensive literature now exists on the rich variety of faults seen in these large unit-cell materials and on the attempts of scientists to deduce the geological and thermal history of minerals from high-resolution electron micrographs. Here electron microscopy provides a unique capability and provides information which can only be obtained in a statistical form by X-ray crystallography. Recent promising developments in high-resolution work in materials science suggest that the next two decades will see this success repeated as the next generation of medium-voltage, high-resolution instruments are applied to problems of defect atomic structure determination using the methods of structure imaging outlined in this book. In particular, much current interest attaches to problems of defect structure determination in semiconductors, whose unit-cell dimensions generally fall between those of minerals and those of the close-packed metals. For the analysis of these defects, however, one usually wishes to resolve detail within the unit cell, thus placing demands at the very limits of the performance of present-day instruments. It must be emphasized that for defect studies it is the microscope *point* resolution which limits the structure determination (see Section 6.2) rather than the comonly quoted 'lattice' or line resolution measured using a perfect crystal. Thus, despite recent successes in semiconductor physics, it may be some few years before really useful results are obtained from specimens of metallurgical interest. Methods for the structural analysis of suitable biological molecules also seem likely to be developed in the near future. The study of surfaces by electron microscopy is also likely to be further developed, both in transmission and scanning electron microscopy.

The improvements in instruments over the last two decades have been the result of a large number of small advances. Taken separately, any one of these seems nugatory. They include the improved thermal stability of specimen stages, the development of stable goniometer tilting stages, the control of contamination, improved electronic stability, and many convenient features which reduce microscope maintenance and downtime. Consequently, an untrained operator using a modern machine can now easily record electron micrographs containing high-resolution detail. The interpretation of this detail does, however, require some care. What has happened in recent years is that the scale of detail seen on micrographs has become smaller than the coherence width of the electron beam used. (An approximate estimate of this is the width of Fresnel fringes.) Under these conditions an interpretation of the image contrast in terms of the product of the specimen density and thickness is no longer valid and wave-optical methods similar to those of classical optics must be used for image interpretation. At high resolution the microscope becomes an electron interferometer and the out-of-focus electron image becomes an electron interference pattern. This book is about the control and interpretation of these interference patterns.

Because of the importance of interference effects at high resolution I have devoted separate chapters to coherence and wave-optics. This has involved some mathematics, which I have tried to keep as simple as possible. Nevertheless, non-mathematical readers may wish to skip Chapters 3 and 4. Two chapters on beam–specimen interactions (Chapters 5 and 6) are also included. Unfortunately, despite the best efforts of our theoreticians, it has not proved possible to find a simple analytical relationship between the high-resolution image of a general specimen and the specimen's structure. So it is necessary to investigate in detail the various experimental conditions which make possible a simple image interpretation in terms of specimen structure. Only by understanding the experimental parameters which affect image artefacts can an intelligent choice of experimental conditions be made for a particular specimen. High-resolution dark-field images, for example, are likely to contain misleading features. The popularity of the simplest theories for imaging, however, is likely to continue owing to advances in specimen preparation (which now make it possible to prepare samples of some specimens a few nanometres thick) and the use of higher accelerating voltages.

I have tried to include all the useful mathematical results relevant to high-resolution microscopy (contained mostly in the first half of the book) together with a good deal of practical information on how to record high-resolution images (mostly in the second half of the book). In the second half of the book I have tried to include as much detailed practical information as possible (much of this was written at the microscope). Unfortunately this information is likely to date rather rapidly, as electron microscope models change and techniques improve. The complexity in the interpretation of images increases rapidly as one seeks finer detail or allows the specimen thickness to increase much beyond about 10 nm, so that it is impossible to avoid a certain amount of mathematics in the interests of succinctness. However, I have tried to avoid the style of a theoretical treatise and to compromise between this and that of a recipe book, which lists procedures.

There is another reason for collecting together most of the expressions useful for high-resolution work. Computing methods in electron microscopy seem certain to be further developed both for image simulation and image analysis. The first technique is useful where one has a reasonable idea of the specimen structure and wishes to examine likely images under various conditions, while the second aims at removing *a posteriori* the electron optical defects introduced by the microscope. It should be a simple matter for the interested reader with some programming experience to plot out the various expressions in this book which give the form of the image of a specimen of known structure, and to investigate the effect of changes in experimental conditions. Thus I hope the book will also be useful as a compilation of expressions in consistent notation on which the computer simulation and analysis of images may be based. With the arrival of cheaper on-line computers and more efficient image-sensors, computer analysis of micrographs may become as routine as the computer analysis of X-ray data. It should be borne in mind, however, that image restoration can at best enable the wave function leaving the specimen's lower surface to be retrieved. The relationship between this and the specimen structure is a separate and equally difficult problem. Object reconstruction, rather than image reconstruction, is the ultimate aim of electron microscopy. The present rather laborious solution to this problem is to examine computer-simulated images of likely structures until one is found which matches the experimental image, as described in Chapter 5.

I have not discussed in detail areas of high-resolution work in which I have not had extensive practical experience. Thus, high-resolution diffraction-contrast techniques (e.g. the 'weak beam' method) are not discussed in this book. They are treated in the new edition of Hirsch, Howie, Nicholson, Pashley, and Whelan (1977), in Loretto and Smallman (1975), and in Edington (1976). There is some discussion of hardware for image simulation and analysis but no detailed mathematical treatment of image analysis methods other than the presentation of the fundamental equations on which this is based. Image analysis has been extensively reviewed by Frank (1973), Misell (1979) and Saxton (1978). I have not discussed the principles of three-dimensional object reconstruction.

The SI system of units has been used throughout this book, together with the centimetre unit. The scattering angle θ referred to is generally that between the central beam and a particular Bragg reflection (that is, twice the Bragg angle); however the symbol is defined in the text in any likely case of confusion. I have used the expression 'optical diffractogram' as shorthand to describe the optical diffraction pattern of an electron micrograph obtained using a laser and optical bench. The sign convention for the phase of the electron wave and subsequent electron optical phase shifts, Fourier transforms, etc. is consistent with the choice of an incident plane wave of the form $\exp(-i\mathbf{k} \cdot \mathbf{r} + wt)$. Section 5.12 contains more details of possible sign conventions.

In the hope that this book will become a useful practical reference volume I have compiled an extensive index, against the current publishing trend.

The most useful companion volume on high-resolution electron microscopy is probably Hayat (1976), which contains a discussion by J. M. Cowley on the principles of high-resolution electron microscopy for non-mathematicians. The books by Hall (1966), Grivet (1972), and Hawkes (1972) provide useful introductions to electron optics. The text by Agar, Alderson, and Chescoe (1974) contains a great deal of essential practical information on the operation of microscopes which also complements the material of this book.

References

Agar, A. W., Alderson, R. H., and Chescoe, D. (1974). Principles and practice of electron microscope operation. In *Practical methods in electron microscopy* (ed. A. M. Glauert). North Holland, Amsterdam.

Edington, J. W. (1976). *Practical electron miscroscopy in materials science.* Van Nostrand Reinhold, New York.

Frank, J. (1973). Computer processing of electron micrographs. In *Advanced techniques in biological electron microscopy* (ed. J. K. Koehler). Springer-Verlag, Berlin.

Grivet, P. (1972). *Electron optics,* 2nd edn. Pergamon, Oxford.

Hall, C. E. (1966). *Introduction to electron microscopy,* 2nd edn. McGraw-Hill, New York.

Hawkes, P. W. (1972). *Electron optics and electron microscopy.* Taylor and Francis, London.

Hayat, M. A. (1976). *Principles and techniques of electron microscopy,* Vol. 6. Van Nostrand Reinhold, New York.

Hirsch, P. B., Howie, A., Nicholson, R. B., Pashley, D. W., and Whelan, M. J. (1977). *Electron microscopy of thin crystals,* 2nd edn. Krieger, New York.

Loretto, M. H. and Smallman, R. E. (1975). *Defect analysis in electron microscopy.* Chapman and Hall: Halsted/Wiley, London and New York.

Misell, D. (1979). Image analysis, enhancement and interpretation. In *Practical methods in electron microscopy* (ed. A. M. Glauert) North Holland, Amsterdam.

Saxton, O. (1978). *Computer techniques for image processing in electron microscopy.* Academic Press, New York.

Osakabe, N., Tanishiro, Y., Yagi, K., and Honjo, G. (1980). Reflection electron microscopy of clean and gold deposited (111) silicon surfaces. *Surface Science* **97,** 393.

1

PRELIMINARIES

Until recently the methods described in this book have been restricted to a few research laboratories fitted with the specialized equipment needed for ultimate-resolution microscopy. These methods are now rapidly increasing in popularity and provided one restricts the image resolution to about 0.20 nm it is now possible to outline a straightforward procedure by which images of many specimens can be recorded which can be readily interpreted in terms of specimen structure. This procedure is described in Section 5.16 and Chapter 10.

The conventional transmission electron microscope bears a close resemblance to an optical microscope in which image contrast (intensity variation) is produced by the variation in optical absorption from point to point on the specimen. Most electron microscopists interpret their images in a similar way, taking 'absorption' to mean the scattering of electrons outside the objective aperture. Electrons are never actually 'absorbed' by a specimen; they can only be lost from the image either by large-angle scattering outside the objective aperture or, as a result of energy loss and wavelength change in the specimen, being brought to a focus on a plane far distant from the viewing screen showing the elastic image. This out-of-focus 'inelastic' or energy-loss image then contributes only a uniform background to the in-focus elastic image. Thus image contrast is popularly understood to arise from the creation of an intensity deficit in regions of large scattering or 'mass thickness' where these large-angle scattered rays are intercepted by the objective aperture. The theory which describes this process is the theory of incoherent imaging (see, for example, Cosslett (1958)).

By comparison, the high-resolution transmission electron microscope is a close analogue of the optical phase-contrast microscope. A thorough treatment of the principles and applications of these instruments can be found in the book by Bennett, Jupnik, Osterberg, and Richards (1951). These optical instruments were developed to meet the needs of biological microscopists who encountered difficulty in obtaining sufficient contrast from their thinnest specimens. Nineteenth-century microscopists were dismayed to find the contrast of their biological specimens falling with decreasing specimen thickness. More than a century later, improvements in specimen preparation technique have allowed us to see exactly the same behaviour in electron microscopy when observing specimens just a few atomic layers thick. (Section 10.5 explains how this lack of contrast at exact focus can be put to good use for electron microscope calibration.) Since the transmission image is necessarily a projection of the specimen structure in the beam direction (the depth of focus being large—see Section 2.2) these low-contrast ultra-thin specimens must be used if one wishes to resolve detail at the atomic level.

Broadly then, while low-resolution biological microscopy is mainly concerned with the electron microscope used in a manner analogous to that for a light microscope, this book is devoted to its use as a phase-contrast instrument for high-resolution studies. It might equally have been titled 'Ultimate resolution microscopy' or perhaps 'Imaging by electron interferometry'.

For the benefit of readers with little mathematical background, a simplified summary of the material in the early theoretical chapters is included at the end of each chapter. These mathematical sections are included both for the benefit of microscopists with some undergraduate mathematics training and because image simulation and analysis by compu-ter are likely to play an increasing role in high-resolution microscopy in the future. As with the Fortran programs of X-ray crystallography, the intelligent use of these programs requires some understanding of basic theory. A simple Fortran program which can be used to analyse the data from an optical diffractogram to give the focus increments and spherical aberration constants of a microscope is given in Appendix 1.

1.1   Elementary principles of phase-contrast microscopy

Fresnel edge fringes, atomic lattice fringes, and images of small molecules are all examples of phase contrast. None of these contrast effects can be explained using the 'mass thickness' model for imaging with which most microscopists are familiar. All are interference effects. The distinction between interference or phase contrast and conventional low-resolution contrast is discussed more fully in Section 6.1. In this section, however, a simple optical bench experiment is described which will give the micros-copist some feel for phase contrast and allow him to anticipate the results of many electron imaging experiments. An expensive optical bench is not needed to repeat this experiment—good results will be obtained with a large-diameter lens (focal length, f_0, about 14 cm and diameter, d, about 5

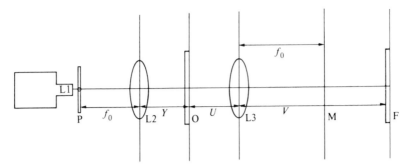

Fig. 1.1 The optical bench arrangement used to record the images shown in Figs. 1.2 to 1.4. Here L1 is a ×40 microscope objective lens at the focus of which is placed a pin-hole aperture P. Lenses L2 and L3 have a focal length of $f_0 = 14$ cm. The object is shown at 0, and film plane at F and the distances $Y = 30$ cm, $U = 17$ cm, and $V = 80$ cm. The pin-hole aperture is used as a spatial filter to provide more uniform illumination. Back-focal plane masks may be inserted at M.

cm) and an inexpensive 1.5 mW He–Ne laser. The object-to-lens and lens-to-image distances U and V could then be about 17 and 80 cm respectively as indicated in Fig. 1.1.

The very thin specimens used in high-resolution electron microscopy can be likened to a piece of glass under an optical microscope. An amorphous carbon film behaves rather like a ground-glass screen for optical wavelengths. In a typical high-resolution electron microscope experiment this 'screen' is used to support a molecule (DNA, for example) of which one wishes to form an image. Figure 1.2 shows the image of a glass microscope slide formed using a single lens and a laser. The laser beam has been expanded (and collimated) using a ×40 optical microscope objective lens. The presence of a small molecule or atom is difficult to simulate exactly. However, the glass slide shown contains a small indentation produced by etching the glass with a small drop of hydrofluoric acid. This indentation will produce a similar contrast effect to that of an atom imaged by electron microscopy. Figure 1.2 shows a 'through-focus series' obtained by moving the lens slightly between exposures. These through-focus series are commonly taken by high-resolution microscopists to allow the best image to be selected (see Chapter 5). In particular, we notice the lack of contrast in Fig. 1.2(b), the image recorded at exact Gaussian focus when the object and image fall on conjugate planes. *Exactly similar effects are seen on the through-focus series of electron micrographs of single atoms shown in Fig. 6.7.*

Figure 1.3 shows the same object imaged using a conventional tungsten lamp-bulb as the source of illumination to provide 'incoherent' illumination conditions. Despite wide changes in focus, little contrast appears.

If this analogy between the indented glass slide and a high-resolution specimen in electron microscopy holds, this simple experiment suggests that our high-resolution specimens will be imaged with strong contrast only if a

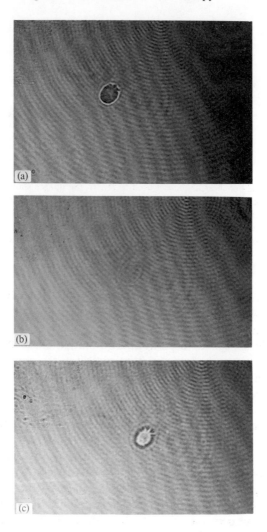

Fig. 1.2 Optical through-focus series showing the effect of focus changes on the image of a small indentation in a glass plate (phase object). The image at (a) was recorded under-focus, that is, with the object too close to the lens L3. It shows a bright fringe surrounding the indentation similar to that seen on electron micrographs of small holes; (b) is recorded at exact focus and shows only very faint contrast; (c) is recorded at an over-focus setting (object too far from L3) and so shows a dark Fresnel fringe outlining the indentation. The background fringes arise in the illuminating system. Compare these with the electron micrographs in Fig. 6.7.

coherent source of illumination is used and if images are recorded slightly out of focus. These conclusions are borne out in practice, and the importance of high-intensity coherent electron sources for high-resolution microscopy is emphasized throughout this book. Two chapters have been devoted to the topic—Chapter 7 deals with experimental aspects while Chapter 4 describes the elementary coherence theory.

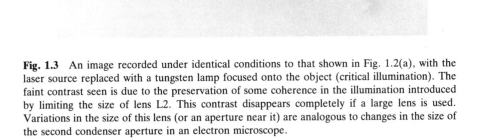

Fig. 1.3 An image recorded under identical conditions to that shown in Fig. 1.2(a), with the laser source replaced with a tungsten lamp focused onto the object (critical illumination). The faint contrast seen is due to the preservation of some coherence in the illumination introduced by limiting the size of lens L2. This contrast disappears completely if a large lens is used. Variations in the size of this lens (or an aperture near it) are analogous to changes in the size of the second condenser aperture in an electron microscope.

To return to our optical analogue. Given coherent illumination the question arises to whether methods other than the introduction of a focusing error can be used to produce image contrast. During the last century many empirical methods were indeed developed to enhance the contrast of images such as Fig. 1.2(b). These included (1) reducing the size of the objective aperture as in Fig. 1.4(a), (2) the introduction of a focusing error as in Fig. 1.2(a); (3) simple interventions in the lens back-focal plane as in Fig. 1.4(b), where Schlieren contrast is shown (The back-focal plane is approximately the plane of the objective aperture for an electron microscope.); (4) the use of back-focal plane phase plates.

All of these methods have their parallel in electron microscopy. The first is not useful, since image resolution is necessarily limited. The second is the usual method of high-resolution electron microscopy and is discussed in more detail in Chapters 3 and 10. Methods based on (3) have appeared from time to time (see, for example, Spence (1974)), while (4) remains a field of active research (for examples, see Unwin (1971) and Downing and Siegel (1973)). The phase-plate method developed by Zernike in the 1930s remain the standard method of optical phase-contrast microscopy.

In both electron and optical microscopy the reasons for the lack of contrast at exact focus are the same—these thin specimens ('phase objects') affect only the phase of the wave transmitted by the specimen and not its amplitude. That is, they behave like a medium of variable refractive index

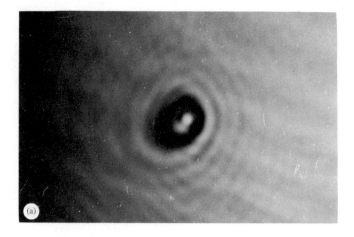

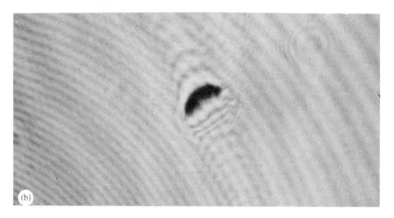

Fig. 1.4 Using a laser source, both these images were recorded at exact focus (exactly as in Fig. 1.2(b). Yet in both, the outline of the indentation can be seen. In (a), a small aperture has been placed on axis at M, severely limiting the image resolution. On removing this aperture the image contrast disappears. The use of a small aperture at M (the back-focal plane) is analogous to the normal low-resolution method of obtaining contrast in biological electron microscopy. In (b) a razor blade has been placed across the beam at M, thus preventing exactly half the diffraction pattern from contributing to the image. The resulting image is approximately proportional to the derivative of the phase shift introduced by the object taken in a direction normal to the razor-blade edge. Notice the fine fringes inside the edge of the indentation arising from multiple reflection within the glass slide.

(see Section 2.1). It is this variation in refractive index from point to point across the specimen (proportional to the specimen's atomic potential in volts for electron microscopy) which must be converted into intensity variations in the image if we are to 'see', for example, atoms in the electron microscope. Through the reciprocity theorem, discussed in Section 5.10, all these considerations apply to the imaging of single atoms in the scanning transmission microscope and we are currently seeing the development of methods analogous to those mentioned above to enhance the contrast of phase objects in scanning microscopy.

We can sharpen these ideas by comparing the simplest mathematical expressions for imaging under coherent and incoherent illumination. For the piece of glass shown in Fig. 1.2, the phase of the wave transmitted through the glass differs from that of an unobstructed reference wave by $2\pi(n-1)/\lambda$ times the thickness of the glass, where n is the refractive index of the glass. If $t(x)$ is the thickness of the glass at the object point x, the amplitude of the optical wave leaving the glass is given from elementary optics (see, for example, Lipson and Lipson (1969)) as proportional to

$$f(x) = \exp(-2\pi i n t(x)/\lambda). \tag{1.1}$$

The corresponding expression for electrons is derived in Section 2.1. Optical imaging theory (see Goodman (1968)) gives the image intensity, proportional to the darkening on a photographic plate, as

$$I(x)_c = |f(x) * S(x)|^2 \tag{1.2}$$

for coherent illumination and

$$I(x)_i = |f(x)|^2 * |S(x)|^2 \tag{1.3}$$

for incoherent illumination. In these equations the function $S(x)$ is known as the instrumental impulse response, and gives the image amplitude which would be observed if one were to form an image of a point object much smaller than the resolution limit of the microscope. Such an object does not exist for electron microscopy; however, the expected form of $S(x)$ for one modern electron microscope can be seen in Fig. 3.6. This function specifies all the instrumental imperfections and parameters such as objective aperture size (which determines the diffraction limit), the amount of spherical aberration, and the magnitude of any focusing error. The asterisk above indicates the mathematical operation of convolution (see Section 3.1), which results in a smearing or broadening of the object function $f(x)$. It is not possible to resolve detail finer than the width of the function $S(x)$. Using eqn (1.1) in eqns (1.2) and (1.3) we see that under incoherent illumination the image intensity from such a phase object does not vary with position in the object, since

$$|\exp\{-2\pi i t(x)n/\lambda\}|^2 = 1, \tag{1.4}$$

Only by using coherent illumination and an 'imperfect' microscope can we hope to obtain contrast variations in the image of a specimen showing only variations in refractive index. *In high-resolution electron microscopy of thin specimens the accurate control of illumination coherence and defect of focus are crucial for success.* The amount of fine detail in a high-resolution micrograph increases dramatically with improved coherence of illumination, while completely misleading detail may be observed in images recorded at the wrong focus setting. The correct choice of these important experimental conditions is discussed in Chapters 5, 6, and 10.

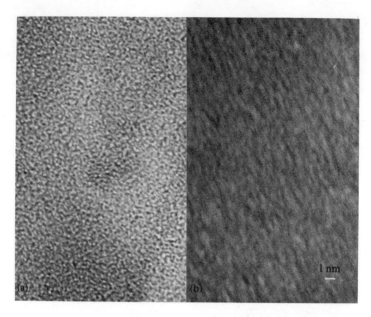

Fig. 1.5 Two electron images of amorphous carbon films recorded at the same focus setting but using different condenser apertures. In (a) a small second condenser aperture has been used, resulting in an image showing high contrast and fine detail. This contrast is lost in (b), where a large aperture has been used. This effect is seen even more clearly in optical diffractograms (see Chapter 4).

To demonstrate the effect of coherence on image quality at high resolution, Fig. 1.5 shows the images of an amorphous carbon film using widely different illumination coherence conditions but otherwise identical conditions. The loss of fine detail in the 'incoherent' image is clear. In practice, for a microscope fitted with a conventional hair-pin filament, the illumination coherence is determined by the size of the second condenser lens aperture, a small aperture producing high coherence. With a sufficiently small electron source (such as a field-emission, pointed-filament, or LaB_6 source) the illumination may be perfectly coherent and independent of the choice of second condenser (illuminating) aperture. For these sources the demagnification settings of both condenser lenses are also important in determining the degree of coherence at the specimen (see Section 4.5). The recent development of higher-brightness sources is of particular benefit to high-resolution transmission electron microscopy since the required choice of a small second condenser aperture frequently provides insufficient intensity on the viewing screen to allow accurate focusing at high magnification. A pointed filament, LaB_6 source or field-emission gun greatly facilitates this procedure.

In most cases of practical interest the imaging is partially coherent. By this we loosely mean that for object detail below a certain size X_c we can use the model of coherent phase contrast imaging (see Fig. 1.2) while for detail much larger than X_c the imaging is incoherent. The distance X_c is given approximately by the electron wavelength divided by the semi-angle subtended by the second condenser aperture at the specimen, when using a hair-pin filament. In Section 8.8 a fuller discussion of methods for measuring X_c in practice is given which includes the effect of the objective lens pre-field. Since the width of Fresnel edge fringes commonly seen on micrographs is approximately equal to X_c, we can adopt the rough rule of thumb that the more complicated through-focus effects of phase contrast described in Section 6.2 will become important whenever one is interested in image detail much finer than the width of these fringes.

To summarize, the main changes in emphasis for microscopists accustomed to conventional imaging and considering a high-resolution project are as follows: in conventional imaging we deal with "thick", possibly stained, specimens using a small objective aperture and a large condenser aperture. Focusing presents no special problems. In high-resolution phase-contrast imaging we deal with the thinnest possible specimens (unstained), using a larger objective aperture of optimum dimensions and a small illuminating aperture. The choice of focus becomes crucial and must be guided either by experience with computed images or by the presence of a small region of specimen of known structure in the neighbourhood of the wanted structure. This image simulation experience indicates that for the vast majority of large-unit-cell crystals imaged at a resolution no higher than 0.20 nm the choice of focus which gives a readily interpretable image containing no false detail is between 20 and 60 nm under-focus (weakened lens) for microscopes with a spherical aberration constant between 1 and 2 mm. A calibrated focus control is therefore needed to set this focus condition. Methods for measuring the focus increments are given in Section 8.1. The limits on specimen thickness are discussed in Chapter 5.

1.2 Instrumental requirements and modifications for high-resolution work

The major manufacturers of electron microscopes offer instruments closely matched in performance. All are capable of giving point resolution better than 0.35 nm for bright-field images. The recent trend toward side-entry eucentric goniometer stages, of the kind used by biologists for many years, for high-resolution work greatly facilitates the procedure of orienting a small crystal, since the lateral movement accompanying tilting of the specimen is minimized by these stages. However, the performance of these stages for long-exposure dark-field work has yet to be assessed. In the past the drift rate due to thermal effects using these stages has been considerably higher than that for top-entry stages.

A laboratory which has recently purchased a transmission microscope and wishes to use it for high-resolution studies should consider the following points. This list is intended as a rough checklist to refer the reader to further discussion of these topics in other chapters. Most of the instrumental modifications needed to adapt an existing machine to high resolution work are also listed.

1. The microscope site must be acceptable. Mechanical vibration, stray magnetic fields, and room temperature must all be within acceptable limits. These factors are discussed in Chapter 9 and elsewhere (Alderson 1974; Agar, Alderson, and Chescoe 1974).

2. A reliable, constant temperature and pressure supply of clean cooling water must be assured (see Chapter 9). In installations where internal re-circulating water is not available and the external supply is impure it is common to find both the specimen drift rate and illumination stability deteriorating over a period of weeks as the water filter clogs up, causing the cooling water pressure to fall and leading to fluctuations in specimen stage and lens temperature. A separate closed-circuit water supply and external heat exchanger is the best solution for high-resolution work, with the water supplied only to the microscope. Vibration from the heat exchanger pumps must be considered; these should be in a separate room from the microscope. Despite the use of distilled water, algae are bound to form in the water supply. Some manufacturers warn against the use of algae inhibitors as these may corrode the lens-cooling jacket, repairs to which are very expensive. About all that can be done is to replace the water periodically. Thermal stage drift is one of the most serious problems for high-resolution work, particularly for long-exposure dark-field images. If external hard water is used *it is essential that the condition of the water filter is checked regularly and replaced if dirty.* If the microscope's inbuilt thermoregulator is used it is equally important to check every few days that it is operating within the temperature limits for which it is designed. *The heat-exchanger compressor cooling fins must be cleaned frequently.*

3. In order to record an image at a specified focus defect it will be necessary to measure the change in focus between focus control steps ('clicks') using the methods of Section 8.1. In addition, the spherical aberration constant C_s must be known for the optimum objective lens excitation (see below). This should be less than 2 mm at 100 kV if high-resolution results are expected. Methods for measuring C_s are given in Section 8.2. Microscope calibration—the measurement of defocus increments, spherical abberation constant, and objective lens current—must therefore be one of the first steps in the conversion of a conventional machine to high-resolution work. A digital voltmeter will be required to read the objective lens current as described in Section 8.4. Once the optimum specimen position in the lens gap has been found, the objective lens current needed to focus the specimen at this position should be noted, and the highest-quality images recorded near this lens current. This is not always possible when using a tilting specimen holder if a non-axial object point is required, since the specimen height (and hence the objective lens current needed for focus) varies across the inclined specimen.

4. The resolution obtained in a transmission image depends amongst other things on the specimen position in the objective lens. The optimum 'specimen height' must be found by trial and error. In reducing the specimen height and so increasing the objective lens current needed for focus, the lens focal length and spherical aberration constant are reduced, leading to improved resolution. In Fig. 2.16 the dependence of spherical aberration on lens excitation is shown. A second effect, however, results from the depth of the objective lens pre-field, which increases as the specimen is immersed more deeply into the lens gap. This increases the overall demagnification of the illumination system and increases its angular magnification (see Section 2.2). The result is an increase in the size of the diffraction spots of a crystalline specimen which, for a fixed illumination aperture, allows the unwanted effect of spherical aberration to act over a larger range of angles (see Section 3.3). Finally, as shown in Fig. 2.17, for many lens designs the chromatic aberration coefficient C_c passes through a minimum as a function of lens excitation and this parameter affects both the contrast and resolution of fine image detail. The complicated interaction between all these factors which depend on specimen position can best be understood using the 'damping envelope' concept described in Section 4.2 and Appendix 3. This 'damping envelope' controls the information resolution limit (loosely referred to by manufacturers as the 'line' or 'lattice' resolution) of the instrument and depends chiefly on the size of the illumination aperture and C_c. The point resolution, however, is determined by spherical aberration. A method has been described which would allow both these important resolution limits to be measured as a function of specimen position in the lens bore through an analysis of optical diffractogram pairs (see Section 8.10). A systematic analysis of this kind, however, represents a sizeable research project, involving many practical difficulties. For example, in such an analysis, diffractograms are required at similar focus defects for a range of specimen positions, yet the 'reference focus' used to establish a known focus defect (see Section 10.5) itself depends on C_s, which varies with specimen position. In addition, both the focal increment corresponding to a single step on the focus control and the chromatic aberration constant (which affects the 'size' of diffractogram ring patterns) depend on the specimen position. For preliminary work, then, a simpler procedure is to record images of a large-unit-cell crystal at the Scherzer focus for a range of specimen heights and to select the highest-quality image, judging this by eye. Experience in comparing computed and experimental images of large-unit-cell crystals shows that a useful judgment of image quality (a combination of contrast, point, and line resolution) can be made by eye. Once this near-optimum specimen position has been found, further refinements can be made using the computer image-matching technique together with diffractogram analysis. To obtain the required range of specimen heights on top-entry machines it is necessary to fit small brass washers above or below the specimen. A shallow threaded cap to fit over the specimen may also be needed to bring the tilted specimen closer to the lower cold-finger. The measured objective lens current (see Section 8.4) needed to bring the image into focus is then used as a measure of specimen height, and must be recorded for each image. The specimen height and corresponding lens current which give images of the highest quality can then be determined and permanently recorded for the particular electron microscope. These large-unit-cell crystals have the useful property that images of highest contrast are usually produced near the Scherzer focus, so that this focus setting can be found routinely by experienced microscopists working with these materials. (This may not be the case for metals and semiconductors, whose lattice images of highest contrast appear at much larger under-focus settings—see Section 5.15.)

5. A vacuum of 0.5×10^{-6} Torr or better is needed (measured in the rear pumping line). Ultra-high-vacuum transmission electron microscopy is discussed in Section 12.8. The simplest way to trace vacuum leaks is to use a partial pressure gauge (see Section 9.5). A microscope which is to be used for high-resolution work should either be fitted permanently with such a gauge, or should be fitted with a blanked-off T-connection on the rear pumping line to allow a gauge available on loan to be fitted at short notice. Inexpensive gauges of the radiofrequency quadrupole type are useful for conventional work. A machine which has been used for routine use for several years will often be found to contain several leaks. Leaks in the region of the gun chamber should be particularly avoided in high-resolution work as these affect high-voltage stability. A plentiful supply of gun chamber O-rings is therefore required. Specimen contamination must be within acceptable limits. Photographic plates, a major source of water vapour in the microscope, must be well out-gassed and the interior of the high-vacuum pumping lines must be clean. These can be out-gassed using flexible heating tape wrapped around the metal pipes and heated for several hours. A sizeable improvement in vacuum is usually possible using this method. In a clean, well-out-gassed microscope with no serious leaks and a vacuum of 0.5×10^{-6} Torr the major source of contamination is frequently the specimen itself. Always compare the contamination rates for several different specimens before concluding that the microscope itself is causing a contamination problem.

6. The microscope high voltage must be sufficiently stable to allow high-resolution images to be obtained. Given a sufficiently stable high-voltage supply (see Sections 9.2 and 2.8.2) this generally requires a scrupulously clean gun chamber and Wehnelt cylinder. Methods for observing high-voltage instability are discussed in Section 10.2.

7. The room containing the microscope must be easily darkened completely), and a room-light dimmer control needs to be fitted within arm's reach of the operator's chair. This is needed in addition to the normal room light switch on the microscope console which should be wired in series with the dimmer rheostat. Specimen changes can then be made in dim lighting so that the user does not lose the slow chemical dark adaptation of his eyes.

8. The room used for photographic processing may requre attention. A common problem in areas where water quality is poor is drying marks. These may be of comparable size to the image detail being sought and can be eliminated by the methods described in Section 10.7. A good-quality optical microscope and calibrated graticule will be required for micrograph examination. A microscope fitted with zoom magnification control is convenient, since one frequently requires a large field of view in order to assess the mean electron and thickness noise level. A high-quality enlarger capable of being operated horizontally to give magnified images of optically dense plates up to 30 times will be required for small-molecule imaging. The newer electron sources (field-emission, LaB_6, and pointed-filament) will allow focusing under adequate coherence conditions up to a magnification of perhaps 700 000, thus reducing the need for subsequent optical enlargement. Using a hair-pin filament it is difficult to focus images of even the thinnest specimens at magnifications much above 500 000, if the beam divergence is limited to a milliradian or less. Under these conditions an optical enlargement of 20 times would be required to give a final scale on the print of 1 cm = 1 nm. Light sources suitable for enlarging high-resolution plates are discussed in Section 10.7.

9. A simple optical bench and laser are extremely useful in high-resolution work, mainly for training microscopists in astigmatism correction and for distinguishing the effect of astigmatism from that of drift (see Section 8.7). Accurate astigmatism correction is probably the most important and difficult skill the high-resolution microscopist must learn (see Section 10.1). The optical diffraction pattern of a micrograph of a thin carbon film gives an immediate indication of the presence of astigmatism. This is sometimes difficult to detect in an image but causes an oval distortion of the rings seen in an optical diffraction pattern. Even the optical bench is not strictly needed for this simple and sensitive test. If lens holders and a micrograph holder can be made locally, these can be aligned with an inexpensive He–Ne laser as shown in Fig. 8.4 on a firm bench to give useful results. Astigmatism and specimen drift account for the vast majority of image defects in high-resolution work. For short-exposure (2–5 s) bright-field work it is common to reject perhaps 30 per cent of all plates taken owing to these defects. For dark-field work, where exposures of half a minute are not uncommon, it may be necessary to take several dozen plates before one is found free of astigmatism, and whose resolution is not limited by specimen movement during the exposure. This, despite all the precautions listed in Chapter 10.

1.3 First experiments

This section outlines a procedure which will enable a microscopist new to high-resolution work to become familiar with some of the techniques of high-resolution transmission microscopy and to practise the required skills of focusing, astigmatism correction, and micrograph examination. The test specimen used consists of small clusters of heavy atoms supported on a thin carbon film. The microscope is assumed to satisfy the conditions of Section 1.2 and the focal increments and spherical aberration constant are presumed known.

1. Specimens can be purchased or made up using an evaporator. With patience a specimen can be made which will give higher contrast than most commercially available specimens by selecting the thinnest possible support film. To prepare the specimens, set up the evaporator as shown in Fig. 1.6, using a glass microscope slide carefully cleaned with distilled water and detergent at A. A second slide is placed above this to shield the lower slide from direct evaporation. The amount of gold wire added to the tungsten coil must be judged by experiment—start with about 1 cm of fine gold wire.

2. Pump down the evaporator bell-jar to 10^{-5} Torr or better and evaporate the carbon for a few seconds. Increase the current in the tungsten heater quickly to well above the metlting point of gold, then switch off and allow the specimen to cool. Break the bell-jar vacuum.

3. Remove the lower glass slide, which should now contain two well-defined carbon shadow edges. Fill a clean evaporating dish with distilled water and gently slide the microscope slide into the still water at a low angle

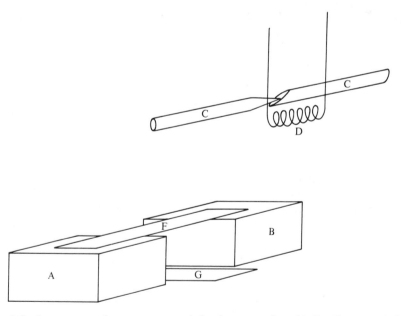

Fig. 1.6 Arrangement of an evaporator unit for the preparation of indirectly evaporated gold atom clusters on amorphous carbon. A and B are supports of material which does not out-gas under vacuum (e.g., clean metal). G and F are clean glass microscope slides—freshly cleaved mica can also be used at G. The tungsten heater loaded with a minute amount of fine gold wire is shown at D and the carbon rods used for the amorphous carbon film are indicated at C. The distance between D and G should be as large as possible and several trial evaporations will be needed to obtain a low density of gold clusters.

to the surface. The carbon should detach from the glass and fragments will be seen floating freely on the water surface. Keep track of a thin fragment from the shadow edge—this must be picked up by drawing a 400-mesh microscope grid (held with tweezers) up from below the surface of the water. If commercial holey-carbon films are used in place of the 400-mesh grids even thinner films can be made.

4. Drying the specimen without damaging the carbon film is difficult. For rough preliminary work with thicker films the water drop enclosing the grid can usually be drawn off by touching a piece of filter paper against the *side* of the grid, or introducing a sliver of filter paper between the tweezer prongs and running this up to the specimen. For the thinnest films the tweezers should be clamped together using an electrician's crocodile clip and the specimen, held in the tweezers, left to dry alone.

5. Examine the specimen in the microscope. If too much gold has been used a continuous or semi-continuous film of gold will be seen. The correct amount of gold produces isolated islands of gold atoms, well separated and of varying size. The carbon support should be as thin as possible and not torn within the grid square of interest.

6. Once a suitable specimen area has been found, bright-field images can be recorded at high magnification (400 000 or more). Chapter 10 gives details of all necessary precautions (in particular Sections 10.1 and 10.2). A microscope which has been switched on in the morning, with all cold-traps kept filled with liquid nitrogen, should be in thermal equilibrium by early afternoon when serious work can commence. The morning is profitably spent examining possible specimens. The cold-traps must not be allowed to boil dry during a lunch break. *The thermal stability and cleanliness of the three components in the objective lens pole-piece gap (objective aperture, cold-finger, and specimen) are of the utmost importance in high resolution work.* If finances allow, the best arrangement is to leave the microscope running through the week, fitted with an automatic liquid-nitrogen pump for the cold-traps. Record images without an objective aperture inserted (so that resolution is limited by incoherent instabilities—see Section 4.2), and using an aperture whose size is given by eqn (6.16). The semi-angle θ_{ap} subtended by the objective aperture can be measured by taking a double exposure of the diffraction pattern of some continuous gold film with and without the aperture in place (see Section 8.2). A check of the following must be made before taking pictures: (1) condenser astigmatism (for maximum image intensity); (2) gun tilt; (3) current or voltage centre (see Section 10.2); (4) high-voltage stability (see Section 9.2); (5) absence of thermal contact between specimen and cold-finger (see Section 10.2); (6) cleanliness of objective aperture (7) specimen drift (see Section 9.4); and (8) contamination (see Section 9.5).

Use the minimum-contrast condition (see Section 10.5) to correct astigmatism and take several bright-field images in the neighbourhood of the focus value given by eqn (6.16). A through-focus series about the minimum contrast focus in steps of, say 20 nm, will show the characteristic change from a dark over-focus Fresnel fringe around an atom cluster to a bright fringe in the under-focus images.

7. Develop the micrographs, noting the precautions mentioned in Section 10.7. After drying they should be examined under an optical microscope (emulsion side uppermost) and those showing obvious signs of drift should be discarded.

8. Optical diffractograms of the remaining images should be taken as described in Section 8.7. The measured diameter of the rings seen in these can be used with the simple computer program given in Appendix 1 to find the focus setting for each micrograph and the microscope's spherical aberration constant. More immediately, these optical diffraction patterns reveal at a glance the presence of astigmatism or specimen movement during the exposure (drift—see Section 8.7). This immediate 'feedback' is essential for a microscopist learning the skills of astigmatism correction and focusing. The optical diffractograms should be examined as soon as possible after processing the micrographs. In this way the microscopist will be able to investigate the through-focus appearance of high-resolution images and confirm experimentally for himself that the atom-cluster images of highest contrast are found near the focus setting given by eqn (6.16). With practice the number of plates which must be discarded will decrease and the microscopist will become adept at finding the minimum-contrast condition, correcting astigmatism, and resetting the focus control a fixed number of 'clicks' toward the under-focus side to obtain images of highest contrast and resolution.

References

Agar, A. W., Alderson, R. H., and Chescoe, D. (1974). Principles and practice of electron microscope operation. In *Practical methods in electron microscopy* (ed. A. M. Glauert). North-Holland, Amsterdam.

Alderson, R. H. (1974). The design of the electron microscope laboratory. In *Practical methods in electron microscopy* (ed. A. M. Glauert). North-Holland, Amsterdam.

Bennett, A. H., Jupnik, H., Osterberg, H., and Richards, O. W. (1951). *Phase microscopy. Principles and applications.* Wiley, New York.

Cosslett, V. E. (1958). Quantitative aspects of electron staining. *J. R. Microsc. Soc.* **78,** 18.

Downing, K. H. and Siegel, B. M. (1973). Phase shift determination in single-sideband holography. *Optik* **38,** 21.

Goodman, J. W. (1968). *Introduction to Fourier optics.* McGraw-Hill, New York.

Lipson, S. G. and Lipson, H. (1969). *Optical physics.* Cambridge University Press, London.

Spence, J. C. H. (1974). Complex image determination in the electron microscope. *Opt. Acta* **21,** 835.

Unwin, N. (1971). Phase contrast and interference microscopy with the electron microscope. *Phil. Trans. R. Soc., Lond.* **B261,** 95.

2

ELECTRON OPTICS

An elementary knowledge of electron optics is important for the intelligent use of an electron microscope, particularly at high resolution. This chapter is intended to give the simplest account of electron optics that will expose the important physical properties of magnetic electron lenses, such as image rotation, aberrations, minimum focal length, and the distinction between projector and objective modes. In high-resolution work it is common practice to experiment with changes in specimen position, the effects of which can be understood from the discussion of Section 2.7. It is also sometimes convenient to increase the overall magnification available on an older machine—the effect of the projector lens pole-piece dimensions on magnification is also mentioned. Some calculations of lens characteristics used for lens design are also given, showing the way in which lens aberrations, of prime importance at high resolution, depend on the lens excitation and geometry. Since the emphasis is on the practical aspects of the behaviour of real lenses, I do not discuss the elegant and considerable contribution of early workers using simple algebraic approximations for the lens field. These workers' investigations, such as those of Lenz, Glaser, Grivet, and Ramberg, can be traced through the electron optics texts included in the references for this chapter. Useful introductory accounts of electron optics can be found in the books by Hall (1966) and Hawkes (1972). Grivet (1965) is a useful general text on electron optics for the solution of practical electron-optical problems. Modern lens designs are guided by a great deal of experience, together with computed solutions of the Laplace equation (see Septier (1967) for a review of methods for doing this) and subsequent numerical solution of the ray equation (see, for example, Mulvey and Wallington (1972) or Kamminga, Verster, and Francken (1968/9)). The accurate measurement of electron-optical para-

25

meters such as the spherical aberration and chromatic aberration constants, which has become increasingly important with developments in image analysis at high resolution, is discussed in Chapter 8. Promising developments in superconducting electron optics are not discussed, as these are fully described in Dietrich (1977).

2.1 The electron wavelength and relativity

As the resolution of electron microscopes improves, the wave properties of electrons become increasingly important. Indeed, it is impossible to interpret high-resolution micrographs of very thin specimens on an incoherent particle model. A particle model has traditionally been used, and is adequate, for thick specimens imaged at moderate and low resolution (see Chapters 5 and 6). This book is primarily concerned with recordings from very thin specimens from thich it is hoped to obtain information on a scale below 1 nm, where questions of coherence, lens aberrations, and Fresnel diffraction become crucial. All these are wave properties which require interference effects to be included in the description of the imaging process.

Rather than solve the Schrödinger equation for the electron microscope as a whole, it is simpler to separate the three problems of beam–specimen interactions, magnetic lens action, and fast electron sources. The first problem is a many-body problem solved by optical-potential methods, while the second has traditionally been treated classically. A wavelength is assigned to the fast electron as follows.

The principle of conservation of energy applied to an electron of charge $-e$ traversing a region in which the potential varies from 0 to V_0 gives

$$eV_0 = p^2/2m = h^2/2m\lambda^2 \tag{2.1}$$

where p is the electron momentum and h is Planck's constant. Thus,

$$\lambda = \frac{h}{\sqrt{2meV_0}} \tag{2.2}$$

where the de Broglie relation $p = mv = h/\lambda$ has been used. An electron leaves the filament with high potential energy and thermal kinetic energy, and arrives at the anode with no potential energy and high kinetic energy. The zero of potential energy is taken at ground potential. If λ is in nanometres and V_0 in volts, then

$$\lambda = 1.22639/\sqrt{V_0} \tag{2.3}$$

At higher energies the relativistic variation of electron mass must be considered. Neglect of this leads to a 5 per cent error in λ at 100 kV. The relativistically corrected mass is

$$m = m_0/(1 - v^2/c^2)^{1/2}$$

and the equation corresponding to (2.1) is

$$eV_0 = (m - m_0)c^2$$

with m_0 the electron rest mass and c is the velocity of light. These equations may be combined to give an expression for the electron momentum mv. Used in the de Broglie relation, this gives the corrected electron wavelength as

$$\lambda = h/(2m_0 eV_r)^{1/2}. \tag{2.4}$$

where

$$V_r = V_0 + \left(\frac{e}{2m_0 c^2}\right) V_0^2$$

is the 'relativistic accelerating voltage', introduced as a convenience. For computer calculations discussed in later chapters the value of λ may be taken as

$$\lambda = 1.22639/(V_0 + 0.97845 \times 10^{-6} V_0^2)^{1/2} \tag{2.5}$$

with V_0 the microscope accelerating voltage in volts and λ in namometres.

The relativistic correction is important for high-voltage electron microscopy (HVEM). If V_0 is expressed in MeV, a good approximation is $V_r = V_0 + V_0^2$, so that $V_r = 6$ MeV for a 2 MeV microscope. The largest instruments currently available operate at 3 MeV. The formal justification for these definitions of a relativistically corrected electron mass and wavelength must be based on the Dirac equation, as first pointed out by A. Howie and K. Fujiwara (see Section 5.7).

A method for measuring the relativistically corrected electron wavelength directly from a diffraction pattern is discussed in Chapter 8. This method requires only a knowledge of a crystal structure and does not require the microscope accelerating voltage or camera length to be known.

The positive electrostatic potential ϕ_0 inside the microscope specimen further accelerates the incident fast electron, resulting in a small reduction in wavelength inside the specimen (Fig. 2.1). Ignoring the periodic variation of specimen potential, which gives rise to diffraction and the dispersion surface construction, the mean value of this inner potential is given by v_0, the zero-order Fourier coefficient of potential (see Section 5.3.2). A typical value of v_0 is 10 volts. The refractive index of a material for electrons is then given by the ratio of wavelength λ in vacuum to that inside the specimen λ'. Applying the principle of conservation of energy with careful regard to sign gives

$$n = \frac{\lambda}{\lambda'} = \left(\frac{1.23}{\sqrt{V_0}}\right)\left(\frac{\sqrt{V_0 + \phi_0}}{1.23}\right) \approx 1 + \frac{\phi_0}{2V_0}$$

The phase shift of a fast electron passing through a specimen of thickness t with respect to that of the vacuum wave is then

$$\theta = 2\pi(n - 1)t/\lambda = \pi\phi_0 t/\lambda V_0 = \sigma\phi_0 t$$

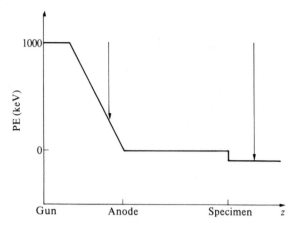

Fig. 2.1 Simplified potential energy diagram for an electron microscope. The length of the vertical arrow is proportional to the kinetic energy of the fast electron, and inversely proportional to the square of its wavelength. The sum of the electron's potential energy (represented by the height of the graph) and its kinetic energy is constant. Electrons leave the filament with low kinetic energy and high potential energy (supplied by the high-voltage set) and exchange this for kinetic energy on their way to the anode, which is at ground potential. As with a ball rolling down a hill, they are further accelerated as they 'fall in' to the specimen. Approximate distance down the microscope column is represented on the abscissa and the potential step at the specimen has been exaggerated.

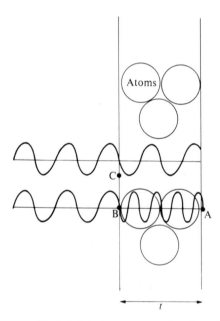

Fig. 2.2 The electron wave illustrated in two cases passing through a specimen. The wave passing through the centre of an atom (where the potential is high) has its wavelength reduced and so suffers a phase advance relative to the wave passing between the atoms which experiences little change in its wavelength. The assumption of this simplified model used in the phase grating approximation is that the amplitude at A can be calculated along the optical path AB with no contribution at A from a point such as C. For thick specimens this approximation is unsatisfactory.

as suggested in Fig. 2.2. Here $\sigma = \pi/\lambda V_0 = 2\pi me\lambda/h$. If the approximation is then made that the exit face wavefunction can be found by computing its phase along a single optical path such as AB in Fig. 2.2, the product $\phi_0 t$ can be replaced by the specimen potential function projected in the direction of the incident beam (see Section 3.4). The neglected contributions from paths such as CA can be included using the Feynman path-integral method as discussed in Jap and Glaeser (1978).

2.2 Simple lens properties

Modern electron microscopes use four or five imaging lenses, of variable focal length, below the specimen with the position of the object and final viewing screen fixed for the purposes of focusing. At the high magnifications usually used for high-resolution work, the lens currents (which determine the focal lengths) of lenses L2, L3, and L4 are used to control the magnification as shown in Fig. 2.3 and Table 2.1. For a fixed magnification setting, focusing is achieved by adjusting the strength of the objective lens

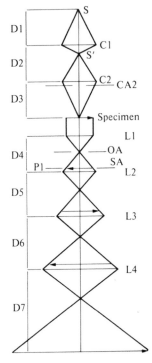

Fig. 2.3 Ray diagram for an electron microscope with two condenser lenses, C1 and C2 and four imaging lenses, L1, L2, L3, and L4, operating at high magnification. A typical set of dimensions for $D1$ to $D7$ is given in Table 2.1 for the JEM-100C, together with the possible range of focal lengths. These values may be used for examples throughout this book. Here OA is the objective aperture, P1 is a fixed plane, and SA is the selected area aperture.

Table 2.1 Electron-optical data for an electron microscope (JEM-100C). (See Fig. 2.3)

Distances Between Lens Centres (approximate)	Focal Length Range
$D1 = 143.6$ mm	$1.65 < f(C1) < 19$ m
$D2 = 94.3$ mm	$30 < f(C2) < 1060$ m
$D3 = 251.4$ mm	$15.4 < f(L2) < 281$ m
$D4 = 215.5$ mm	$3.1 < f(L3) < 99.5$ m
$D5 = 44.9$ mm	$2.06 < f(L4) < 16.4$ m
$D6 = 73.6$ mm	
$D7 = 345.6$ mm	

For magnifications greater than 100 000 the magnification is controlled adjusting the focal length of L3 with $f(L2) = 15.4$ mm fixed and $f(L4) = 2.1$ mm fixed. The focal length of L3 is set as follows: $f(L3) = 9.8, 7.0, 5.0, 3.1$ mm for $M = 150$ K, 200 K, 400 K, and 750 K respectively.

L1 until the fixed plane P1 is conjugate to the exit face of the specimen. Some of the properties of these lenses can be understood from the equations describing ideal lens behaviour given below, which also provide results used in later chapters.

The study of electron optics seeks to determine the conditions under which the electron wavefield passing through an electron lens satisfies the requirements for perfect image formation. For comparison purposes, it is convenient to set up the model of the ideal lens. The ideal lens is a mathematical abstraction which provides perfect imaging given by a projective transformation between the object and image space. The constants appearing in this transformation specify the positions of the cardinal planes of the lens. The six important cardinal planes are the two focus planes, the two principal planes, and the two nodal planes as shown in Fig. 2.4. For magnetic lenses the nodal planes coincide with the principal planes. The points where the axis crosses the nodal planes are called nodal points, N1 and N2. Principal planes are planes of unit lateral magnification, while nodal planes are planes of unit angular magnification. For an axially symmetric lens, the projective transformation for perfect imaging simplifies to

$$\frac{y_i}{y_0} = \frac{f_i}{x_0} = \frac{x_i}{f_0} \tag{2.6}$$

where the symbols are as defined in Fig. 2.4. The second of these equations is Newton's lens equation. Figure 2.4 provides a convenient graphical construction for eqn (2.6). Here F1 and F2 are known as the object and image focus respectively with H1 and H2 the object and image principal planes. The determination of the positions of these planes is the key problem of electron optics—once they are known the rules for graphical construction of figures satisfying eqn (2.6) can be used to find the image of an arbitrary object. The rule for a construction which gives the conjugate

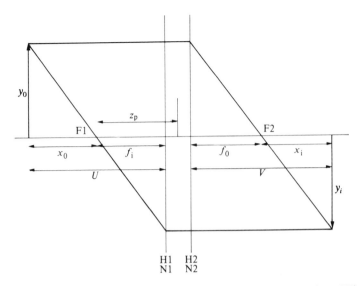

Fig. 2.4 The thick lens. The nodal planes (N1, N2), principal planes (H1, H2), and focal planes (F1, F2) are shown together with the lens focal lengths (f_i, f_0) and the object focal distance z_p. For a magnetic electron lens the principal planes are crossed.

image point of a known object point P is:

1. Draw a ray through P and F1, intersecting H1 at Q. Through Q draw a ray YQ parallel to the axis extending into both object and image spaces.
2. Draw a ray parallel to the axis through P to intersect H2. From this intersection draw a ray through F2 to intersect the ray YQ at P′. P′ is the image of P.

As an example of this construction, Fig. 2.5 shows these rules applied to the objective lens of a modern electron microscope operating at moderate magnification (about 40 000). Note that the image formed by the objective lens is virtual, and that the principal planes are crossed, as they are for all magnetic electron lenses. The use of this mode on a four-lens instrument has advantages for biological specimens where radiation damage must be minimized (see Section 6.9). At this moderate magnification lens L3 is switched off. At high magnification all lenses are used. An extensive discussion of the methods of graphical construction can be found in Conrady (1957). Modern lens designers use the methods of matrix optics; an elementary introduction to these techniques can be found in the book by Nussbaum (1968).

The simple thin-lens formula can still be used if the object and image distances U and V are measured from the lens principal planes H1 and H2.

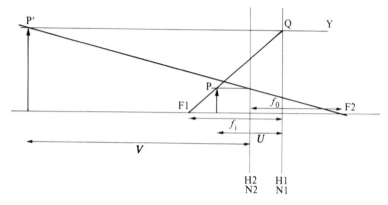

Fig. 2.5 Ray diagram for the objective lens of a microscope operating at moderate magnification. The image is virtual and the principal planes are crossed. Object and image focal lengths are equal for magnetic lenses. A typical value for f_2 is 2 mm, and the magnification $M = V/U$ may be about 20. This diagram would apply to the Phillips EM 300 microscope in the middle magnification range.

Equation (2.6) becomes

$$\frac{f_i}{U} + \frac{f_0}{V} = 1$$

If the refractive indices in the object and image space are equal, as they are for magnetic electron lenses, then

$$f_i = f_0 = f \tag{2.7}$$

and so

$$\frac{1}{U} + \frac{1}{V} = \frac{1}{f} \tag{2.8}$$

A construction can also be given to enable the continuation of a ray segment in the image space to be found if it is known in the object space. Note that a ray from P arriving at N1 at an arbitrary angle leaves N2 at the same angle. For a thin lens the principal planes coincide, so this is the ray through the lens origin. Equation (2.8) is quite general if the following sign convention is obeyed: U is positive (negative) when the object is to the left (right) of H1, V is positive (negative) when the image is to the right (left) of H2. Both focal lengths are positive for a convergent lens, and all magnetic lenses are convergent for electrons and positrons. From eqn (2.8) and the definition of magnification (eqn (2.9)) three cases emerge:

1. $U < f$. Image is virtual, erect, and magnified.
2. $f < U < 2f$. Image is real, inverted, and magnified.
3. $U > 2f$. Image is real, inverted, and reduced.

Some additional terms, commonly used in electron optics, are defined below.

1. The lateral magnification M is given by

$$M = \frac{y_i}{y_0} = -\frac{V}{U} \tag{2.9}$$

Using eqn (2.8) we have

$$M - 1 = -\frac{V}{f} \tag{2.10}$$

so that if V is fixed and the magnification is large, as it is for the objective lens of an electron microscope, the magnification is inversely proportional to the objective lens focal length. For high magnification, U must be slightly greater than f—both are about a millimetre for a high-resolution objective lens.

2. The angular magnification m is, for small angles,

$$m = \frac{\tan \theta_i}{\tan \theta_0} \approx \frac{\theta_i}{\theta_0} = \left| \frac{1}{M} \right| \tag{2.11a}$$

as shown in Fig. 2.6.

3. The entrance and exit pupil of an optical system are important in limiting its resolution and light-gathering power. These concepts also simplify the analysis of complex optical systems. The entrance pupil is defined as the image of that aperture, formed by the optical system which precedes it, which subtends the smallest angle at the object. The image of the entrance pupil formed by the whole system is known as the exit pupil. The 'aperture stop' is the physical aperture whose image forms the entrance pupil, as shown in Fig. 2.7.

4. The Gaussian reference sphere for an image point P is defined as the sphere, centred on P, which passes through the intersection of the optic axis

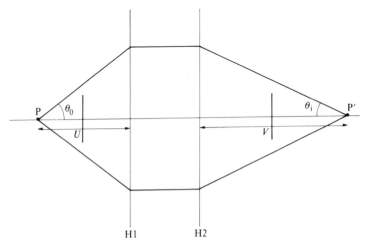

Fig. 2.6 Angular magnification. The image P' of a point P is shown together with the angles which a ray makes with these points.

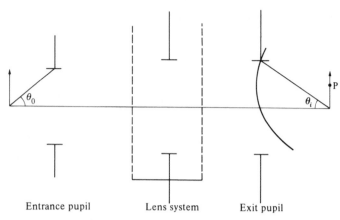

Fig. 2.7 The entrance and exit pupils of an optic system. A complicated optical system consisting of many lenses can be treated as a 'black box' and specified by its entrance and exit pupils and a complex transfer function. A Huygens spherical wavefront is shown converging to an image point P.

with the exit pupil (see Fig. 2.7). For an unaberrated optical system, the surface of constant phase for a Huygens spherical wavelet converging toward P coincides with this reference sphere. The deviation of the wavefront from the Gaussian reference sphere specifies the aberrations of the system (see Section 3.3), while the diffraction limit is imposed by the finite size of the exit pupil, or, equivalently, the entrance pupil. This is discussed further in Chapter 3.

5. The longitudinal magnification, M_z, can be used to relate depth of field to depth of focus (see below). Differentiation of eqn (2.8) gives

$$\frac{\Delta V}{\Delta U} = -M^2 = M_z \tag{2.11b}$$

Thus an object displacement ΔU causes a displacement ΔV of conjugate image planes given by this equation. For example the image planes conjugate to the upper and lower surfaces of an atom 0.3 nm 'thick' are separated by three metres if the lateral magnification M is 100 000.

6. Incoherent imaging theory gives the depth of field or range of focus values (referred to the object plane) over which an object point can be considered 'in focus' as $2d/\theta$, where θ is the objective aperture semi-angle and d is the microscope resolution. This result cannot, however, be accurately applied to the coherent high-resolution imaging of phase objects (see Section 3.4).

The methods used by lens designers to determine the position of the cardinal planes of electron lenses are discussed in the next sections. From Fig. 2.4 it is seen that the axis crossing of a ray entering (leaving) parallel to the axis defines the image (object) focus. In electron optics the trajectory of an electron entering the lens field parallel to the axis can similarly be used to find the lens focus once the equation of motion for the electron can be solved for a particular magnetic field distribution. Real electron trajectories follow smooth curves within the lens magnetic field. In order to use the ideal lens model it may be necessary to use the virtual extensions of a ray from a point well outside the influence of the field to define the lens focus.

2.3 The paraxial ray equation

The focusing action of an axially symmetric magnetic field can be understood as follows. Figure 2.8 shows a simplified diagram of an electron lens, including a typical line of magnetic flux. The actual arrangement used for one instrument (Siemens 102) in the important region of the lens pole-pieces is shown in Fig. 2.9. The dimensions of the pole-piece are S, R_1, and R_2 as shown in Fig. 2.8. The magnetic field B is confined to the pole-piece gap, where an electron of charge $-e$ and velocity v experience a force

$$F = -ev \times B = m\frac{\mathrm{d}^2r}{\mathrm{d}t^2} \tag{2.12}$$

The direction of F is given by the left-hand rule (current flow opposite to electron flow), which from Fig. 2.8 is seen to be into the page as the electron enters the field on the left (assuming the upper pole-piece is a North pole). An electron entering on the right side experiences a force out

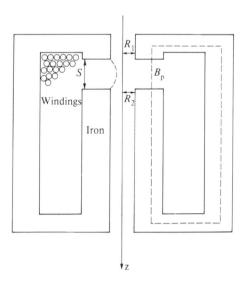

Fig. 2.8 Simplified diagram of a magnetic electron lens. The dimensions of the pole-piece are indicated and the path taken for Ampère's law is shown as a broken line. Also broken is shown a line of magnetic flux which gives a qualitative indication of the focusing action of the lens (see text). The field strength in the gap far from the optic axis is B_p. In practice the windings are water cooled and the pole-piece is removable.

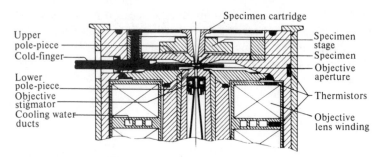

Fig. 2.9 Detail of an actual high-resolution pole-piece (Siemens 102). The upper pole-piece bore diameter $2R_1$ is 9 mm, the lower bore diameter $2R_2$ is 3 mm, and the gap S is 5 mm in this commercial design. (See Fig. 2.8 for the definitions of R_1, R_2, and S.)

of the page. These forces result in a helical rotation of the electron trajectory. The rotational velocity component r_θ imparted interacts with the z component of the field $B_z(r)$ to produce a force toward the axis, again given by the left-hand rule. This force is responsible for the focusing action of the lens.

Using certain approximations described in most optics texts, eqn (2.12) can be simplified for meridional rays in a cylindrical coordinate system. These are rays in a plane containing the axis before reaching the lens. The simplifications include the neglect of terms which would lead to imperfect imaging (lens aberrations). The paraxial ray equation which results contains only the z component of the magnetic field evaluated on the optic axis $B_z(z)$. This is

$$\frac{dr}{dz^2} + \frac{e}{8mV_r} B_z^2(z)r = 0 \qquad (2.13)$$

where r is the radial distance of an electron from the optic axis and V_r is the relativistically corrected accelerating voltage. Once the field $B_z(z)$ on the axis is known, computed solutions of eqn (2.13) can be used to trace the trajectory of an electron entering the field. For a symmetrical lens, a ray entering parallel to the axis will define all the electron lens parameters discussed in later sections. Note that the approximation has been made that the z component of the field does not depend on r.

Equation (2.13) is a linear differential equation of second order with two linearly independent solutions. It can be shown that these solutions describe rays which satisfy the conditions for perfect lens action. Using a modern computer one can also solve eqn (2.12) to find the electron trajectories for a particular magnetic field. By comparing these true electron trajectories with the idealized trajectories satisfying eqn (2.13) (and producing perfect imaging) it is possible to determine the aberrations of a magnetic lens. Alternatively, it is possible to use simple expressions for the various aberration coefficients which are given as functions of the solution to eqn (2.13) (see Section 2.8.1).

Similarly, the helical rotation of meridional rays can be found. The rotation of this plane is given by

$$\theta_0 = \left(\frac{e}{8mV_r}\right)^{1/2} \int_{z_1}^{z_2} B_z(z)\, dz. \tag{2.14}$$

Rays entering the lens in a given meridional plane remain in that plane which rotates through θ_0 as the rays traverse the lens field. Note that the total image rotation is $(180 + \theta_0)$ degrees on account of the image inversion (M is negative).

It can be seen that the lens shown in Fig. 2.8 is very inefficient since most of the power dissipated supports a field in the z direction which produces no force on the electron entering parallel to the axis. An octopole lens (see Section 2.8.3) while far more efficient, appears to have intractable practical problems associated with accurate alignment.

2.4   The constant-field approximation

A review of computing methods used in the solution of eqn (2.13) is given in Mulvey and Wallington (1972). However, eqn (2.13) is easily solved if the z dependence of the field can be neglected. It then resembles the differential equation for harmonic motion. The accuracy of this 'constant field' approximation has been investigated by Dugas, Durandeau, and Fert (1961), who found it to give good agreement with experiment for the focal lengths of projector lenses at moderate and weak excitation. It is included here for the physical insight it allows into magnetic lens action and to clarify the definition of the lens focal length. Since the expression for C_s (eqn (2.32)) involves derivatives of the field, we cannot expect to understand the influence of aberrations using such a crude model.

If the origin of coordinates is taken on a plane midway between the pole-piece gap (length S), then a field constant in the z direction is given by

$$B_z(z) = B_p \quad \text{for} \quad -S/2 \leqslant z \leqslant S/2$$
$$= 0 \quad \text{elsewhere} \tag{2.15}$$

Equation (2.13) becomes

$$\frac{d^2r}{dz^2} + k^2 r = 0 \tag{2.16}$$

with

$$k^2 = \left(\frac{e}{8m_0 V_r}\right) B_p^2 \tag{2.17}$$

that is,

$$k = 1.4827 \times 10^5 B_p / V_r^{1/2}$$

with B_p in teslas and V_r in volts. The solution to eqn (2.16) is

$$r = A \cos kz + B \sin kz \qquad (2.18)$$

where A and B are constants to be determined from the boundary conditions. Matching the slope and ordinate r_0 at $z = S/2$ for a ray which leaves the lens parallel to the axis gives

$$r = r_0 \cos k(z - S/2) \qquad (2.19)$$

The 'constant field' strength B_p can be related to the number of turns N and the lens current I using Ampère's circuital law. If the bore of the lens D is sufficiently small that it does not disturb the magnetic circuit and the reluctance of the iron is considered negligible compared to that of the gap S, then we have, for the circuit indicated in Fig. 2.8,

$$B_p = \frac{\mu_0 N I}{S} = 4\pi \times 10^{-7} \left(\frac{NI}{S} \right) \qquad (2.20)$$

with I in amperes and B_p in teslas. Note that B_p gives the flux density in any lens gap if measured sufficiently far from the optic axis.

2.5 Projector lenses

Lenses which use the image formed by a preceding lens as object, such as lenses L2, L3, and L4 of Fig. 2.3, are known as projector lenses. 'Intermediate' lenses fall in this category, to distinguish them from lenses which use a physical specimen as object.

It may happen that the image formed by, say, L2 falls within the lens field of L3. The image formed by L3 can nevertheless be found by the constructions of Section 2.2 if a virtual object is used for L3. This virtual object is the image formed by L2 with L3 removed. A similar procedure applies if the L2 image falls beyond the centre of L3.

The behaviour of real electron trajectories within the lens is given on a simple model by eqn (2.19). This equation is now used to give the focal length and principal plane position of an equivalent ideal lens. The image formed by a system of lenses can then be found by successive applications of the ideal lens construction.

The projector object focus f_p may fall inside or outside the lens field. The two cases are indicated in Figs 2.10(a) and 2.10(b). Notice that the extension of the asymptotic ray direction has been used for $z \to -\infty$ to define the 'virtual' or asymptotic projector focal length f_p. The distance between the principal plane H1 and the object focus in Fig. 2.10(a) is

$$f_p = r_0/\tan \theta = r_0 \Big/ \left(\frac{dr}{dz} \right)_{-\infty} = r_0 \Big/ \left(\frac{dr}{dz} \right)_{-S/2}$$

Using eqn (2.19) gives

$$f_p/S = [Sk \sin(Sk)]^{-1} \qquad (2.21)$$

(a)

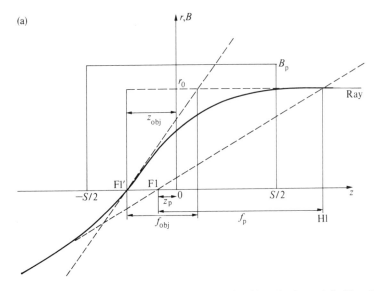

Fig. 2.10 (a) Definition of the objective and projector focal lengths f_{obj} and f_p. The objective and projector focal distances z_{obj} and z_p are also shown. The specimen for an objective lens is placed near F1'. This diagram shows the lens used both as an objective (with focus at F1') and as a projector (with focus F1) if the focus lies inside the lens field.

(b)

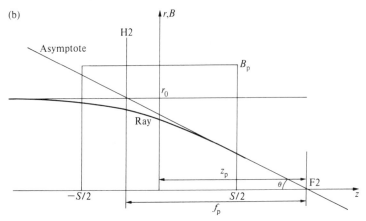

Fig. 2.10 (b) Ray diagram for a projector lens if the focus lies outside the lens field, which extends from $-S/2$ to $S/2$. The ordinate represents both the field strength and the distance of a ray from the optic axis. The image focus occurs at F2.

This function is plotted in Fig. 2.11, and shows the minimum focal length characteristic of projector lenses. Lenses are generally operated in the region $0 < Sk < 2$, though the properties of 'second zone' lenses with $Sk > 2$ have been investigated for high resolution (von Ardenne 1941). Equation (2.21) gives a minimum for $f_p = 0.56S$.

Using eqn (2.20) the abscissa for Fig. 2.11 is, under the constant-field

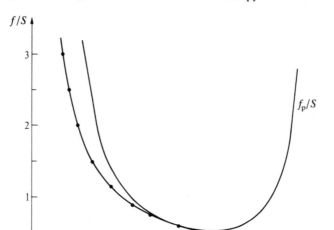

Fig. 2.11 Relative focal lengths of objective and projector lenses (f_{obj} and f_p) as a function of lens excitation (proportional to the lens current) for the constant-field model. The abscissa is defined in the text. The projector minimum focal length occurs at $f_p = 0.56S$, where S is the pole-piece gap size.

approximation,

$$Sk = 0.1862\left(\frac{NI}{\sqrt{V_r}}\right) \tag{2.22}$$

The object focal distance z_p is defined as the distance between the centre of the lens and the object focus. From Fig. 2.10(a) it is seen that, together with f_p this specifies the position of the principal plane. From the figure,

$$z_p = S/2 - r(-S/2)\bigg/\left(\frac{dr}{dz}\right)_{-S/2}$$

Using eqn (2.19) gives

$$\frac{z_p}{S} = \left[\frac{1}{2} - \frac{1}{kS \tan kS}\right] \tag{2.23}$$

for a projector lens. The sign is reversed for Fig. 2.10(b).

If the focal length is much larger than the lens gap ($f_p \gg S$) the lens is a weak lens such as C2 of Fig. 2.3. With $kS < 2$, eqn (2.21) gives

$$S/f_P = (Sk)^2 \tag{2.24}$$

which is the formula for a 'thin' magnetic lens when the distance between principal planes is neglected. A slightly more general form of this result was obtained by Busch (1926) in one of the first papers on electron optics. His result is obtained by integration of eqn (2.13) assuming r constant within the lens field. Then

$$\frac{1}{f_p} = \frac{e}{8mV_r} \int_{-\infty}^{\infty} H_z^2(z) \, dz \qquad (2.25)$$

which agrees with eqn (2.24) for a 'top-hat' field.

In an attempt to express the focal properties of all magnetic lenses on a single universal curve, Liebman (1955a) found that lenses of finite bore D could be described by a modified form of eqn (2.25). His result is

$$\frac{1}{f} = \frac{A_0(NI)^2}{V_r(S + D)} \qquad (2.26)$$

where A_0 is approximately constant. This is a good approximation for weakly excited lenses and can be extended to include asymmetrical lenses by taking

$$D = R_1 + R_2$$

Unfortunately, the properties of strong lenses become increasingly sensitive to the details of pole-piece geometry, making eqn (2.26) unreliable for many modern lenses. In particular, since the focal length of a modern objective lens is not large compared with the pole-piece dimensions, these lenses cannot be treated as 'thin' lenses.

2.6 The objective lens

The objective lens, which immediately follows the specimen, is the heart of the microscope. Since angular magnification is inversely proportional to lateral magnification (eqn (2.11a)), the magnification provided by this lens ensures that rays travel at very small angles to the optic axis in all subsequent lenses. We shall see that lens aberrations increase sharply with angle, so that it is the objective lens, in which rays make the largest angle with the optic axis, which determines the final quality of the image. Lens aberrations are specified by aberration constants C_c and C_s, which are very approximately equal to the lens focal length (Liebman 1955a). For high-resolution work, where the reduction of aberrations is important, the focal length and hence lens aberrations can be reduced by introducing the specimen into the field of the objective lens. This is known as an immersion lens. The field maintained on the illuminating side of the object is known as the pre-field (see Section 2.9) and plays no part in image formation. The extent of the remaining field available for image formation depends on the object position. Thus the position of the lens cardinal planes and aberration coefficients depend on the object position z_{obj}, which becomes an important electron-optical parameter for these lenses. It is measured from the centre of the lens. By convention, z_{obj} and the objective lens focal length f_{obj} are specified for a lens producing an image at infinity (infinite magnification) so that the specimen is at the exact object focus. Thus z_{obj} is equal to the object focal distance.

For the simple lens described by eqn (2.17) it is easy to show from symmetry that the image and object focal distances are equal. These are defined by rays entering and leaving the lens parallel to the axis respectively. The object focal distance can consequently be found from Fig. 2.10(a) as indicated on the figure. From the figure and eqn (2.19), the ordinate is zero when the cosine argument is $-\pi/2$. That is

$$\frac{z_{obj}}{S} = -\left(\frac{\pi}{2kS} - \frac{1}{2}\right) \tag{2.27}$$

The 'real' focal length f_{obj} is given by

$$f_{obj} = r_0 \bigg/ \left(\frac{dr}{dz}\right)_{z_{obj}}$$

that is

$$\frac{f_{obj}}{S} = (Sk)^{-1} \tag{2.28}$$

as shown in Fig. 2.11.

A comparison with eqn (2.21) shows that the focal length of a lens used as an objective may be shorter than its focal length when used as a projector lens. In practice a lower limit to the focal length is set by saturation of the lens pole-pieces.

Objective lens data such as those given in Section 2.7 include a plot of lens excitation against z_{obj}, both in suitable units. This is interpreted as follows. For a specified excitation, the ordinate gives the object position required to produce an image at infinity. This will be the approximate specimen position for a lens operating at high magnification. The object is then separated from the object principal plane by the focal length and is usually situated on the illuminating system side of the lens.

For an asymmetrical objective $(R_1 \neq R_2)$ operated at high excitation, the focal length depends on which of R_1 or R_2 is in the image space. For weaker lenses this is not so. Highly asymmetrical objectives are popular in modern microscopes since, for a top-entry stage in which the specimen is introduced from above, the specimen-change mechanism must be accommodated. A typical value of $2R_1$ is about 20 mm for a tilt stage, while $2R_2$ may be as small as 2 mm. The effect of pole-piece asymmetry is to increase the maximum value of $B_z(z)$, (R_1 and S held constant) and to shift the position of the maximum toward the smaller pole-piece (R_2). For $R_2 \to 0$ this maximum coincides with the vanishing pole-piece face. For real lenses $B_z(z)$ is a smooth peaked function and the important refractive effect of the lens occurs near the maximum of the field. Liebman (1955b) was able to show from analogue computations that this maximum value for an asymmetrical lens agrees within 1 per cent of the maximum for an equivalent symmetrical lens whose pole-pieces have diameter $D = R_1 + R_2$. The focal properties of a highly asymmetrical lens can be obtained by tracing two rays through the lens, one entering parallel to the axis and one leaving parallel to the axis. In practice a single ray traced through an equivalent symmetrical lens gives a good approximation to the lens focal length, but the equivalent lens should not be used for calculation of the spherical aberration constant C_s (see Section 2.8.1).

2.7 Practical lens design

Figure 2.12(a) shows the variation of measured aberration constants C_s and C_c for the JEOL UHP pole piece ($C_s \approx 0.7$ mm) as a function of specimen position. It can be seen that there is an optimum specimen position to minimize C_c, but that C_s steadily decreases with z_0, the specimen height. Here b_1 is the upper pole-piece diameter (the lower diameter is 2 mm).

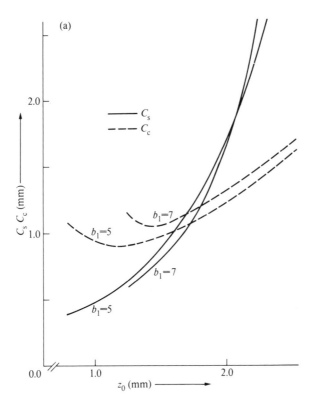

Fig. 2.12   (a) The variation of C_c and C_s with specimen position for the JEOL UHP pole-piece ($C_s = 0.7$ mm). Here z_0, the specimen position, is measured from the top of the lower pole-piece. The values of C_c and C_s are experimentally measured values, courtesy of Drs Shirota, Yonezawa, and Yanaka.

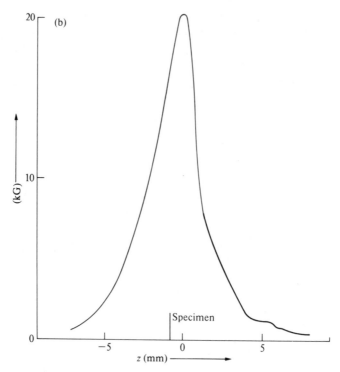

Fig. 2.12 (b) The measured lens field distribution, obtained by the image-rotation method. The position of the specimen, where most of the refractive effect of the lens occurs, is shown. The ordinate indicates the z component of the axial field strength in kilogauss.

Figure 2.12(b) shows the lens-field distribution, measured by observing the shadow-image rotation as an edge is moved down the optic axis.

Some computed curves, accurate to about 1 per cent, on which the design of real lenses has been based, are also shown in Figs 2.13–2.17. These are derived from both analogue (Liebman 1950) and digital (Mulvey and Wallington 1969) computations of the lens magnetic field. These figures give a useful impression of the way in which focal lengths and aberration constants vary with excitation and specimen position. A better approximation to the behaviour of real lenses can only be obtained if a more realistic form of $B_z(z)$ can be obtained, say from experimental measurements (see, for example, Engler and Parsons (1974)). These aberration coefficients were computed by a recurrence method described by Liebman (1949) in a form well suited to a modern computer program or programable calculator (equations 18(b) and 18(e) of Liebman's paper are incorrect and should be rederived from the earlier ones given).

The lens excitation NI in these curves is expressed in units of the excitation NI_A required for absolute minimum focal length. This minimum focal length determines the maximum magnification possible in a microscope. The unit NI_A is given in Table 2.2 for particular values of S/D, the

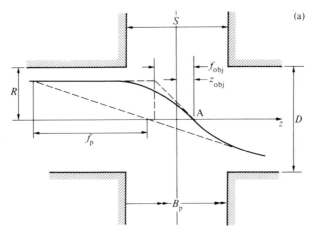

(a)

Fig. 2.13 (a) Definition of the lens dimensions for the data shown in the following figures.

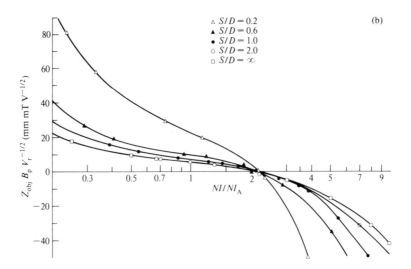

(b)

Fig. 2.13 (b) The object focal distance z_{obj} as a function of lens excitation. The excitation is given in units of the excitation needed for minimum projector focal length shown in Table 2.2. Here B_p is the saturation flux density. The object focal distance gives the specimen position in the lens needed to form a high (strictly infinite) magnification image.

ratio of lens gap to average bore size. For $S/D < 2$, the absolute minimum projector focal length is given by

$$f_p(\text{min}) = 0.55(S^2 + 0.56D^2)^{1/2}. \tag{2.29}$$

The ordinate for the figures is given in terms of B_p, the actual flux density in the lens gap. An upper limit to this is set by saturation of the pole-piece iron—a typical saturation flux density is 2.4 tesla (24 000 gauss) for Permendur, a commonly used type of iron.

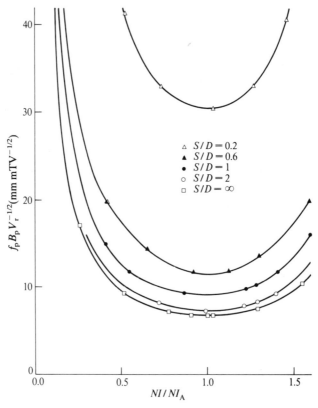

Fig. 2.14 The variation of projector focal length with lens excitation for various pole-piece proportions. The minimum projector focal length is seen to be almost independent of the ratio of the pole-piece gap to bore diameter. The units for the abscissa and ordinate are the same as those in Fig. 2.13.

From the curves we see that the spherical and chromatic aberration constants are of the same order as the lens focal length, and vary with lens excitation in a similar way. The best specimen position for minimum chromatic aberration is seen to be well to the illumination side of the centre of the lens, while spherical aberration is minimized by placing the specimen close to the centre of the lens. From eqns (2.29), (2.20), and (2.26) we see that a reduction in the lens gap increases the flux in the gap and reduces the lens focal length, unless it is otherwise limited. The overall instrumental magnification of an older microscope can be increased in this way by reducing the projector lens pole-piece gap or by increasing the current in the lens if the additional heat dissipated can be absorbed by the lens-cooling system, thus making the instrument more suitable for high-resolution work. The very large depth of focus at high magnification means that the overall magnification can be increased by reducing the projector focal length without altering the strength of other lenses. (See note on p. 59.)

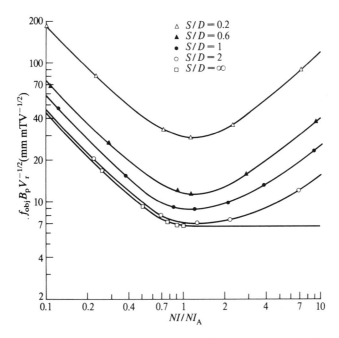

Fig. 2.15 Objective lens focal length as a function of lens excitation NI. Here I is the lens current in amps and N is the number of turns on the lens winding. NI_A is obtained from Table 2.2.

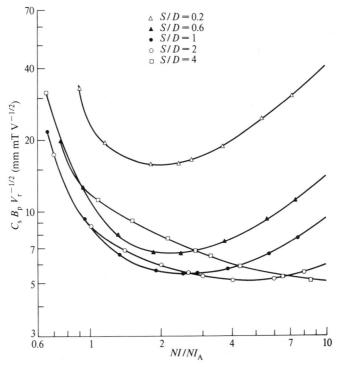

Fig. 2.16 The variation of the spherical aberration coefficient C_s with lens excitation for various pole-piece geometries. Once the excitation needed for minimum spherical aberration has been found from this figure, Fig. 2.13 may be used to determine the specimen position in the pole-piece.

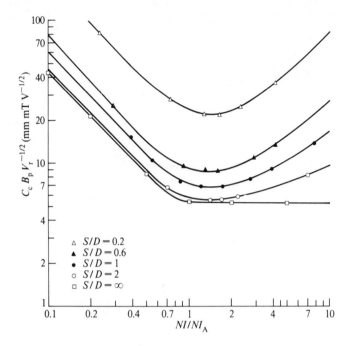

Fig. 2.17 Chromatic aberration coefficient C_c of various objective lenses as a function of lens excitation. The main source of chromatic aberration in high-resolution images of thin specimens ($t < 10$ nm) arises from the thermal energy spread of electrons leaving the filament (see Section 2.8.2).

Table 2.2 The unit of excitation NI_A used in Figs 2.13–2.17 for various ratios of lens gap to average bore size S/D

S/D	0	0.5	1.0	1.5	2.0	3.0	4	5	10	100	∞
$NI_A V_r^{-1/2}$	11.2	10.85	10.55	10.30	10.00	9.63	9.40	9.25	8.9	8.44	8.44

2.8   Aberrations

The consequences of including higher-order terms in the derivation of the ray equatin have been discussed by many workers (see, for example, Zworykin, Morton, Ramberg, Hillier, and Vance (1945)). The departure from perfect imaging which results from the inclusion of these terms has components which are classified by analogy with the corresponding aberrations in light optics. Retention of the term in r^3 is analogous to the inclusion of the term in θ^3 in the expansion of sin θ used in Snell's law applied to a glass lens. The result is that the third-order aberrations are generated, only one of which, spherical aberration, is important at high resolution.

Most high-resolution images are recorded at high magnification ($M >$

80 000). As discussed in Chapter 6, the use of moderate magnification ($15\,000 < M < 80\,000$) may have advantages for high-resolution experiments where radiation damage must be minimized. The magnification dependence of the aberrations of a system of lenses is not discussed here (see Hawkes 1970) and we restrict ourselves to the case in which the object is near the objective lens object focus. At high magnification, then, the quality of the final image is determined by the aberrations of the first lens (objective lens), since here the scattering angle (θ_0 of Fig. 2.4) is largest and the spherical aberration (ray aberration) depends on the cube of this angle. Since the images formed by subsequent lenses are larger, their angular spectra (diffraction patterns) are smaller.

The distinction between ray and wave aberrations is discussed in Chapter 3; the following sections deal only with the geometrical-optics treatment of aberrations. The power of θ_0 is increased by 1 for an aberration expressed as a wave aberration or phase shift.

The only other image defect important for high resolution is due to both electronic instabilities and beam–specimen interactions and is known as chromatic aberration. It is only on the most recent generation of instruments that other instrumental instabilities and aberrations can be made negligible, allowing these two image imperfections to limit resolution and contrast. The aberrations affecting the projector and intermediate lenses (the distortions and chromatic aberration of magnification) are unimportant for a small axial object at high resolution and are not discussed here.

An elegant treatment of aberrations makes use of Fermat's principle applied to the optical path $W(P_1, P_2)$ between two points P_1 and P_2. This is, with n the refractive index,

$$W(P_1, P_2) = \int_{P_1}^{P_2} n \, \mathrm{d}s$$

which, according to Fermat's principle, is a minimum or takes a stationary value. The function $W(P_1, P_2)$ is also known as Hamilton's point characteristic function. The expansion of this function as a power series allows terms to be identified which describe perfect imaging. Deviations from perfect imaging are expressed by terms of higher order. This theory goes far beyond the interests of high-resolution imaging; however, the following sections express the important results of aberration theory in a comprehensible way. Fuller treatments in the English language can be found in Septier (1967), Zworykin et al. (1945), Cosslett (1946), and Grivet (1965). A summary of the various methods which have been proposed or tested for eliminating aberrations using multipole lenses can be found in Reimer (1984). The use of one conventional lens and two sextupole lenses has been proposed by Crewe and Salzman (1982) as a means of eliminating spherical aberration. The performance of a STEM instrument based on this design would then be limited only by the fifth-order aperture aberrations, energy spread, diffraction, and tip vibration effects (Crewe 1984).

2.8.1 *Spherical aberration*

The effect of spherical aberration is indicated in Fig. 2.18. Rays leaving an axial object point at a large angle θ_0 are refracted too strongly by the outer zones of the lens and brought to a focus (the marginal focus for a ray which just strikes the aperture) before the Gaussian image plane. This is the plane satisfying eqn (2.8), where paraxial rays, obtained as solutions of the paraxial ray equation (2.13), are focused. The trajectories of aberrated rays can be obtained by computed solution of the full-ray equation. If all rays from an axial object point are considered, the image disc of smallest diameter is known as the circle of least confusion. The distance Δr_i in the image plane at high modification is found to be proportional to θ_0^3, the constant of proportionality being the third order spherical aberration constant C_s if Δr_i is referred to object space. That is

$$\Delta r_0 = C_s \theta_0^3 \qquad (2.30)$$

The radius of the circle of least confusion is $\frac{1}{4} C_s \theta_0^3$. The value of C_s depends on lens excitation and object position, but is usually quoted for an object close to the object focal plane, corresponding to the case of high magnification. On modern instruments, C_s ranges between about 0.5 and 2.5 mm and sets the limit to 'interpretable' resolving power. A value for C_s can be obtained either from computed solutions of the full (non-paraxial) ray equation or through the use of an expression given by Glaser (1956). Simple recurrence relations have been given by Liebman (1949) for the first method, expressing the deviation of the true trajectory from the paraxial trajectory at increments along the ray path. If the deviation at the point where a paraxial ray, which left the object parallel to the axis, crosses the axis is Δ, then the spherical aberration is given by

$$C_s = \Delta / \beta^3 \qquad (2.31)$$

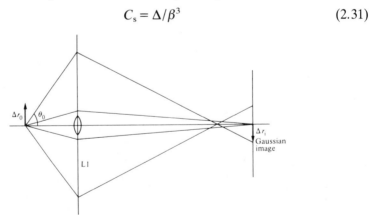

Fig. 2.18 The effect of spherical aberration. Rays passing through the outer zones of the lens (far from the axis) are refracted more strongly than those paraxial rays passing close to the axis. These outer rays meet the axis before the Gaussian image plane, and meet that plane a distance Δr_i from the axis. Spherical aberration (or aperture defect) is the most important defect affecting the quality of high-resolution images.

where β is the slope of the paraxial ray at the image focus. Once the deviations for two independent trajectories are known, all five Seidel aberrations are easily calculated.

The integral expression of Glaser (1956) is

$$C_s = \frac{e}{16 m_0 V_r} \int_{z_1}^{z_2} \left[\left\{ \frac{\partial B_z(z)}{\partial z} \right\}^2 + \frac{3e}{8 m V_r} B_z^4(z) - B_z^2(z) \left\{ \frac{h'(z)}{h(z)} \right\}^2 \right] h^4(z) \, dz \quad (2.32)$$

Here $h(z)$ is the *paraxial* trajectory of a ray which leaves an axial object point with unit slope. The occurrence of derivatives in eqn (2.32) makes the evaluation of C_s particularly sensitive to the shape of the field $B_z(z)$. For high resolution, where resolution is limited by C_s (see Chapter 3), eqn (2.32) shows that the detailed form of $B_z(z)$, and hence the shape of the pole-piece, sensitively determine the instrumental resolution. It is for this reason that the utmost care must be taken when handling the pole-piece. An expression similar to eqn (2.32) has been given by Scherzer (1936). The fifth-order spherical aberration term C_5 may, however, also be important for very high-resolution work. For example, it has been found that at 100 kV, with $C_s = 0.5$ mm, the effects of C_5 cannot be neglected for spacings smaller than 0.1 nm (Uchida and Fujimoto 1986).

Simple techniques for *in situ* experimental measurement of C_s for the objective lens of an electron microscope are given in Chapter 8. Spherical aberration is discussed from the point of view of wave optics in Chapter 3.

2.8.2 Chromatic aberration

Chromatic aberration arises from the dependence of the lens focal length on the wavelength of the radiation used, and hence on the electron energy. With polychromatic illumination, in-focus images are formed on a set of planes, one for each wavelength present in the illuminating radiation. There are three important sources of wavelength fluctuation in modern instruments. These are, in decreasing order of importance at 100 kV, for the thin specimens ($t < 10$ nm) used at high resolution:

1. The energy spread of electrons leaving the filament (see Chapter 7). At a high gun-bias setting, $\Delta E / E_0 = 10^{-5}$.
2. High-voltage instabilities. A typical fluctuation specification is $\Delta V_0 / V_0 = 2 \times 10^{-6}$/min with V_0 the microscope high voltage.
3. Energy losses in the specimen.

It is convenient to include fluctuations both in the high-voltage supply and in the objective lens current in the definition of the chromatic aberration constant C_c. Differentiation of eqn (2.26) gives

$$\frac{\Delta f}{f} = \frac{\Delta V_0}{V_0} - \frac{2 \Delta B}{B} = \frac{\Delta V_0}{V_0} - \frac{2 \Delta I}{I} \quad (2.33)$$

To allow for deviations from the thin-lens law (eqn (2.8)), a constant of proportionality K is introduced. The chromatic aberration constant C_c is defined as $C_c = Kf$, so that

$$\Delta f = C_c\left(\frac{\Delta V_0}{V_0} - \frac{2\,\Delta B}{B}\right) = C_c\left(\frac{\Delta V_0}{V_0} - \frac{2\,\Delta I}{I}\right) \tag{2.34}$$

In practice the fluctuations in lens current and high voltage are unlikely to be correlated, so that the random fluctuations in focal length should be obtained from the rules given for manipulation of variance in statistical theory. Adding the fluctuations in quadrature gives the variance

$$\frac{\sigma^2(f)}{f^2} = \frac{\sigma^2(V_0)}{V_0^2} + \frac{4\sigma^2(I)}{I^2} \tag{2.35}$$

The fluctuation in focal length, expressed by the standard deviation $\sigma(f)$, is given by

$$\frac{\sigma(f)}{f} = \left[\frac{\sigma^2(V_0)}{V_0^2} + \frac{4\sigma^2(I)}{I^2}\right]^{1/2} = \Delta/C_c \tag{2.36}$$

The variances in I and V_0 must be obtained by measurement, or their values may be deduced either by treating Δ as a fitting parameter for the image matching experiments described in Section 5.5, or from diffractogram analysis (see Section 8.10).

The effect on the image of a small change in focal length is now determined using the methods of geometric optics as a first approximation. As shown in Fig. 2.19, a ray at slightly higher energy, for which the lens has a longer focal length, is brought to a focus distance ΔV from the Gaussian image plane where

$$\Delta r_i = \Delta V \theta_i$$

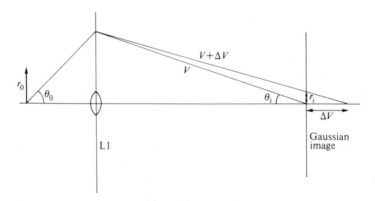

Fig. 2.19 The effect of chromatic aberration. The faster electrons which have been accelerated through a potential $V + \Delta V$ are less strongly refracted than lower electrons accelerated through a potential V. These higher energy electrons are thus brought to a focus beyond the Gaussian image plane, which they pass at a distance r_i from the axis.

Differentiating the lens equation gives

$$\frac{dV}{df} = \left(\frac{V}{f}\right)^2 \approx M^2$$

so that

$$\Delta V = M^2 \Delta f \qquad (2.37)$$

thus

$$\Delta r_i = M^2 \Delta f \theta_i$$

Referred to the object space, the extended disc image of an axial object point has radius

$$\Delta r_0 = M \Delta f \theta_i$$

Using eqn (2.11) this is

$$\Delta r_0 = \theta_0 \Delta f$$

From eqn (2.34) we have

$$\Delta r_0 = \theta_0 C_c \left(\frac{\Delta V_0}{V_0} - \frac{2 \Delta I}{I}\right) \qquad (2.38)$$

As an example, consider the image of a specimen which introduces a discrete energy loss of 25 eV. Many biological specimens show a rather broad peak at this energy. The loss image will be out of focus by an amount ΔU referred to the object plane. At high magnification, the lens law gives

$$\Delta U \approx \Delta f \qquad (2.39)$$

so that the focus defect for the loss image can be obtained from eqn (2.34) with the fluctuation in lens current neglected. For $C_c = 1.6$ mm and $\Delta E = 25$ eV, this gives $\Delta U = 400$ nm at 100 keV. Since the lens has a shorter focal length for the loss electrons, with V fixed an objective lens focused on the elastic image will have to be weakened to bring the inelastic image into focus. Thus inelastic loss images appear on the under-focused side of an image where the first Fresnel fringe (see Chapter 3) appears bright and the lens current is too weak. Unfortunately the optimum focus condition for high-resolution phase contrast is also on the under-focus side (about 90 nm at 100 kV) so that any inelastic contribution appears as an out-of-focus background blur in the elastic image. For many specimens the total inelastic image contribution may exceed the elastic contribution (see Chapter 6).

In practice C_c can be evaluated once the field distribution is known (Glaser 1952) using

$$C_c = \frac{e}{8mV_r} \int_{z_1}^{z_2} B_z^2(z) h^2(z) \, dz \qquad (2.40)$$

with $h(z)$ the function in eqn (2.32). The chromatic difference of magnification and chromatic difference of rotation are given by similar formulae; however, these are of no consequence at high resolution and are dealt with in many texts on electron optics. Chromatic aberration can be eliminated in principle by combining a magnetic lens with an electrostatic mirror, whose chromatic aberration is of the opposite sign.

The constant C_c can be measured experimentally from measurements of the focal length change with high voltage or lens current. A method is described in Chapter 8. A discussion of the effect of inelastic scattering on high-resolution images can be found in Chapters 5 and 6. Chromatic aberration is absent from STEM instruments if no lenses are used after the specimen. The representation of chromatic aberration in terms of wave optics is discussed in Chapter 3, and it should be stressed that eqn (2.38) is a geometrical approximation, accurate for the spatially incoherent imaging of point objects at moderate defocus, that is, for ΔV not too small. For small ΔV or coherent illumination (or both) the wave-optical methods of Chapter 3 must be used.

The effect of variations in C_c or Δ on high-resolution image detail is shown in Fig. 4.3. (See also Appendix 3.)

2.8.3   Astigmatism

Severe astigmatism is caused by an asymmetric magnetic field. A departure from perfect symmetry in the lens field can be represented by superimposing a weak cylindrical lens on a perfectly symmetrical lens. Figure 2.20 shows the image of a point formed by an astigmatic lens as two line foci at different points on the optic axis. Astigmatism can be thought of as an azimuthally dependent focal length from the form of eqn (3.22). *In situ* methods for measuring the astigmatism constant z_a are discussed in Chapter

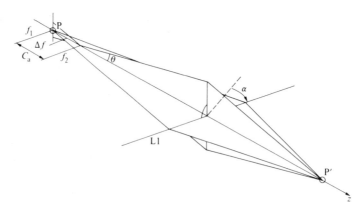

Fig. 2.20 Astigmatism. The focal length of the lens depends on the azimuthal angle α of a ray leaving the object. These rays are in a plane containing the optic axis. Planes at right angles for the maximum and minimum focal length are shown, with a mean focus Δf. The difference between the maximum and minimum focus is the astigmatism constant C_a.

8. An experienced operator using a modern instrument fitted with a stigmator for correcting astigmatism can reduce this aberration so that it does not limit resolution; however, this requires considerable practice (see Chapter 10). A stigmator is a weak quadruple lens whose excitation is controlled electronically to allow correction of field asymmetry in any direction by introducing a compensating weak cylindrical lens of the correct strength and orientation.

A machining tolerance of a few micrometres in the ovality of the objective lens pole-piece is required to bring astigmatism within easily correctable limits. The main bearing of the lathe used to cut the pole-piece is specially selected for this job. In consequence, a pole-piece which has been dropped is completely useless for high-resolution work. Generous compressible packing should be used when transporting a pole-piece. The lower surface is equally important, since scratches on the soft iron will disturb the magnetic circuit and introduce stray fields. The homogeneity of the iron used is also important. Since it is almost impossible to avoid scratching the lower face of a pole-piece if it is removed frequently, multi-purpose instruments should be avoided for high-resolution work. When removing a pole-piece, place it carefully on its side (and prevent it from rolling) or upside-down on lint-free cloth on the bench. Cover it with an inverted glass beaker so that it can be clearly seen.

Astigmatism is discussed from the point of view of wave optics in Chapter 3. The two forms of astigmatism present in a perfectly symmetric field can be neglected in high-resolution electron microscopy.

2.9   The pre-field

A comparison of Figs 2.13 and 2.16 shows that for an objective lens of minimum spherical aberration, the specimen will be placed well within the lens field. The field on the illuminating side of the specimen is known as the pre-field and has the effect of a weak lens placed before the specimen. Figure 2.21 shows a ray entering a 'constant-field' objective parallel to the axis with the lens excitation to the right of the specimen arranged to produce an image at infinity. These rays define both the pre-field focal length and the objective focal length. Expressions for the pre-field focal length, focal distance, and the demagnification of the incident beam by the pre-field have been given by Mulvey and Wallington (1972) using an analysis similar to that of Section 2.6. The significance of this for high resolution is that the incident-beam divergence, and hence the coherence length, may be affected by the strength of the pre-field lens. In practice, pre-field effects are included in a measurement of transverse coherence width if this is obtained from measurements of diffraction spot sizes as discussed in Section 8.8. It follows that for an immersion lens the illumination conditions are affected by specimen position and objective lens focusing. The modern trend has been toward the use of increasingly

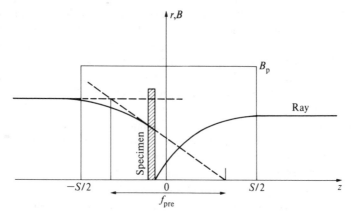

Fig. 2.21 The effect of the objective lens pre-field. The specimen is immersed in the lens field and a ray entering the field parallel to the axis is bent toward the axis, providing partially converged illumination. Diffraction spots appear larger than they would in the absence of the pre-field to the left of the specimen. The focal length of the pre-field f_{pre} is shown, which depends on the specimen position. A large pre-field is commonly used for high-resolution work since the shortest possible focal length is required to reduce aberrations.

symmetrical 'condenser-objective' lenses. These allow convenient switching from the HREM to the probe-forming mode for CBED work on the same specimen region (see Section 11.4). A ray diagram, showing in highly simplified form the operation of the symmetrical lens, can be found in Spence and Carpenter (1986).

2.10 Summary

This chapter has described the elementary properties of electron lenses. Since we are primarily concerned with electron interference effects similar to those revealed by Young's slit experiment (see Fig. 4.4), the chapter commences by giving a simple formula for the electron wavelength in terms of the electron microscope high voltage to emphasize that it is the wave properties rather than the particle properties of electrons which are of interest at high resolution. A technique for measuring this wavelength which does not require the high voltage to be known is outlined in Section 8.9. Just as light rays are bent by glass or water, so an electron wave feels the effect of the refractive index of a specimen. A simple formula for the electron refractive index of a specimen is also given in terms of its inner potential. In regions of high atomic potential—for example areas of heavy-atom stain—the refractive index is high, causing a reduction in the fast electron's wavelength and a phase advance relative to a wave passing through regions of lower atomic number or density (see Fig. 2.2). The trick of phase-contrast microscopy is to convert these phase variations into image-intensity variations (and so expose the specimen's inner potential) by a judicious choice of focusing 'error'.

In Section 2.2 the properties of ideal thin lenses are outlined. These give an immediate qualitative impression of the effects of changing lens currents, or the specimen's position. The ray-tracing methods of geometrical optics are shown applied to a modern electron microscope. The lens law and the definition of magnification are used to show the variation of magnification with objective lens focal length in the practical case of interest at high resolution where objective lens focal length changes are common. Using the lens-current metering described in Section 8.4, this magnification change with specimen height can be allowed for as described in Section 8.3. The reason for the very large depth of focus in an electron microscope is also explained.

In Sections 2.3–2.6 a simple explanation for the operation of electron lenses is given. The non-mathematical reader may wish to skip these sections; however, they do expain the reason for several lens properties of practical importance—their focusing effect, the distinction between projector and objective modes of operation, image rotation, the presence of a minimum focal length, and the possibility of 'second-zone' operation. Second-zone lenses are the subject of active research for high resolution. The focal length of an electron lens is shown to be inversely proportional to the square of the lens current and proportional to the electron accelerating voltage. Some lens curves are given in Section 2.7 to indicate the various compromises which must be made in designing electron microscopes for high resolution. For example, these curves show that a trade-off commonly exists between spherical aberration and chromatic aberration, so that the best specimen position and lens excitation for imaging ultra-thin specimens (where electron energy losses are small) may not also be the optimum conditions for imaging thicker specimens where the minimization of chromatic aberration may be paramount. The modification of an older instrument for high-resolution work by increasing the projector lens magnification is also mentioned.

Finally, the aberrations of electron lenses which are important at high resolution are discussed. The point is made that the aberration coefficients, which determine the quality of the final image, depend sensitively on the form of the lens magnetic field and so on the detailed shape of the pole-pieces. Any distortion of the pole-piece (caused, for example, by its being dropped) can, therefore, have a catastrophic effect on the performance of the microscope. The images formed by electrons which have lost energy in passage through the specimen are seen to occur on the under-focus side of exact focus—a lens focused on the 'elastic' electron image must therefore be weakened to bring them into focus. Practical methods for measuring aberration coefficients are discussed in Chapter 8. Finally, a ray diagram showing the effect of the objective lens pre-field is given. The presence of this field has important consequences for high-resolution imaging as described in Section 8.8.

References

Ardenne, M. von (1941). Zur Prufung von kurzbrennweitigen Elektronenlinsen. *Z. Phys.* **117**, 602.

Busch, H. (1926). Berechnung der Bahn von Kathodenstrahlen im axia symmetrischen elektromagnetischen Felde. *Ann. Phys.* **81**, 974.

Conrady, A. E. (1957). *Applied optics and optical design.* Dover, New York.

Cosslett, V. E. (1946). *Introduction to electron optics.* Oxford University Press.

Crewe, A. V. (1984). The sextupole corrector. 1: Algebraic calculations. *Optik* **69**, 24.

Crewe, A. and Salzman, D. B. (1982). On the optimum resolution for a corrected STEM. *Ultramicroscopy* **9**, 373.

Dietrich, I. (1977). *Superconducting electron optic devices.* Plenum, New York.

Dugas, J., Durandeau, P., and Fert, C. (1961). Lentilles electroniques magnetiques symetriques et dyssymetriques. *Rev. Opt.* **40**, 277.

Engler, P. E. and Parsons, D. F. (1974). The accurate determination of spherical aberration and focal length of an objective pole piece using a miniature magnetic field probe. *Optik* **41**, 309.

Glaser, W. (1952). *Grundlagen der Elektronenoptik.* Springer-Verlag, Berlin.

Glaser, W. (1956). In *Handbuch der Physik,* Vol. 33. Springer-Verlag, Berlin.

Grivet, P. (1965). *Electron optics.* Pergamon, London.

Hall, C. E. (1966). *Introduction to electron microscopy,* 2nd edn. McGraw-Hill, New York.

Hawkes, P. W. (1970). The addition of round lens aberrations. *Optik* **31**, 592.

Hawkes, P. W. (1972). *Electron optics and electron microscopy.* Taylor and Francis, London.

Jap, B. and Glaeser, R. (1978). The scattering of high energy electrons. I. Feynman path integral formulation. *Acta Crystallogr.* **A34**, 94.

Kamminga, W., Verster, J. L., and Francken, J. C. (1968/9). Design considerations for magnetic objective lenses with unsaturated pole pieces. *Optik* **28**, 442.

Liebmann, G. (1949). An improved method of numerical ray tracing through electron lenses. *Proc. Phys. Soc.* **B62**, 753.

Liebmann, G. (1950). Solution of partial differential equations with a resistance network analogue. *Br. J. Appl. Phys.* **1**, 92.

Liebmann, G. (1955a). A unified representation of magnetic electron lens properties. *Proc. Phys. Soc.* **B68**, 737.

Liebmann, G. (1955b). The field distribution in asymmetrical magnetic electron lenses. *Proc. Phys. Soc.* **B68**, 679.

Mulvey, T. and Wallington, M. J. (1969). The focal properties and aberrations of magnetic electron lenses. *J. Phys. E* **2**, 446.

Mulvey, T. and Wallington, M. J. (1972). Electron lenses. *Rep. Prog. Phys.* **36**, 348.

Nussbaum, A. (1968). *Geometric optics: an introduction.* Addison-Wesley, Reading, Mass.

Reimer, L. (1984). *Transmission electron microscopy.* Springer-Verlag, Berlin.

Scherzer, O. (1936). Uber einige Fehler von Elektronenlinsen. *Z. Phys.* **101**, 593.

Septier, A. (1967). *Focusing of charged particles,* Vol. 1. Academic Press, New York.

Spence, J. C. H. and Carpenter, R. A. (1986). Electron microdiffraction. In *Elements of analytical electron microscopy* (D. Joy, A. Romig, J. Hren, and H. Goldstein) Chap. 3, Plenum, New York).

Uchida, Y. and Fujimoto, F. (1986). Effect of higher order spherical aberration term on transfer function in electron microscopy. *Japan J. Appl. Phys.* **25**, 644.

Zworykin, V. K., Morton, G. A., Ramberg, E. G., Hillier, J., and Vance, A. W. (1945). *Electron optics and the electron microscope.* Wiley, New York; Chapman and Hall, London.

Bibliography

Perhaps the most useful introduction to electron optics, and one which stresses the practical considerations involved in the design of an instrument, is the book by M. E. Haine *The electron microscope* (Spon, London (1961)). The books by Hall and Grivet referenced above provide additional theoretical and practical information at a similar level.

Note

A complete set of computer programs developed by E. Munro for calculating the properties of electron lenses are available on request from the University of Cambridge (UK) Engineering Department librarian.

3

WAVE OPTICS

Recent improvements in instrumental stability and the development of brighter electron sources have enabled high-resolution images to be formed under highly coherent conditions. The microscope then becomes an interferometer and the out-of-focus image must be interpreted as a coherent interference pattern. The particle model of the fast electron, useful for the classical ray-tracing techniques described in the last chapter to determine simple lens properties, must then be replaced by a wave-optical theory taking account of the finite electron wavelength. A quantum-mechanical theory is required to do this correctly; however, this shows (Komrska and Lenc 1972) that the simple wave-optical methods give accurate results for electron interference and imaging experiments with a suitable change of wavelength and allowance for image rotation. Consequently, in this chapter, for the purposes of discussing interference effects the electron lens is simply replaced by an equivalent ideal thin lens whose wave-optical properties are discussed. The squared modulus of the complex wave amplitude $\psi(x, y)$ is then interpreted as a probability density (with dimensions Length to the power -3), rather than giving the time average of the amplitude of the electric field vector as in optics. The image current density in an electron microscope, formed from non-interacting fast electrons, is then proportional to $\psi(x, y)\psi^*(x, y)$, the electron 'intensity'.

For complicated optical systems it is convenient to use the concepts of entrance and exit pupil to account for the diffraction limit of the system. Aberrations are introduced as a phase shift across the exit pupil added to the spherical wavefront which would converge to an ideal image point. In electron microscopy the important aberrations for high resolution are confined to the objective lens; consequently the following sections treat a single lens only. Then the objective aperture forms the exit pupil and the intensity distribution across the objective lens back-focal plane is the intensity recorded in an electron diffraction pattern. In this chapter we are interested in finding both the back-focal plane amplitude and the image amplitude and to relate these to the wave leaving the object. The relationship between this complex amplitude and the specimen structure is a more difficult problem. Most of the work on image analysis has concentrated on retrieving an improved version of the object exit-face wavefunction, rather than relating this to the specimen structure. This second problem of structure analysis is discussed in Chapter 5, and in this chapter we are mainly concerned with understanding the way in which the microscope aberrations contribute to the image complex amplitude. The emphasis in this chapter is on interference effects, so that perfectly coherent illumination is assumed throughout.

The diffraction limit, an important 'aberration' for high-resolution images, is also discussed. The simplified scalar theories of Kirchhoff together with Sommerfeld's work showed that the early physical models for diffraction phenomena due to Huygens and Young are a useful basis for understanding diffraction. That is, that diffraction can be thought of as arising from interference between the wave transmitted through the aperture unobstructed and an 'edge wave' originating as a line source around the aperture rim. A similar model is sometimes applied to Fresnel diffraction at an edge. Huygens saw diffraction arising from the interference between an infinite number of fictitious secondary spherical waves constructed along a wavefront. The mathematical description of one of these spherical waves is the Green's function for the wave equation.

A more modern view of the diffraction limit is expressed by the Fourier optics approach and based on the Abbe theory and concepts taken from communications engineering in the early post-war years. In the jargon of electrical engineering, image formation under coherent illumination is band-limited, that is, the image is synthesized from a truncated Fourier series. A scattering angle θ is associated with each object periodicity d through Bragg's law and the image is synthesized from a Fourier summation of those beams which are included within the objective aperture. Only for the very thinnest specimens can these Fourier coefficients be used to represent a simple object property such as its projected electric potential or charge density. On the Fourier optics interpretation, the image formation consists of two stages—a Fourier transform from the specimen exit-face wavefunction giving the back-focal plane amplitude, followed by a second transform from this plane to reconstruct the image. An accessible account of Fourier electron optics can be found in the book by Hawkes (1972).

The subject matter of this chapter has been treated from slightly different points of view by many workers, including Lenz (1965), Hanszen (1971), and Misell (1973). The properties and wider applications of Fourier transforms and convolutions are also discussed in many books, including that by Bracewell (1965). A comprehensive modern account of optical aberration theory can be found in the book by Welford (1974).

3.1 Propagation and Fresnel diffraction

At high resolution the best images are not usually those obtained at exact Gaussian focus. Figure 3.1 shows an out-of-focus plane conjugate to the viewing phosphor which is imaged by an electron lens. In order to interpret out-of-focus images it is important to understand the propagation of the electron wave across the defocus distance Δf. The mathematical development of Huygen's principle, due to Fresnel and Kirchhoff gives a good approximation to the wave amplitude at P′ as

$$
\psi(x, y) = \frac{i}{\lambda \, \Delta f} \int\limits_{-\infty}^{\infty}\!\!\int \psi_e(X, Y) \exp\!\left(-\frac{2\pi i \, |r|}{\lambda}\right) \mathrm{d}X \, \mathrm{d}Y
$$

$$
\approx \frac{i \exp(-2\pi i \, \Delta f / \lambda)}{\lambda \, \Delta f} \int\limits_{-\infty}^{\infty}\!\!\int \psi_e(X, Y)
$$

$$
\times \exp\!\left\{-\frac{i\pi}{\lambda \, \Delta f}\left[(x - X)^2 + (y - Y)^2\right]\right\} \mathrm{d}X \, \mathrm{d}Y \qquad (3.1)
$$

The essential features of this important expression can be understood from Fig. 3.2, and a more detailed account of its derivation can be found in many optics texts (see, for example, Goodman (1968)). Equation (3.1) expresses the amplitude at P′ as the sum of the amplitudes due to an infinite number of fictitious secondary sources along $\psi_e(X, Y)$, each emitting a coherent spherical wave. The justification for using this expression in electron microscopy and the approximations involved (e.g., use of a scalar theory, neglect of backscattering) are discussed elsewhere (Cowley 1975). The scattering angles are all much smaller for electrons than for light. In eqn (3.1) $\psi_e(X, Y)$ is the two-dimensional complex amplitude across the exit face of the specimen. It is related to the incident wave $\psi(X, Y)$ by a

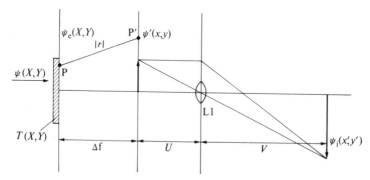

Fig. 3.1 The out-of-focus image ψ_i of an object with transmission function T(X, Y). The wave amplitude conjugate to the image plane is $\psi(x, y)$ and the incident wave $\psi(X, Y)$. For coherent imaging this is taken to have unit amplitude with the phase origin at the object.

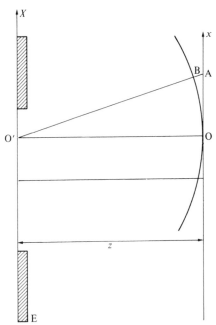

Fig. 3.2 A spherical wave (Huygens 'secondary source') emitted by a single point O' in the plane of a diffracting aperture E. The phase is constant along OB and the path difference $AB = (z^2 + (x - X)^2)^{1/2} - z \approx (x - X)^2/2z$ where x is the coordinate of A and X that of O'. The phase variation across Ox is thus $\psi(x) = \exp(-2\pi i AB/\lambda) = \exp(-i\pi(x - X)^2/\lambda z)$ due to a single point source at O'. Equation (3.1) expresses the total wavefunction across Ox as the sum of many such contributions from an object such as a transparency in the plane O'X, each weighted by the wavefunction $\psi_e(X, Y)$ leaving the object.

transmission function $T(X, Y)$, where

$$\psi_e(X, Y) = T(X, Y) \cdot \psi(X, Y) \tag{3.2}$$

Equation (3.1) can be used to compute the out-of-focus image of a specimen if the in-focus wave amplitude $\psi_e(X, Y)$ is known. For example, the Fresnel fringes seen at the edge of a partially absorbing specimen with finite refractive index have been calculated using this equation by Fukushima, Kawakatsu, and Fukami (1974). The anomalous difference in contrast between the over-focus and under-focus fringes, a subject of considerable research in the past, is investigated in this way. For an edge, $\psi_e(X, Y) = \psi_e(X)$ and so, performing the Y integration,

$$\psi'(x) = \frac{1 + i}{\sqrt{2}} \frac{\exp(-2\pi i \, \Delta f/\lambda)}{\sqrt{\lambda \, \Delta f}} \int_{-\infty}^{\infty} \psi_e(X) \exp\left\{-\frac{i\pi}{\lambda \, \Delta f}(x - X)^2\right\} dX \tag{3.3}$$

It is the real and imaginary parts of this integral, with the integration limits taken from 0 to a, say, which are represented by the Cornu spiral. While eqn (3.3) has been used to predict the form of Fresnel fringes in electron microscopy, recent work clearly shows that at high resolution both the properties of the edge-scattering material and the effect of spherial aberration must be considered. Dynamical calculations which include all these effects can be found in Wilson, Spargo, and Smith (1982). Similar problems arise in the simulation of HREM 'profile' images, as described in Section 12.8, and in the simulation of images of planar defects such as nitrogen platelets in diamond (see Section 12.2).

A convenient notation results if eqn (3.1) is written as a convolution. Then

$$\psi'(x, y) = A\psi_e(x, y) * \mathscr{P}_z(x, y)$$

$$= A \int_{-\infty}^{\infty} \psi_e(X, Y)\mathscr{P}_z(x - X, y - Y)\, dX\, dY \qquad (3.4)$$

Table 3.1 Properties of the Fresnel propagator $\mathscr{P}_z = (i/\lambda z)\exp(-i\pi(x^2 + y^2)/\lambda z)$. The wave amplitude a distance z from a plane on which it is known is obtained by convoluting the wave across that plane with $\mathscr{P}_z$. Notice that transmission through a thin lens is given by multiplication with $\mathscr{P}_{-f} = \mathscr{P}_f^*$. Property 1 shows that the effect of choosing an arbitrary intermediate plane A for the construction of Huygens' sources produces the same wave amplitude at B as would be obtained by a single propagation from 0 to B. Property 2 shows the effect of Fresnel propagation on a plane wave, while the third property expresses the spherical wave emitted by a point source. The final properties 4 and 5 can be used to show how the focal lengths of thin lenses in contact combine. These results are all obtained using standard integrals (a scaling constant has been omitted in 4 and 5)

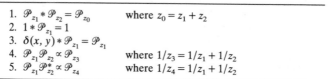

1. $\mathscr{P}_{z_1} * \mathscr{P}_{z_2} = \mathscr{P}_{z_0}$ where $z_0 = z_1 + z_2$
2. $1 * \mathscr{P}_{z_1} = 1$
3. $\delta(x, y) * \mathscr{P}_{z_1} = \mathscr{P}_{z_1}$
4. $\mathscr{P}_{z_1}\mathscr{P}_{z_2} \propto \mathscr{P}_{z_3}$ where $1/z_3 = 1/z_1 + 1/z_2$
5. $\mathscr{P}_{z_1}\mathscr{P}_{z_2}^* \propto \mathscr{P}_{z_4}$ where $1/z_4 = 1/z_1 + 1/z_2$

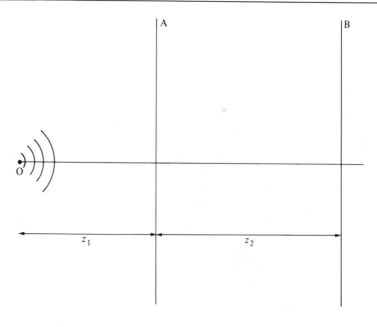

where A is a complex constant and

$$\mathcal{P}_z(x, y) = \exp\left(-\frac{i\pi(x^2 + y^2)}{\lambda \,\Delta f}\right) \tag{3.5}$$

is called the Fresnel propagator.

Thus the evolution of the electron wave amplitude in the near-field region important for out-of-focus images is described qualitatively by Huygens' principle and quantitatively, within approximations accurate for electrons, by eqn (3.4). A list of the properties of the propagation factor $\mathcal{P}_z(x)$ is given in Table 3.1. Note that two successive applications of the propagator leaves the wavefield unchanged, in support of Huygen's original physical intuition.

The next section discusses the image of a point formed by a lens of finite aperture. The aperture limitation results in a blurred image amplitude, known as the lens impulse response. From eqn (3.4) the amplitude at the entrance surface of a lens distance U from a point object is, in one dimension,

$$\psi'(x) = A\delta(x) * \mathcal{P}_u(x) = A \exp\left(-\frac{i\pi x^2}{\lambda U}\right) \tag{3.6}$$

which is a quadratic approximation to the wavefield of a spherical wave diverging from the object point. This is shown in Fig. 3.2, taking $X = 0$. For a converging spherical wave, the sign in the exponential is reversed.

3.2 Lens action and the diffraction limit

The analogy between electron and light optics suggests that the action of a magnetic lens is equivalent to the introduction of a 'focusing' phase shift $[2\pi/\lambda](x^2/2f)$, where f is the lens focal length and x the radial coordinate in the plane of the lens. The varying thickness of an ideal thin glass lens of suitable refractive index introduces such a phase shift. Thus, for a plane wave incident on such a lens (with the phase origin taken at the entrance to the lens) the emerging wave would be proportional to

$$\psi_e = \exp\left(\frac{\pi i x^2}{\lambda f}\right) \tag{3.7}$$

From Fig. 3.2 we see that this is a quadratic approximation to a spherical wave converging to the optic axis at $z = f$. Thus a lens which introduces the phase shift of eqn (3.7) produces a focusing action. For the diverging spherical wave given by eqn (3.6), the wave immediately behind the lens will be, in one dimension, (for a lens of infinite aperture),

$$\psi_e = A \exp\left(\frac{-i\pi x^2}{\lambda U}\right) \exp\left(\frac{i\pi x^2}{\lambda f}\right) = A \exp\left(\frac{i\pi x^2}{\lambda}\left[\frac{1}{f} - \frac{1}{U}\right]\right)$$

Now from the lens law, $1/f - 1/U = 1/V$, so that the wave emerging from this lens is a spherical wave converging to the conjugate image point P' at $z = V$, that is,

$$\psi_e = A \exp\left(\frac{i\pi x^2}{\lambda V}\right)$$

The effect of a limiting aperture on such a wave is now considered, as shown in Fig. 3.3. This aperture may be due to the lens itself or, as in electron optics, to the objective aperture situated near the back-focal plane. In electron microscopy the objective aperture is rarely exactly in the back-focal plane since the aperture position is fixed whereas the objective lens focal length is variable. Applying the Fresnel propagator from the plane of the aperture to the image gives

$$\psi(X) = AP(x) \exp\left(\frac{i\pi x^2}{\lambda d}\right) * \exp\left(-\frac{i\pi x^2}{\lambda d}\right)$$

$$= A' \int_{-\infty}^{\infty} P(x) \exp(2\pi i u x)\, dx \qquad (3.8)$$

where A' is a quadratic phase factor, $u = X/\lambda d$, and $P(x)$ is the exit pupil function, equal to unity within the aperture and zero elsewhere. Thus, regardless of the position of the aperture, the image of a point formed by an ideal lens is given by the Fourier transform of the limiting aperture pupil function. Fourier transforms are also used to describe far-field or Fraunhofer scattering. In this special case of an aperture illuminated by a coherent converging spherical wave, the intensity of the Fresnel diffraction pattern equals that of the Fraunhofer pattern. The image intensity for a point object using a circular aperture is shown in most optics texts. The phase contrast impulse response (point image) for an electron lens is further discussed in Section 3.4 and shown in Fig. 3.6. In passing, it is interesting to

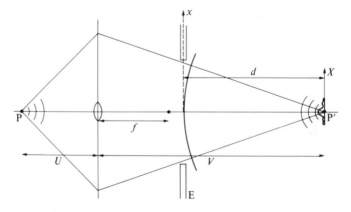

Fig. 3.3 The conjugate image P' of a point P. The image is broadened by the effect of the diffraction limit set by the aperture E. The amplitude at P' is proportional to the Fourier transform of the pupil function $P(x)$ describing the aperture.

note that the condition for Fraunhofer diffraction $(z < d^2/\lambda)$ with an object of size d would require a viewing screen to be placed more than 27 metres from a 10-micrometre objective aperture in order to observe the Fraunhofer pattern using 100 kV electron illumination and no lenses.

The imaging properties of simple lenses can also be understood using the Fresnel propagator to trace the progress of a wavefunction through the lens system. With unit amplitude coherent illumination on an object with transmission function $T(x, y)$, the wavefunction incident on the lens becomes

$$\psi_1(x_1, y_1) = \frac{i}{\lambda U} \psi_0(x, y) * \mathcal{P}_U(x, y) \qquad \text{where} \qquad \psi_0(x, y) = T(x, y)$$

while the wavefunction immediately beyond the lens is

$$\psi_2(x_2, y_2) = \psi_1(x_2, y_2) \exp\left\{\frac{i\pi(x_2^2 + y_2^2)}{\lambda f}\right\}$$

and the complex amplitude in the back-focal plane is given by

$$\psi_d(X, Y) = \frac{i}{\lambda f} \psi_2(x_2, y_2) * \mathcal{P}_f(x_2, y_2)$$

Evaluating these integrals and using eqn (3.5) gives

$$\psi_d(X, Y) = \frac{i}{\lambda f} \exp\left(-\frac{i\pi(X^2 + Y^2)}{\lambda} \left\{\frac{1}{d} - \frac{U}{f^2}\right\}\right)$$

$$\times \int\!\!\!\int_{-\infty}^{\infty} \psi_0(x, y) \exp\left(2\pi i\left\{\left(\frac{X}{f\lambda}\right)x + \left(\frac{Y}{f\lambda}\right)y\right\}\right) dx \, dy \qquad (3.9)$$

The quadratic phase factor disappears if the object is placed at $U = f$, as is approximately the case in electron microscopy. This important result shows that a function proportional to the Fourier transform of the object exit-face wave amplitude (the wave leaving the object) is found in the lens back-focal plane, in accordance with the Abbe interpretation of coherent imaging. All parallel rays leaving the object are gathered to a focus in the back-focal plane at a point distance X from the optic axis where $X = f\theta$, where θ is the angle at which rays leave the object. Figure 3.4 illustrates the Abbe interpretation for a periodic specimen where the diffraction pattern consists of a set of point amplitudes, if the illumination is coherent (small condenser aperture). These points can be thought of as a set of secondary sources like the holes in Young's pin-hole experiment which produce waves which interfere in the image plane to form fringes. The image is a Fourier synthesis of all these fringe systems. To confirm this mathematically, the back-focal plane amplitude of eqn (3.9) is written

$$\psi_d(X, Y) = \frac{Ai}{\lambda f} F(u, v)$$

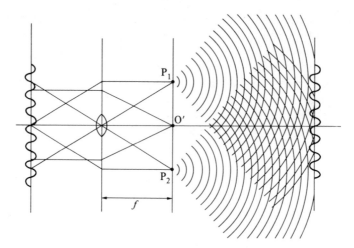

Fig. 3.4 The Abbe interpretation of imaging. An object with sinusoidal amplitude transmittance produces two diffracted orders in directions given by the Bragg law, together with the unscattered beam. These are focused to points in the back-focal plane. The image is formed from the interference between spherical waves from each of the three 'sources' in the back-focal plane.

where

$$F(u, v) = \iint\limits_{-\infty}^{\infty} \psi_0(x, y) \exp\{2\pi i(ux + vy)\} \, dx \, dy = \mathscr{F}\{\psi_0(x, y)\} \quad (3.10)$$

In order to use published tables of Fourier transforms it is convenient to introduce the variables $u = X/\lambda f = \theta/\lambda$ and $v = Y/\lambda f = \beta/\lambda$, where θ and β are angles made with the optic axis. An imperfect lens with an aberration phase shift $\chi(u, v)$ can be incorporated into this analysis by defining the modified back-focal plane amplitude

$$\psi_d'(X, Y) = \frac{Ai}{\lambda f} F(u, v) P(u, v) \exp(i\chi(u, v)) \quad (3.11)$$

where $P(u, v)$ is the objective aperture pupil function. Then the image amplitude is

$$\psi_i(x', y') = \frac{i}{\lambda z_0} \psi_d'(X, Y) * P_{z_0}(X, Y) \quad (3.12)$$

where $z_0 = V - f$ is the propagation distance between the back-focal plane and the image. From the lens law for conjugate planes this is $z_0 = Mf$, where M is the lateral magnification. Evaluating eqn (3.12) gives the image amplitude as

$$\psi_i(x', y') = -\frac{1}{M} \exp\left(-\frac{i\pi(x'^2 + y'^2)}{\lambda M f}\right) \iint\limits_{-\infty}^{\infty} F(u, v) P(u, v) \exp(i\chi(u, v))$$

$$\times \exp\left(2\pi i\left\{u\frac{x'}{M} + v\frac{y'}{M}\right\}\right) du \, dv \quad (3.13)$$

It is convenient to assume unit magnification in most image calculations. Then the image amplitude, aside from a quadratic phase factor, is seen to be given by the Fourier transform of the product of the back-focal plane amplitude $F(u, v)$ and the transfer function $P(u, v) \exp(i\chi(u, v))$. The back-focal plane amplitude is proportional to the Fourier transform of the object exit-face wave amplitude. These two Fourier transforms produce an inverted image, the sign of M being taken as positive.

Inserting the expression for $F(u, v)$ in eqn (3.13) gives finally for the image amplitude in an ideal lens with $P(u, v) \exp(i\chi(u, v)) = 1$

$$\psi_i(x', y') = \frac{1}{M} \exp\left(-\frac{i\pi(x'^2 + y'^2)}{\lambda f M}\right) \psi_0\left(-\frac{x'}{M}, -\frac{y'}{M}\right) \tag{3.14}$$

where the properties of the Dirac delta function have been used.

Equation (3.13) can be written symbolically at unit magnification,

$$\psi_i(x', y') = \mathscr{F}\{F(u, v)A(u, v)\} \tag{3.13}$$

where $A(u, v) = P(u, v) \exp(i\chi(u, v))$ is the microscope transfer function. Using the fourier transform convolution theorem this is

$$\psi_i(x', y') = \mathscr{F}\{F(u, v)\} * \mathscr{F}\{A(u, v)\} = \psi_0(-x', -y') * \mathscr{F}\{A(u, v)\} \tag{3.15}$$

showing that each point of the ideal image is smeared or broadened by the impulse response function $\mathscr{F}(A(u, v))$. This result applies only in a small isoplanatic patch near the optic axis where the form of $\mathscr{F}(A(u, v))$ does not depend on image position. The imaging is seen also to be linear in complex amplitude from the distributive properties of convolution (convolution is also commutative and associative). More generally $\psi_0(x', y')$ can be taken as the image predicted by geometrical optics, taking into account image inversions, magnification, and rotation.

3.3   Wave and ray aberrations

In this section the aberration phase shift $\chi(u, v)$ is investigated, commencing with the term arising from a small error in focusing.

Aside from a constant term, the wavefield imaged in Fig. 3.1 is, using eqn (3.1),

$$\psi'_{\Delta f}(x, y) = \psi_e(x, y) * \mathscr{P}_{\Delta f}(x, y)$$

with $\mathscr{P}_{\Delta f}(x, y)$ given by eqn (3.5). Using the convolution theorem for Fourier transforms the back-focal plane amplitude can be written

$$\psi_d(u, v) = \mathscr{F}\{\psi_e(x, y)\} \cdot \mathscr{F}\{\mathscr{P}_{\Delta f}(x, y)\}$$

A table of Fourier transforms then gives the transform of the Fresnel propagator as

$$\mathscr{F}\{\mathscr{P}_{\Delta f}(u, v)\} = \frac{i}{\lambda \, \Delta f} \iint \exp\left(-\frac{i\pi x^2}{\lambda \, \Delta f}\right) \exp\left(-\frac{i\pi y^2}{\lambda \, \Delta f}\right)$$

$$\times \exp(2\pi i(ux + vy)) \, \mathrm{d}x \, \mathrm{d}y = \exp(i\pi \, \Delta f \, \lambda(u^2 + v^2)) \tag{3.16}$$

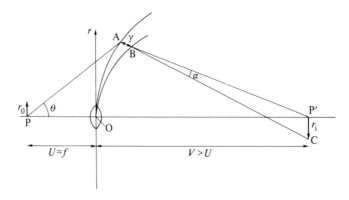

Fig. 3.5 Determination of the phase shift due to spherical aberration (see text). The curvature of the wavefronts has been exaggerated. The rays AP' and BC run normal to the surfaces OA and OB respectively.

which is a radially symmetric function of the scattering angle $\theta = (u^2 + v^2)^{1/2}\lambda$ and can be written $\exp(i\chi_1(\theta))$, where

$$\chi_1(\theta) = \frac{2\pi}{\lambda}(\Delta f\,\theta^2/2) \tag{3.17}$$

Spherical aberration can also be incorporated into the complex transfer function as follows. Figure 3.5 shows an aberrated ray whose deviation r_i in the Gaussian image plane has been obtained as a function of θ by, say, numerical solution of the full ray equation (see Section 2.8). Only the case of high magnification is of interest and we ignore any dependence of C_s on M. Then, from Section 2.8.1

$$r_i = MC_s\theta^3 \tag{3.18}$$

The rays drawn run normal to the surfaces of constant phase OA and OB. The surface OA is spherical, corresponding to the spherical wave converging from the exit pupil to an ideal image point P'. We require an expression for

$$\chi_2(\theta) = \frac{2\pi}{\lambda}\gamma$$

Now

$$r_i = \left(\frac{V}{f}\right)C_s\theta^3 = \frac{VC_s r^3}{f^4}$$

and

$$\alpha = \frac{d\gamma}{dr} \quad \text{with} \quad \alpha = \frac{r_i}{V} \tag{3.19}$$

so that

$$\gamma = \frac{1}{V}\int r_i\,dr = C_s\theta^4/4$$

giving $\chi_2(\theta) = (2\pi/\lambda)(C_s\theta^4/4)$. Combining this result with that obtained for defocus gives

$$A(\theta) = P(\theta) \exp\left(\frac{2\pi}{\lambda}(C_s\theta^4/4 + \Delta f\theta^2/2)\right) \qquad (3.20)$$

where a positive value of Δf implies an over-focused lens.

The focusing error Δf due to image formation by radiation of energy $V_0 - \Delta V_0$ must be considered with respect to the Gaussian image formed by radiation of energy V_0. The focusing action of a magnetic lens is greater for electrons of lower energy and longer wavelength. At high magnification $\Delta U \approx \Delta f$, so that eqns (2.34) and (3.17) give

$$\chi_3(\theta) = \frac{2\pi}{\lambda}\left[C_c\left(\frac{\Delta V_0}{V_0}\right)\theta^2/2\right] \qquad (3.21a)$$

where C_c is the chromatic aberration constant. To a good approximation the intensities of the various chromatically aberrated images can be added to form the final image, weighted by the normalized distribution function of incident electron energies (Misell 1973). Only for recording times less than $\Delta\tau = h/\Delta V_0$ should a coherent addition of image amplitudes be considered. A simple method of incorporating the effect of chromatic aberration into the electron microscope transfer function has been given by Fejes (1977). By assuming a Gaussian distribution for the spread in focus given above (there will also be a contribution from instabilities in the lens excitation current), he obtains a damping envelope applied to the transfer function of the form

$$\exp(-\tfrac{1}{2}\pi^2 \Delta^2\theta^4/\lambda^2) \qquad (3.21b)$$

This approximation can only be made if the central 'unscattered' beam is much stronger than any other. Here Δ is the standard deviation for the distribution of focus values, defined in eqn (2.36). The general case in which the central beam is not strong involves a convolution in reciprocal space. A typical value for Δ found to give good agreement between computed and experimental images is $\Delta = 1.2$ nm (JEM-100B).

A similar damping envelope has been derived to describe the way in which the use of a finite illumination angle θ_c attenuates the transmission of high-resolution detail (Spence, O'Keefe, and Kolar 1977; Wade and Frank 1977). For a uniformly filled illumination disc, Anstis and O'Keefe (unpublished) have obtained a damping function of the form

$$\frac{2J_1 |2\pi\theta_c(\Delta f\,\mathbf{K} + \lambda(\lambda C_s - i\pi\,\Delta^2)\mathbf{K}^3)|}{|2\pi\theta_c(\Delta f\mathbf{K} + \lambda(\lambda C_s - i\pi\,\Delta^2)\mathbf{K}^3|} = \frac{2J_1(\mathbf{q})}{\mathbf{q}} \qquad (3.22a)$$

where $J_1(x)$ is a Bessel function of the first order and kind and $\mathbf{K}$ is a vector in the back-focal plane. Again this function can only be used if the central beam is stronger than any other. Note that there is an interaction between the effects of the focus instability due to the incident electron energy spread and the illumination angle θ_c. This effect is discussed in a simpler way in Section 5.2, and a fuller discussion of damping envelopes can be found in Section 4.2 and in Appendix 3.

The azimuthally dependent focusing effect of astigmatism is similarly found to contribute a term (Fig. 2.20)

$$\chi_4(\theta) = \frac{2\pi}{\lambda}(\tfrac{1}{4}C_a\theta^2 \sin 2\phi) \tag{3.22b}$$

where ϕ is the azimuthal scattering angle. Collecting all these terms together gives a transfer function

$$A(\theta) = P(\theta)\exp\left(\frac{2\pi i}{\lambda}(\Delta f\,\theta^2/2 + C_s\theta^4/4 + \tfrac{1}{4}C_a\theta^2 \sin 2\phi)\right)$$

$$\times \exp(-\tfrac{1}{2}\pi^2\,\Delta^2\theta^4/\lambda^2)\frac{2J_1(\mathbf{q})}{\mathbf{q}} \tag{3.23}$$

Thus the effects of aberrations and instabilities in the microscope can be understood by imagining the scattering (diffraction pattern) observed in the object lens back-focal plane to be multiplied by this function. The last two terms attenuate the high-angle (large θ) scattering, which corresponds to the fine detail in the image. Thus, even if no objective aperture is used for lattice imaging, these terms impose a 'virtual aperture' as shown in Figs 4.11, 4.3, and A3.2.

In terms of the back-focal plane coordinates X and Y with $\mu = X/\lambda f$ and $v = Y/\lambda f$ the transfer function including defocus and spherical aberration effects only becomes

$$A(u, v) = P(u, v)\exp(2\pi i[\Delta f\lambda(u^2 + v^2)/2 + C_s\lambda^3(u^2 + v^2)^2/4])$$
$$= P(u, v)\exp(i\chi(u, v)) \tag{3.24}$$

3.4 Strong-phase and weak-phase objects

A highly simplified theory based on the weak-phase object and useful only for low- and medium-weight molecules and ultra-thin unstained biological specimens ($t < 5$ nm) has become popular owing to its convenience for *a posteriori* image analysis. The neglect of Fresnel diffraction (focus variation) within the specimen (but not of multiple scattering) allows the exit-face complex wave amplitude to be written

$$\psi_e(x, y) = \exp(-i\sigma\phi_p(x, y)) \tag{3.25}$$

as suggested by Fig. 2.2. Here $\sigma = 2\pi me\lambda_r/h^2$ (a positive quantity, with relativistically corrected values of λ_r and m) and $\phi_p = \int_{-t/2}^{t/2} \phi(x, y, z)\,\mathrm{d}z$ is the projected specimen potential in volt-nanometres. At 100 kV, $\sigma = 0.009244$. This strong-phase object takes the Ewald sphere to be a plane normal to the incident beam direction defined by the z-axis. It represents the complete N-beam solution to the dynamical scattering problem in the limit of infinite voltage (Moodie 1972).

The further approximation

$$\psi_e(x, y) \approx 1 - i\sigma\phi_p(x, y) \tag{3.26}$$

is known as the weak-phase object approximation and assumes kinematic scattering within the specimen. This requires that the central 'unscattered' beam be much stronger than the diffracted intensity. Like the strong-phase object, this approximation is a projection approximation—atoms could be moved vertically within the specimen without affecting the computed image. The potential $\phi_p(x, y)$ may be made complex with an imaginary component used to represent either depletion of the elastic wavefield by inelastic scattering or the presence of an objective aperture, as described in Appendix 3.

Leaving aside constants, the complex amplitude in the back-focal plane of the objective lens is given by the Fourier transform of eqn (3.26), that is, with $\phi_p(x, y)$ a real function,

$$\psi_d(u, v) = \delta(u, v) - i\sigma\mathscr{F}\{\phi_p(x, y)\} \tag{3.27a}$$

Introducing the microscope transfer function for a bright-field image formed with a central objective aperture then gives

$$\psi_d'(u, v) = \delta(u, v) - i\sigma\mathscr{F}\{\phi_p(x, y)\}P(u, v)\exp(i\chi(u, v)) \tag{3.27b}$$

A further Fourier transform then gives the image complex amplitude as

$$\psi_i(x, y) = 1 - i\sigma\phi_p(-x, -y) * \mathscr{F}\{P(u, v)\exp(i\chi(u, v))\} \tag{3.28}$$

Using the fact that since the sine and cosine of $\chi(u, v)$ are even functions, their transforms are both real and the image intensity can be found to first order as

$$I(x, y) = \psi_i(x, y)\psi_i^*(x, y) \approx 1 + 2\sigma\phi_p(-x, -y) * \mathscr{F}\{\sin \chi(u, v)P(u, v)\} \tag{3.29}$$

The function $\mathscr{F}\{\sin \chi(u, v)P(u, v)\}$ is negative and sharply peaked as shown in Fig. 3.6, so the bright-field image of a weak-phase object consists of dark detail on a bright background at the optimum value of Δf and correctly chosen $P(u, v)$, determine by the objective aperture size (see Section 6.2). On this simplest theory of image contrast, the contrast is proportional to the projected specimen potential, convoluted with the impulse response of the instrument. Where this is radially symmetric (no astigmatism), it can be written

$$\sigma\mathscr{F}\{\sin \chi(u, v)P(u, v)\} = \frac{2\pi}{\lambda^2} \int_0^{\theta_{ap}} \sin \chi(\theta) J_0\left(\frac{2\pi\theta r}{\lambda}\right)\theta \, d\theta \tag{3.30}$$

as shown in Fig. 3.6 for a modern high-resolution instrument. This function is laid down at every point in the phase-contrast image, and so limits the

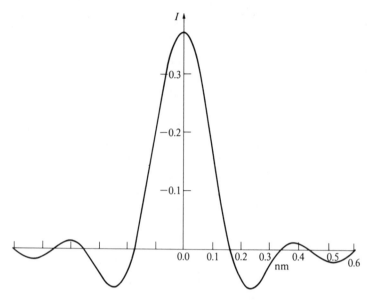

Fig. 3.6 The impulse response of an electron microscope for phase contrast. The parameters are $C_s = 0.7$ mm, 100 kV, $\theta_{ap} = 12.8$ mrad, $\Delta f = -61$ mm. This function, added to the constant bright-field background, gives the image of an ideal point object under the optimum imaging conditions for phase contrast. The full width at half maximum height of this function is 0.204 nm, which gives an indication of the resolution. A more accurate estimate of resolution would be based on the coherent superposition of two adjacent functions and take into account the scattering properties of the specimen and instrumental instabilities.

instrumental resolution. The parameters used for Fig. 3.6 are the optimum values of θ_{ap} and Δf for phase contrast on an instrument with $C_s = 0.7$ mm (Scherzer conditions—see Section 6.2).

Higher-order terms in the expansion of eqn (3.25) correspond to the multiple-scattering terms of the Born series. Their importance has been investigated by Erickson (1974) and detailed numerical calculations comparing these two approximations with the correct N-beam dynamical solution have also been given by Lynch, Moodie, and O'Keefe (1975). These workers found serious discrepancies for middle-weight atoms between the amplitudes and phases of diffracted beams calculated using the kinematic equation (3.26) and those calculated from the complete N-beam solution for specimens as thin as 1 nm. For a lighter element such as carbon, eqn (3.26) may hold for perhaps 6 nm. As has been pointed out, the validity of the weak-phase object approximation depends on the specimen structure and orientation. For crystalline specimens seen in projection with all atoms aligned in the beam direction, the phase change between an image point below a column of heavy atoms (see Fig. 2.2) and a neighbouring empty tunnel will increase rapidly with thickness. For a light amorphous specimen such as an unstained biological specimen, the local deviation of phase from the mean phase shift will increase less rapidly with thickness.

3.5 The optical bench

The use of coherent optical diffraction patterns of electron micrographs for biological image processing and electron microscope testing (see Section 8.7) has become increasingly popular. Since there are some dangers in the interpretation of these optical 'diffractograms', the relationship between the electron microscope specimen and a diffractogram of its electron image is outlined in this section.

We are concerned only with low-contrast specimens imaged in bright field. Then the electron image optical density recorded is

$$D(x, y) = D_0 + D_0 f(x, y) \tag{3.31}$$

where D_0 represents the density of the bright-field background, and $f(x, y)$ is the contrast term given either by

$$f(x, y) = 2\sigma\phi_p * \mathscr{F}\{\sin \chi(u, v) P(u, v)\}, \tag{3.32}$$

or

$$f(x, y) = \frac{\Delta f \lambda \sigma}{2\pi} \nabla^2 \phi_p(x, y), \tag{3.33}$$

depending on the precise experimental conditions and provided the specimen is sufficiently thin (see Chapter 5). The intensity of laser light transmitted by the micrograph in an optical bench is, from the definition of optical density,

$$I_e(x, y) = I_0 \exp(-D(x, y))$$

so that the amplitude immediately beyond the micrograph is

$$\begin{aligned} \psi_e(x, y) &= \sqrt{I_0} \exp(-\tfrac{1}{2}D(x, y)) \\ &= \sqrt{I_0} \exp(-D_0/2) \exp(-f(x, y)D_0/2) \\ &= C \exp(-f(x, y)D_0/2) \end{aligned}$$

where C is a constant.

For a dark-field micrograph $D_0 \approx 2$, and if the contrast is low, $f(x) < 1$ so that

$$\psi_e(x, y) \approx C - \tfrac{1}{2}D_0 C f(x, y)$$

The complex amplitude in the optical lens back-focal plane is given, aside from phase factors, by the Fourier transform of this,

$$\psi_d(u, v) = C\delta(u, v) - \tfrac{1}{2}D_0 C \mathscr{F}\{f(x, y)\} \tag{3.34}$$

with

$$u = \theta/\lambda_e, \qquad v = \phi/\lambda_e$$

where λ_e is the laser light wavelength. The intensity recorded on film (the diffractogram) is

$$I(u, v) = \psi_d \psi_d^* = B\delta(u, v) + \tfrac{1}{4}D_0^2 C^2 |\mathscr{F}\{f(x, y)\}|^2 \tag{3.35}$$

where B is a constant. For the weak-phase object (eqn (3.31))

$$|\mathscr{F}\{f(x, y)\}|^2 = 4\sigma^2 |\mathscr{F}\{\phi_p(x, y)\}|^2 \sin^2 \chi(u, v) P(u, v) \qquad (3.36)$$

If $|\mathscr{F}\{\phi_p(x, y)\}|^2$ is sufficiently slowly varying, the important structure seen will be the term $\sin^2 \chi(u, v)$, that is, the square of the transfer function shown in Fig. 6.1. This is the basis of the interpretation given to Thon's diffractograms (Section 8.7). The interpretation depends on two important conditions—low bright-field contrast and a monotonic form for $|\mathscr{F}\{\phi_p(x, y)\}|^2$.

Notice that the intensity of the electron diffraction pattern, given by the squared modulus of eqn (3.27), does not reveal the transfer function $\sin \chi(u, v)$. It does, however, reveal the form of $|\mathscr{F}\{\phi_p(x, y)\}|^2$, which for thin carbon films shows appreciable structure indicating the degree of order in these 'amorphous' specimens.

For thin, low-atomic-number periodic specimens (e.g., unstained catalase) we see from eqn (3.35) that the optical diffraction spots will appear in the same places (though with incorrectly weighted intensities) as the corresponding electron diffraction spots. This allows the possibility of optical filtering of electron micrographs under these conditions. Masks are used to exclude particular diffraction orders or the diffuse scattering between orders, thereby enhancing the contrast of the optically re-constructed image. A bright-field micrograph can be used to form a high-contrast dark-field optical image by excluding the central optical diffraction spot. The crucial requirement is that the specimen is thin enough for the central electron diffraction spot to be much more intense than any diffracted orders (kinematic conditions). In particular it is incorrect to interpret the optical diffraction spot intensities as if they were proportional to the corresponding electron diffraction pattern spot intensities. However, by carrying the analysis through to the optical image plane it can be shown that this image intensity is proportional to the electron micrograph density. Thus the image intensity recorded with an optical mask in place will be similar to the micrograph apart from the effects of the mask. The use of a mask corresponds to the introduction of *a priori* information, and results should be treated accordingly. For example it should be borne in mind that the optical image periodicity is determined by the positions of the holes in the mask. For a particular mask, an image will be formed whose periodicity is completely independent of the micrograph used, so long as it provides some scattering to illuminate the mask.

If an optical image of an 'almost periodic' micrograph (such as a damaged biological crystal) is formed using a back-focal plane mask which allows only the sharp Bragg spots to contribute to the image, a 'periodically averaged' image of the micrograph will be formed. This periodically averaged image is perfectly periodic, and each unit cell in this image is proportional to the sum of all the individual damaged unit cell images in the original micrograph, placed in registry. The reasoning behind this result, and a discussion of the effect of enlarging the mask holes, can be found in Section 6.9 and Fig.

6.19. A fuller discussion of the use of optical diffractograms is given in Beeston, Horne, and Markham (1972).

For dark-field electron micrographs, a simple interpretation of the optical diffraction pattern is not generally possible, since the density of the micrograph is proportional to the square of the deviation of the potential from the mean potential on a kinematic model (see Section 6.5).

The use of optical diffractions for electron microscope testing is described in Section 8.7.

3.6 Summary

The factors which complicate image interpretation at high resolution can be divided into two groups—those associated with the specimen (such as multiple scattering), and those arising from electron-optical imperfections. This chapter concerns this second group of resolution-limiting effects. Image processing is concerned with the removal of these effects by computer. Many of the equations on which this data processing is based are included in this chaper. Image processing is likely to occupy a larger place in high-resolution work in the future with the development of more efficient image sensors and inexpensive, fast, on-line digital computers.

The early sections are devoted to explaining why it is that the best images are recorded slightly out-of-focus at high resolution, and how the size of the objective aperture and the effect of spherical aberration limit image resolution. To illustrate this, the image of a point object is shown for a modern microscope as it would appear if such an object were available. The important point is made that a straightforward interpretation is possible for a greater thickness of light amorphous material than for crystalline specimens. Finally, some notes on the use of the optical bench are given. For image analysis, this technique is usually restricted to low-contrast bright-field images for which one has some certain *a priori* structural information.

References

Beeston, B. E. P., Horne, R. W., and Markham, R. (1972). Electron diffraction and optical diffraction techniques. *Practical methods in electron microscopy* (ed. A. M. Glauert), Vol. 1. North-Holland, Amsterdam.
Bracewell, R. (1965). *The Fourier transform and its applications*. McGraw-Hill, New York.
Cowley, J. M. (1975). *Diffraction physics*. North-Holland, Amsterdam.
Erickson, H. P. (1974). The Fourier transform of an electron micrograph—first order and second order theory of image formation. *Adv. Opt. Electron Microsc.* **5**, 163.
Fejes, P. L. (1973). Approximations for the calculation of high resolution electron microscope images of thin films. *Acta Crystallogr.* **A33**, 109.
Fukushima, K., Kawakatsu, H., and Fukami, A. (1974). Fresnel fringes in electron microscope images. *J. Phys. D* **7**, 257.
Goodman, J. W. (1968). *Introduction to Fourier optics*. McGraw-Hill, New York.

Hanszen, K. J. (1971). The optical transfer theory of the electron microscope: fundamental principles and applications. *Adv. Opt. Electron Microsc.* **4**, 1.

Hawkes, P. (1972). *Electron optics and electron microscopy.* Taylor and Francis, London.

Komrska, J. and Lenc, M. (1972). Wave mechanical approach to magnetic lenses. In *Proc. 5th Eur. Congress. Electron Microsc.*, p. 78. Institute of Physics, Bristol.

Lenz, F. (1965). The influence of lens imperfections on image formation. *Lab. Invest.* **14**, 70.

Lynch, D. F., Moodie, A. F. and O'Keefe, M. (1975). N-beam lattice images. V. The use of the charge density approximation in the interpretation of lattice images. *Acta Crystallogr.* **A31**, 300.

Misell, D. L. (1973). Image formation in the electron microscope with particular reference to the defects in electron-optical images. *Adv. Electron. Electron Phys.* **32**, 63.

Moodie, A. F. (1972). Reciprocity and shape functions in multiple scattering diagrams. *Z. Naturforsch.* **27a**, 437.

Spence, J. C. H. O'Keefe, M. A., and Kolar, H. (1977). High resolution image interpretation in crystalline germanium. *Optik* **49**, 307.

Wade, R. H. and Frank, J. (1977). Electron microscope transfer functions for partially coherent axial illumination. *Optik* **49**, 81.

Welford, W. T. (1974). Aberrations of the symmetrical optical system. Academic Press, London.

Wilson, A. R., Spargo, A. E. C., and Smith, D. J. (1982). The characterisation of instrumental parameters in the high resolution electron microscope. *Optik* **61**, 63.

Bibliography

Two excellent texts on imaging theory which emphasize ideas important for high-resolution electron microscopy are:

Martin, L. C. (1966). *The theory of the microscope.* American Elsevier/Blackie, London.

Goodman, J. W. (1968). *Introduction to Fourier optics.* McGraw-Hill, New York.

Sections of J. M. Cowley's book referenced above also deal with the material of this chapter. Treatments of Fourier electron optics can be found in the articles by Lenz, Hanszen, and Misell cited above.

4

COHERENCE

The coherence of a wavefield refers to its ability to produce interference effects. The high-resolution detail in an electron micrograph arises from coherent interference. In a bright-field image, for example, it is the interference between the central beam and the various waves scattered by the specimen which forms the image. So long as the resolution of the electron microscope was limited by electronic and mechanical instabilities to distances much larger than that coherently illuminated, the question of coherence remained unimportant. These incoherent instabilities no longer limit the resolution of modern electron microscopes and wave-optical interference controls the fine structure of a modern electron image.

Some of the important ideas of optical coherence theory are described below. This theory was developed in optics, but much of it has been found useful in electron optics and the validity of a fundamental optical coherence theorem (the van Cittert–Zernike theorem) has now been tested experimentally for electrons (Burge, Dainty, and Thom 1975). In optics, the waves emanating from different atomic oscillators in a light source are treated as incoherent. The total intensity in the interference patterns due to each atomic oscillator must be added together. Similarly, the wavefields of successive fast electrons emitted from the filament in the electron microscope are incoherent. In the words of Paul Dirac 'each electron interferes only with itself'. Now the image theory outlined in Chapter 3 was developed for a specimen illuminated by an idealized infinite electron wavetrain. Where many electrons, arriving from slightly different directions, are used to illuminate the specimen the image intensities due to each fast electron must be added together.

Some further qualitative ideas, described in more mathematical detail in the sections which follow, are set out below.

1. An effective source can be defined for an electron microscope. It lies in the exit pupil of the second condenser lens which is usually taken as coincident with the illuminating aperture. The effective source is an imaginary electron emitter filling the illuminating aperture. A mathematical definition is given in Hopkins (1957). Each point within the aperture is supposed to represent a point source of electrons. The emerging spherical wave is approximately plane at the specimen and this is focused to a point in the lens back-focal plane (Fig. 4.1). At the specimen, each electron can be specified by the direction of an incident plane wave. Increasing the size of the illuminating aperture increases the size of the central diffraction spot accordingly. The conditions under which the illuminating aperture may not be incoherently filled are discussed in Section 4.5.

2. A second important concept is that of spatial coherence width, also known as lateral or transverse coherence width. The term coherence length should be reserved for temporal or longitudinal coherence. The coherence width is the distance at the object over which the illuminating radiation may be treated as perfectly coherent. Thus the scattered waves from a specimen consisting of two atoms separated by less than this distance X_c will interfere and it is the complex amplitudes of these waves which must be added, in this case to produce a cosine-modulated atomic scattering factor. Atoms separated by distances much greater than X_c scatter incoherently and the intensities of their scattered radiation must be added. The intermediate range is described by the theory of partial coherence (see Fig. 4.2).

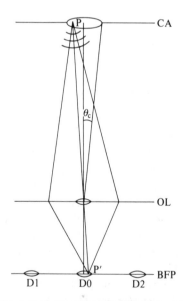

Fig. 4.1 Formation of the central 'unscattered' diffraction spot in an electron microscope. Each point P in the illuminating aperture CA is focused to a point P' in the central diffraction spot DO in the back-focal plane BFP of the objective lens OL. D1 and D2 are two other Bragg reflections. The beam divergence θ_c is also shown.

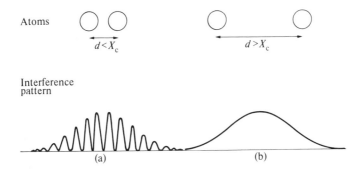

Fig. 4.2 The intensity of scattering recorded at a large distance from two atoms separated by (a) less than the coherence width X_c and (b) a distance much greater than X_c. There is no interference in the second case. The unscattered beam is not shown.

Under normal operating conditions there is a simple relationship between the coherence width X_c and the semi-angle θ_c subtended by the illuminating aperture at the specimen (the beam divergence). This result is given in Section 4.3 as

$$X_c = \lambda/2\pi\theta_c$$

The objective lens pre-field must also be considered (see Section 2.9).

3. The effect of coherence width or beam divergence on the contrast of phase-contrast images is discussed in Section 4.2. This is an important question since it enables the experimentalist to make the best choice of illuminating aperture size for a particular experiment. A strong phase object (one which shows only refractive index variation) such as the ultra-thin biological specimens used for high resolution produces little contrast unless coherent illumination is used. At the other extreme, source coherence becomes unimportant for the contrast of thick 'amorphous' specimens (such as thick biological specimens) since the phase relationship between scattered waves soon becomes indeterminate with increasing multiple scattering. An approach to the difficult problem of determining the effect of an almost randomly distributed set of atoms on the coherence of a plane wave is described in Sellar (1976). In the intermediate region the contrast of specimens of moderate thickness may be due to interference effects (phase contrast) at high resolution, but incoherent for the coarser image detail. The division between these two ranges of image detail is specified very approximately by the coherence width. Fresnel diffraction, lattice fringes, and single-atom images are three examples of high-resolution phase-contrast detail. To obtain this type of contrast a choice of θ_c must be made which keeps X_c larger than the coarsest detail of interest. Only a range of specimen detail smaller than X_c will produce phase contrast of the type described in Chapter 3 which can be enhanced by the Scherzer optimum focus technique, since this relies on Fresnel interference.

4.1 Independent electrons and computed images

The elastic scattering of a fast electron by a thin specimen is generally treated as two-body problem in which the specimen is described by a suitable complex optical potential and a solution is obtained for the wavefunction $\psi_0(\mathbf{r}_0, \mathbf{k}_i)$ of the fast electron (incident wavevector $\mathbf{k}_i$) on the specimen exit face. Successive fast electrons are assumed independent and any interaction between them (such as the Boersch effect) is neglected. We assign a separate wavevector and direction to each incident electron. Two electrons with the same wavevector would arrive at the specimen at different times. For an extended source, the intensity at a point in the final image $I(\mathbf{r}_i)$ can be obtained by summing the intensities of the images due to each fast electron. Thus

$$I(\mathbf{r}_i) = \int_{-\infty}^{\infty} \int_{-\infty}^{\infty} |\psi_i(\mathbf{r}_i, \Delta f, \mathbf{K}_0)|^2 \, F(\mathbf{K}_0) B(\Delta f) \, d\mathbf{K}_0 \, d\Delta f \qquad (4.1)$$

where $\mathbf{r}$ and $\mathbf{K}_0$ are two-dimensional vectors $(\mathbf{K}_0 = u\hat{\mathbf{i}} + v\hat{\mathbf{j}})$ and $F(\mathbf{K}_0)$ describes the normalized distribution of electron wavevectors. Thus $F(\mathbf{K}_0) \, d\mathbf{K}_0$ is the probability that the incident electron has a wavevector with $\hat{\mathbf{i}}$ and $\hat{\mathbf{j}}$ components in the range $\mathbf{K}_0$ to $\mathbf{K}_0 + d\mathbf{K}_0$. Here $\mathbf{k}_i = \mathbf{K}_0 + w\hat{\mathbf{k}}$ and $|\mathbf{k}_i| = 1/\lambda_r$ with $\hat{\mathbf{i}}$, $\hat{\mathbf{j}}$, and $\hat{\mathbf{k}}$ orthogonal unit vectors. $B(\Delta f)$ describes the distribution of energy present in the electron beam, and may also include effects due to fluctuations in the objective lens current. All these effects can be represented as time-dependent variations in the focus setting Δf.

It is shown in a later section (and can be demonstrated experimentally using an electron biprism) that to a good approximation the filled final illuminating aperture can be treated as a perfectly incoherent source of electrons. It is thus not necessary to trace each electron back to its source at the filament in using eqn (4.1). For practical computations, the illuminating cone of radiation under focused illumination may be taken as uniformly filled, corresponding to the choice of a 'top-hat' function for $F(\mathbf{K}_0)$ with $|\mathbf{K}_0|_{max} = |\mathbf{k}_i| \sin \theta_c$, where θ_c is the beam divergence. The exact profile of $F(\mathbf{K}_0)$ can be measured from a densitometer trace taken across the central diffraction spot. A slow emulsion must be used to avoid film saturation.

There are now three possible approaches to the problem of understanding and simulating partial coherence effects in HREM.

1. The images may be computed exactly, using eqn (4.1) and the result of a multiple-scattering computer calculation for $|\psi_i(\mathbf{r}_i, \mathbf{K})|^2$ (see Chapter 5). This method makes no approximation but requires a separate dynamical calculation for each component wavevector $\mathbf{K}_0$ in the incident cone of illumination. The results of such calculations can be found in O'Keefe and Sanders (1975) for lattice images (see Fig. 4.9), and in Holmes, Cockayne, and Ray (1974) for single-beam images.
2. To avoid the need for many dynamical calculations, an approximation

valid for small beam divergence may be adopted. This requires a single dynamical calculation for $\psi_i(\mathbf{r}_i, \mathbf{K}_0)$, and is described in Section 5.8.

3. In addition to assuming small θ_c, we may make the further weak-phase object approximation, in order to obtain a result in the form of a multiplicative transfer function. This is done below.

Under rather unusual experimental conditions (see Section 4.5) it may happen that the illuminating aperture is coherently filled. Then the complex image amplitudes for each incident direction must be summed, rather than their intensities. This occurs if a field-emission electron source is used for HREM work. The appropriate transfer function for this case has been derived by Humphreys and Spence (1981), and compared with the incoherent illumination case.

4.2   Coherent and incoherent images and the damping envelopes

The labour of detailed image simulation can be avoided and a simple understanding of the effect of coherence on contrast can be obtained for specimens sufficiently thin that the approximation

$$\psi_0(\mathbf{r}_0, \mathbf{K}_0) = \psi_0(\mathbf{r}_0, 0) \exp(2\pi i \mathbf{K}_0, \mathbf{r}_0) \qquad (4.2)$$

can be made. This approximation is satisfied for both the strong- and weak-phase object. Here $\psi_0(\mathbf{r}_0, 0)$ is the specimen exit-face wave for normally incident illumination. The approximation neglects the orientation dependence of scattering within the specimen, that is the rotation of the Ewald sphere with respect to the crystal lattice. The neglect of excitation error effects is equivalent to the neglect of Fresnel diffraction (describing Huygens wavelets) within the specimen if refractive index and propagation effects are taken as separable (see Section 6.3).

For specimens sufficiently thin that eqn (4.2) applies ($t < 5$ nm for low atomic number amorphous specimens), we now consider the two extremes of spatially coherent and incoherent illumination. For the present we ignore chromatic aberration effects and take $B(\Delta f) = \delta(\Delta f)$. The transfer equation for imaging is (Section 3.2)

$$\psi_i(\mathbf{r}_i, \mathbf{K}_0) = \int \psi_0(\mathbf{r}_0, \mathbf{K}_0) \bar{A}(\mathbf{r}_i - \mathbf{r}_0)\, d r_0 \qquad (4.3)$$

Using eqns (4.1), (4.2), and (4.3) gives

$$I(\mathbf{r}_i) = \int\int \psi_0(\mathbf{r}_0, 0)\psi_0^*(\mathbf{r}_0', 0)\bar{A}(\mathbf{r}_i - \mathbf{r}_0)\bar{A}^*(\mathbf{r}_i - \mathbf{r}_0')\gamma(\mathbf{r}_0' + \mathbf{r}_0)\, d r_0\, d r_0' \qquad (4.4)$$

where

$$\gamma(\mathbf{r}_0) = \int F(\mathbf{K}_0) \exp(-\pi i \mathbf{K}_0 \cdot \mathbf{r}_0)\, dk \qquad (4.5)$$

The function $\gamma(\mathbf{r}_0)$, if normalized, is known as the complex degree of coherence as will be discussed in Section 4.3.

For coherent illumination, $F(\mathbf{K}_0) = \delta(\mathbf{K}_0)$ and $\gamma(\mathbf{r}_0) = 1$, so that eqn (4.4) becomes

$$I(\mathbf{r}_i) = \left| \int \psi_0(\mathbf{r}_0, 0) \bar{A}(\mathbf{r}_i - \mathbf{r}_0) \, d r_0 \right|^2 \tag{4.6}$$

in agreement with the image intensity given by eqn (4.3) for normal plane-wave illumination. For perfectly incoherent illumination, $F(\mathbf{K}_0)$ is constant and $\gamma(\mathbf{r}_0) = \delta(\mathbf{r}_0)$ and we obtain

$$I(\mathbf{r}_i) = \int |\psi_0(\mathbf{r}_0, 0)|^2 \, |\bar{A}(\mathbf{r}_i - \mathbf{r}_0)|^2 \, d r_0 \tag{4.7}$$

For a pure-phase object, as discussed in Section 1.1 and used as a model for ultra-thin biological specimens, we have

$$\psi_0(\mathbf{r}_0, 0) = \exp(-i\sigma\phi_p(\mathbf{r}_0))$$

where $\phi_p(\mathbf{r})$ is real. Since the squared modulus of this function is unity, eqn (4.7) indicates that no contrast is possible from such a specimen using perfectly incoherent illumination (see also Section 1.2). In practice one is always dealing with partially coherent illumination, since a perfectly incoherent imaging system would require an illumination aperture of infinite diameter.

Loosely speaking then, using a very small condenser aperture in the electron microscope is rather like using a laser source in optics, while a large aperture corresponds to a tungsten lamp source.

The contrast of thin specimens is found to depend sensitively on the illumination coherence conditions. The idea of using this effect to advantage has recently been tried by Japanese workers (Nagata, Matsuda, and Komoda 1975). The distracting grainy background in thin amorphous carbon films is due to Fresnel interference and makes the identification of heavy atoms supported on the film difficult (see Section 10.6 for a discussion of substrates). This 'Fresnel noise' disappears under incoherent illumination. The hope is that a degree of coherence can be found which maximizes the contrast between heavy atoms supported by an amorphous carbon film and the background due to the film itself.

For specimens satisfying eqn (4.2) the transfer of information in the electron microscope is thus linear in complex amplitude under coherent illumination and linear in intensity under incoherent illumination.

We now consider the important intermediate case of partial coherence with which microscopists are chiefly concerned in practice. The incorporation of the effects of partial spatial and temporal coherence into the transfer function described earlier (eqn (3.24)) has been described by many workers in recent years. Results for an incoherently filled disc-shaped effective source have been briefly described in Section 3.3 (see also Appendix 3). We will rely mainly on the work of Frank (1973) and Fejes (1977) which gives a useful qualitative estimate of the likely effects of partial coherence on images for the simpler case of a Gaussian distribution of intensity across the effective source. For small effective source widths and a central zero-order diffracted beam much stronger than any other, these workers have shown that the combined effects of partial coherence and electronic instabilities lead to a transfer function of the form

$$A(\mathbf{K}) = P(\mathbf{K}) \exp\{i\chi(\mathbf{K}) \exp\{-\pi^2\Delta^2\lambda^2\mathbf{K}^4/2\} \gamma(\nabla\chi/2\pi)$$
$$= P(\mathbf{K}) \exp\{i\chi(\mathbf{K})\} \exp\{-\pi^2\Delta^2\lambda^2\mathbf{K}^4/2\} \exp\{-\pi^2 u_0^2\mathbf{Q}\} \qquad (4.8a)$$

in the absence of astigmatism. As discussed earlier in Sections 3.3 and 2.8.2, the quantities $\chi(\mathbf{K})$ and Δ are defined by

$$\chi(\mathbf{K}) = \pi \, \Delta f\lambda\mathbf{K}^2 + \pi C_s\lambda^3\mathbf{K}^4/2 \qquad (4.8b)$$

with $\mathbf{K}$ the vector $u\hat{\mathbf{i}} + v\hat{\mathbf{j}}$ where $(u^2 + v^2)^{1/2} = |\mathbf{K}| = \theta/\lambda$, and

$$\Delta = C_c Q = C_c \left[\frac{\sigma^2(V_0)}{V_0^2} + \frac{4\sigma^2(I)}{I_0^2} + \frac{\sigma^2(E_0)}{E_0^2} \right]^{1/2} \qquad (4.9)$$

where $\sigma^2(V_0)$ and $\sigma^2(I)$ are the variances in the statistically independent fluctuations of accelerating voltage V_0 and objective lens current I_0 respectively. The root-mean-square value of the high-voltage fluctuation is thus equal to the standard deviation $\sigma(V_0) = [\sigma^2(V_0)]^{1/2}$. A term has also been added to account for the energy distribution of electrons leaving the filament. The full width at half maximum height of the energy distribution of electrons leaving the filament is

$$\Delta E = 2.345\sigma(E_0) = 2.345[\sigma^2(E_0)]^{1/2}$$

The normalized Gaussian distribution of intensity assumed for the incoherent effective electron source has the form

$$F(\mathbf{K}_0) = \left(\frac{1}{\pi u_0^2} \right) \exp\left(-\frac{\mathbf{K}_0^2}{u_0^2} \right)$$

If the beam divergence is chosen as the angular half-width θ_c for which this distribution falls to half its maximum value, then u_0 is defined by

$$\theta_c = \lambda u_0 (\ln 2)^{1/2}$$

The quantity $\mathbf{q}$ in eqn (4.8) is

$$\mathbf{q} = (C_s\lambda^3\mathbf{K}^3 + \Delta\lambda f\mathbf{K})^2 + (\pi^2\lambda^4\Delta^4\mathbf{K}^6 - 2\pi^4 i\lambda^3\Delta^2\mathbf{K}^3) \qquad (4.10)$$

A clear understanding of the properties of eqn (4.8) is essential for the microscopist seriously interested in ultra-high-resolution work, since it expresses all the resolution-limiting factors of practical importance (except specimen movement). The relative importance of these factors is discussed in more detail in Appendix 3. For the present, we note the following features of eqn (4.8).

1. A crucial approximation needed to obtain it is the assumption of a strong zero-order diffracted beam. This condition is satisfied both in very thin crystals and in thicker areas of wedge-shaped crystals showing crystals showing strong Pendellösung, as described in Section 5.6. Note that terms such as those containing $\Phi_h\Phi_{-h}$ in eqn (5.9) which lead to the appearance of 'half-period fringes' are neglected in this analysis. These fringes have been used in the past to give a misleading impression of high-resolution detail (see Section 8.10).

2. The last bracketed complex term in eqn (4.10) expresses a coupling between the effects of using a finite incident beam divergence angle θ_c (partial spatial coherence) and the consequences of using a non-monochromatic electron beam (partial temporal coherence, $\Delta \neq 0$). The magnitude of this coupling term has been investigated in detail by Wade and Frank (1977), who find that, under high-resolution conditions (e.g., $\theta_c < 0.001$ rad, $\Delta < 20$ nm at 100 kV—their paper should be consulted for other conditions) this term can frequently be neglected. Then eqn (4.8) contains three multiplicative factors, each of which imposes a resolution limit by attenuating high-order spatial frequencies. The first term $P(\mathbf{K})$ expresses the diffraction limit imposed by the objective aperture. The third term describes a damping envelope more severe than Gaussian attenuation with a width

$$u_0(\Delta) = [2/(\pi\lambda\Delta)]^{1/2} \tag{4.11}$$

which will *always* be present even if the objective aperture is removed. This resolution limit $d \approx 1/u_0(\Delta)$ is independent of the illumination conditions used and depends on the existence of instabilities in the objective lens and high-voltage supplies and on the thermal spread of electron energies. The last term in eqn (4.8) shows an apparently complicated dependence on illumination semi-angle θ_c, focus Δf, spherical aberration constant C_s, and wavelength λ. Its behaviour can be given a simple interpretation, however, since the function $\gamma(\nabla\chi/2\pi)$ is just the Fourier transform of the source intensity distribution evaluated with the function's argument equal to the local slope of the aberration function $\chi(\mathbf{K})$. For a Gaussian source, $\gamma(\mathbf{K})$ is also Gaussian and has a width which is inversely proportional to the width of the source. Thus $\gamma(\nabla\chi/2\pi)$ is small in regions where the slope of $\chi(\mathbf{K})$ is large, resulting in severe attenuation of these spatial frequencies. Conversely, in the neighbourhood of regions where the slope of $\chi(\mathbf{K})$ is small, all spatial frequencies are well transmitted by the microscope with high contrast.

Extended regions over which the slope of $\chi(\mathbf{K})$ is small are called passbands or contrast transfer intervals and these can be found for many focus settings, given by

$$\Delta f_n = [C_s\lambda(8n + 3)/2]^{1/2} \tag{4.12}$$

This result may be obtained as follows. By differentiation, it is easily shown that the slope of $\chi(\mathbf{K})$ is zero at K_1 for the corresponding 'stationary phase' focus $\Delta f_0 = -C_s \lambda^2 K_1^2$ (see Fig. A3.1). We require that K_1 lie at the centre of the passband, in order to minimize the damping effects of limited spatial coherence. As a separate condition, however, we also require $\chi = n\pi/2$, with $n = -1, -5, -9, -13$, etc., for good phase contrast (see Section 3.4). Then both the scattering phase shift $\exp(-i\pi/2) = -i$ in eqn (3.26) and the lens phase shift $\exp(-i\pi/2)$ (see eqn (3.27b)) have the same sign, as needed to obtain a high-contrast image which is darker in regions of high potential. We might therefore impose the additional condition that

$$\chi(\Delta f_0) = -\frac{\pi}{2}(1, 5, 9, 13, \ldots)$$

in order to select only negative maxima in $\sin \chi$ for the centre of the passband. However, the passband can be made broader if the value of $\sin \chi(K_1)$ is allowed to decrease slightly as shown by the dip in the passband of Fig. 4.3(b). This is achieved by taking

$$\chi(\Delta f_0) = -\frac{\pi}{2}\left(\frac{8n+3}{2}\right) = -\pi C_s \lambda^3 K^4/2$$

Solving this for K and using this value for K_1 in the stationary phase focus expression gives eqn (4.12). This procedure guarantees both that the slope of $\chi(\mathbf{K})$ is zero (as sketched in Fig. A3.1) and that $\sin \chi = -1$ in the middle of the passband. The zero-order passband ($n = 0$) is commonly known as the 'Scherzer focus' and is the optimum choice of focus for images of defects or single molecules for which a straightforward interpretation in terms of object structure is required (see Section 6.2). Some examples of these passbands are shown in Fig. 4.3. They are seen to move out toward higher spatial frequencies with increasing n and, as discussed in Appendix 3 in more detail, to become narrower with increasing n and C_s. Once the slope of $\chi(\mathbf{K})$ exceeds a certain value beyond these passbands, all spatial frequencies are severely attenuated, and this attenuation is the major consequence of using a cone of illumination to illuminate the specimen from an extended incoherent source. By collecting several images at, say, the $n = 0, 1, 2, 3$ focus values specified by eqn (4.12) and processing these by computer, a composite image can be built up using only the well-transmitted spatial frequencies within the passband from each image, and this idea is the basis of many image-processing schemes. These passbands cannot, however, be moved out beyond the resolution limit set by electronic instabilities (eqn (4.11)). Figure 4.3 shows the transfer function drawn out for a typical modern instrument with $C_s = 2.2$ mm, $\Delta = 120$ Å, $\theta_c = 0.9$ mrad at an operating voltage of 100 kV. To a good approximation the functions shown can be taken to be the last two terms of eqn (4.8) multiplied by $\sin \chi(|\mathbf{K}|)$ with the second, bracketed term in eqn (4.10) set equal to zero (see Appendix 3).

There are, therefore, two resolution limits which can be quoted for an electron microscope. The first, generally called the point-resolution of the

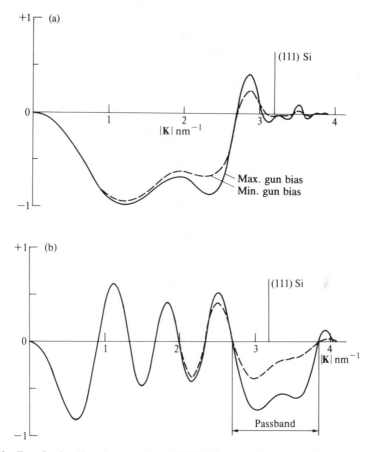

Fig. 4.3 Transfer functions for a modern 100 kV electron microscope with $C_s = 2.2$ mm and beam divergence $\theta_c = 0.9$ mrad. The cases $n = 0$ (Scherzer focus, $\Delta f = -110.4$ nm) and $n = 3$ ($\Delta f = -331.5$ nm) of eqn (4.12) are shown in (a) and (b) respectively. In (a) the 'passband' extends from $u = 0$ out to the point-resolution limit of the instrument, in (b) the passband has moved out to the position indicated. The solid curves are drawn for maximum gun-bias setting ($\Delta = 5.4$ nm) and the dotted curves show the effect of using the minimum gun-bias setting (maximum beam current $\Delta = 12.0$ nm) resulting in increased attenuation of the higher spatial frequencies. Note that the value of n is equal to the number of minima which precede the passband. The position of the (111) Bragg reflection is indicated and is seen to fall beyond the point-resolution limit (see Section 5.8). The imaginary part of eqn (4.8) has been plotted.

instrument, is set by the first zero crossing of the transfer function at the Scherzer focus ($n = 0$ in eqn (4.12)). This is the useful resolution limit of the instrument for the analysis of defects and other non-periodic specimens. On high-voltage machines, the stability-resolution limit (eqn (4.11)) may occur at a lower spatial frequency than the Scherzer cut-off (see eqn (6.17)), in which case eqn (4.11) would determine the point-resolution of the machine (see Appendix 3).

The second resolution specification for an instrument might be called the information-resolution limit. It is set by electronic instabilities and given by eqn (4.11). For the current generation of 100 kV machines this resolution limit generously exceeds the point-resolution, which is chiefly limited by spherical aberration. The information-resolution limit expresses the highest-resolution detail which could be extracted from a micrograph by the methods of image processing (leaving aside problems of electron noise) and can be measured by the Young's fringe diffractogram technique of Frank (see Section 8.7) or by finding the finest three-beam lattice fringes from a perfect crystal which the instrument is capable of recording under axial, *kinematic* conditions (but see Section 5.8). If there is no diffuse scattering between the Bragg reflections, a focus setting can then be found which places one of the passbands of eqn (4.11) across the Bragg reflection of interest. Defects and non-periodic detail in such an image cannot usually be simply interpreted (see Sections 5.8 and 8.10).

4.3 The characterization of coherence

The extent to which the wavefield at neighbouring points on the object vibrates in unison is expressed naturally by the correlation between wave amplitudes at points r_1 and r_2 and is given by the cross-correlation function

$$\Gamma(|r_1 - r_2|, T) = \lim_{\tau \to \infty} \int_{-\tau}^{\tau} \psi^*(r_1, t)\psi(r_2, t + T)\, dt \qquad (4.13)$$

A spatially stationary field has been assumed. When normalized, this function is called the complex degree of coherence $\gamma(x_{1,2}, T)$. Here $x_{1,2} = |r_1 - r_2|$. The function contains a spatial dependence expressing lateral or transverse coherence and a time dependence expressing temporal or longitudinal coherence. In electron microscopy, the temporal coherence is large and we are chiefly concerned with $\gamma(x_{1,2}, 0) = \gamma(x_{1,2})$. In order to obtain strong interference effects such as Bragg scattering from adjacent scattering centres we require the wavefield at these points to be well correlated. That is, that $\gamma(x_{1,2})$ is large for this value of $x_{1,2}$.

The van Cittert–Zernike theorem relates $\gamma(x_{1,2})$ through a Fourier transform to the function $F(k)$ used in Sections 4.1 and 4.2. Despite differences in the nature of the particles (photons are bosons, electrons are fermions) and differing interpretations of the wavefunction, the results of electron interference experiments suggest that this important optical theorem may be taken over into electron optics. It will be seen that the range of object spacings which can be considered coherently illuminated is proportional to the width of $\gamma(x_{1,2})$, so that a narrow source (for which $\gamma(x_{1,2})$ is a broad function) produces more coherent radiation than does a larger source. The theorem only applies to perfectly incoherent sources.

While the effects of partial coherence are important for images, they are seen most dramatically in interference experiments. A familiar example of a near-field interference experiment is the observation of Fresnel fringes at an edge, as discussed in Chapter 8. Note that questions of partial coherence only arise when more than one idealized point source of radiation is used. An interference experiment which may be used to measure $\gamma(x_{1,2})$ is Young's slit experiment. This experiment gives an important physical interpretation to $\gamma(x_{1,2})$—it is the contrast of the interference fringes (if the pin-holes are sufficiently small). Figure 4.4 shows the experimental arrangement used in optics. The relationship between the fringe contrast and the source size is described in most optics texts (e.g., Born and Wolf (1975)). Sharp fringes are obtained from a single point-source P_1. Moving this source to P_2 translates the fringes in the opposite direction. The incoherent superposition of many sets of fringes, slightly out of register and arising from a set of sources along P_1P_2 results in a fringe pattern of reduced contrast. An important point is that a small increase in the width of the

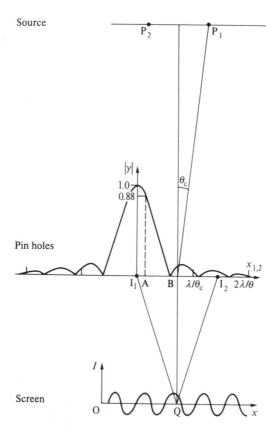

Fig. 4.4 Young's slit experiment in optics. Interference fringes are formed with a single point source P_1 because of the path difference between each of the pin-holes I_1 and I_2 and an image point Q. The visibility of these fringes when an extended source P_1P_2 is used is equal to the ordinate of $|\gamma(x_{1,2})|$ evaluated at the pin-hole I_2. The ordinate has the value $\lambda/2\pi\theta_c$ at A and $0.61\lambda/\theta_c$ at B.

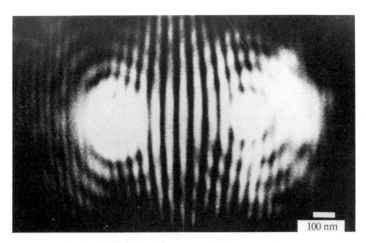

Fig. 4.5 Young's slit experiment performed with electron waves. The 'pin-holes' in this case are two small holes in a thin carbon film. (Courtesy of E. Zeitler. Similar images have been obtained by A. Tonomura using the Hitachi field-emission instrument.)

source will have more effect on the contrast of fine fringes than on coarse fringes. A similar result holds for the effect of source size (condenser aperture) on high-resolution phase-contrast images.

Figure 4.5 shows the result of performing Young's slit experiment with electrons. If the pin-holes are sufficiently small in the optical case, the Michelson visibility (contrast) of these fringes is given by

$$V = \frac{I_{\max} - I_{\min}}{I_{\max} + I_{\min}} = |\gamma(x_{1,2})| \qquad (4.14)$$

for a symmetrical source whose temporal coherence is large. The use of larger pin-holes (as in the electron case) introduces an additional modulating envelope.

The function $|\gamma(x_{1,2})|$ for a uniformly filled incoherent disc source is sketched in Fig. 4.4, centred about the pin-hole I_1. The contrast of the fringes formed in this experiment is then equal to the value of the function evaluated at the second pin-hole I_2. The form of $\gamma(x_{1,2})$ is given by

$$\gamma(x_{1,2}) = \frac{2J_1(u)}{u}; \qquad u = \frac{2\pi\theta_c x_{1,2}}{\lambda} \qquad (4.15)$$

for a circular disc source. Here θ_c is the semi-angle subtended by the source at the pin-holes. By convention, distances smaller than that for which the ordinate has fallen by 12 per cent are said to by coherently illuminated. This occurs when $u = 1$ and so allows the coherence width X_c to be defined for

$$x_{1,2} = \lambda/2\pi\theta_c = X_c \qquad (4.16)$$

A reasonable criterion for incoherent illumination of two points is that

$u = 2\pi$ or larger. Thus we may define an 'incoherence' width

$$X_i = \lambda/\theta_c \qquad (4.17)$$

Points separated by distances greater than X_i are incoherently illuminated. The range of spacings between X_c and X_i is described by the theory of partial coherence.

To summarize, coherence is characterized by the function $\gamma(x_{1,2})$ (the Fourier transform of the source intensity function) whose width gives a measure of the maximum separation between points which can be considered coherently illuminated. The contrast of a pure-phase object decreases as this function become narrower. Physically, the function gives the contrast of fringes in a certain interference experiment. A fuller discussion of these points is given in Barnett (1974).

4.4 Spatial coherence using hollow-cone illumination

As a further example of the application of the van Cittert–Zernike Theorem to electron microscopy, we consider the use of a condenser aperture containing an annular gap. This is taken to be uniformly filled with quasimonochromatic illumination. The resulting 'conical' or 'hollow-cone' illumination has been used in attempts to image single atoms in transmission electron microscopy (Thon and Willasch 1972) and may find further application in high-resolution biological microscopy. Practical aspects of the use of these apertures are discussed at the end of this section. This imaging mode also has an important parallel in scanning microscopy, through the reciprocity theorem (see Section 5.10).

Take the outer radius of the annulus to be r_0 and the inner radius βr_0 as shown in Fig. 4.6. If a function $g(a, x)$ is defined such that

$$g(a, x) = 1 \quad \text{for} \quad x < a$$
$$= 0 \quad \text{elsewhere}$$

then the intensity across an incoherently filled annular gap can be described by

$$f(x) = g(r_0, x) - g(\beta r_0, x)$$

where x is the radial coordinate in the condenser aperture plane. The normalized Fourier transform of $f(x)$ with argument $u = \sin \alpha/\lambda$ gives the complex degree of coherence whose behaviour can be studied as a function of β, the ratio of the inner to outer aperture radii. The normalized transform of $f(x)$ is

$$F(X) = \frac{2}{(1 - \beta^2)} \left[\frac{J_1(X)}{X} - \beta \frac{J_1(\beta X)}{X} \right] \qquad (4.18)$$

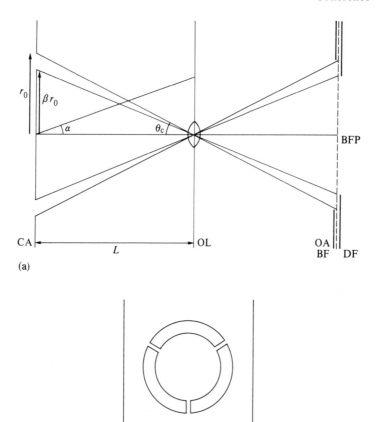

(a)

(b)

Fig. 4.6 (a) Ray diagram for illumination using an annular condenser aperture. The annular gap in the second condenser aperture CA is filled with incoherent radiation and this is focused to a bright ring in the objective back-focal plane (BFP). Figure 4.6(b) shows the shape of a practical aperture whose dimensions must match a particular instrument.

where $J_1(X)$ is a Bessel function of the first order and kind and

$$x = 2\pi r_0 u/\lambda \approx 2\pi r_0 \alpha/\lambda = 2\pi x_{1,2}\theta_c/\lambda \qquad (4.19)$$

For $\beta = 0$ (the normal circular aperture) the transverse coherence width X_c is defined as the specimen spacing $x_{1,2}$ corresponding to the abscissa value for which $F(X) = 0.88$. Two points in the specimen plane are considered 'coherently' illuminated if their spacing is less than X_c. For $F(X) = 0.88$, $X = 1$ and the commonly used coherence criterion is obtained. From eqn (4.19), for $X = 1$,

$$x_{1,2} = X_c = \lambda/2\pi\theta_c$$

Note that $x_{1,2}$ refers to spacings in the object so than any object point can be taken as origin. As β increases, $F(X)$ becomes narrower (the coherence

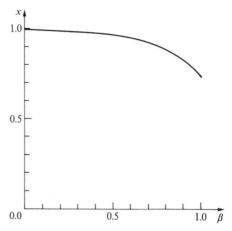

Fig. 4.7 The width of the modulus of the complex degree of coherence (for $F(X) = 0.88$) for an annular illumination aperture as a function of β, the ratio of the inner to outer annular radii. For a particular value of β the coherence width is $X_c = \lambda X / 2\pi\theta_c$.

deteriorates) but retains a form somewhat similar to the Airy's disc function until

$$\lim_{\beta \to 1} F(X) = J_0(X)$$

corresponding to the case of a circular line source. Figure 4.7 shows that the coherence under conical illumination is always less than that obtained using a conventional circular aperture equal in size to the outer diameter of the annular aperture ($\beta = 0$).

The illuminating hollow cone of rays incident on the specimen when using an annular condenser aperture forms a bright ring of unscattered intensity in the objective lens back-focal plane (Fig. 4.6). By using a matching objective aperture of optimum size for phase contrast which either excludes or includes this ring, a dark- or bright-field image may be formed. For matching apertures, $\theta_{ap} = \theta_c$, so that $X_c = \lambda/2\pi\theta_{ap}$ where θ_{ap} is the semi-angle subtended by the objective aperture. Thus for a 10 mrad objective aperture, about the size commonly used for phase contrast, we find $X_c = 0.05$ nm. This would appear to restrict the use of conical illumination to very high-resolution detail for phase contrast. Non-uniform illumination of the condenser aperture and the partially coherent image contribution may account for the limited success of this method at high resolution.

To construct an annular aperature, a diagram similar to Fig. 4.6 must be drawn for the particular instrument which will give the approximate dimensions of the annulus. Depending on the strength of the objective lens (Section 2.9), some deviation from this simple ray diagram must be expected and a certain amount of trial and error is required to obtain matching apertures. Since it is reasonably large (r_0 typically about 2.5 mm), the annular aperture can be made by the photoetching methods used for electronic circuit boards. Some manufacturers, including Siemens, JEOL, and Philips supply these apertures ready-made for their instruments. Because L is much larger than the objective lens focal length, the scale of a modified condenser aperture is much larger than that of an objective aperture, making interventions in the illuminating system easier than in the objective back-focal plane. This is an important advantage of the conical illumination technique over other dark-field methods such as the use of a wire beam stop in the objective lens back-focal plane, where the minute scale of the aperture and contamination of the beam stop are important practical problems. Another advantage is that, as for high-resolution tilted-illumination bright-field imaging, the range of spatial frequencies included within the aperture is about twice that obtained with untilted illumination and a central aperture is about twice that obtained with untilted illumination and a central aperture. The asymmetrical 'Schlieren' distortion (see Section 6.5) which accompanies tilted-illumination dark-field images is also not present using conical illumination. A wave-optical analysis of hollow-cone illumination has been given by Niehrs (1973), and applications to amorphous materials are discussed in Saxton, Jenkins, Freeman, and Smith (1978), and Gibson and Howie (1978/9).

4.5 The effect of source size on coherence

The preceding discussion has been based on the assumption that the illuminating system can be replaced by an effective source filling the final condenser aperture. This makes the degree of coherence at the specimen dependent only on the size of the condenser aperture used (eqn (4.16)), and not on the excitation of the condenser lenses or the source size. In this section the conditions under which source size may be important are investigated.

Equation (4.16) applies only if the illuminating aperture is incoherently filled. This will be so to a good approximation if the coherence width X_a in the plane of the illuminating aperture is small compared with the size of that aperture ($2R_a$). Using the van Cittert–Zernike theorem an expression can be obtained for X_a to test this, as described in Born and Wolf (1975). The result is

$$X_a = \lambda / 2\pi\theta_s$$

where θ_s is now the semi-angle subtended by the focused spot at the aperture (see Fig. 4.8). This angle depends both on the excitation of the lenses and on the source size.

Consider first a hair-pin filament with a tip radius of 15 micrometres, using the illuminating system shown in Fig. 7.1. For this microscope (JEM-100C), leaving aside the effect of the objective pre-field, the total demagnification of the two lenses C1 and C2 varies between 0.54 and 0.03

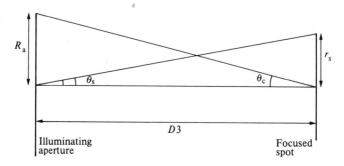

Fig. 4.8 Formation of a focused spot of radius r_s beyond a condenser aperture of radius R_a. The quantities defined enable the degree of coherence in the illuminating aperture plane to be found. In high-resolution scanning electron microscopy where a field-emission source is used the illuminating aperture may be coherently filled.

for the minimum and maximum settings of C1 ('spot size'). This would give maximum and minimum focused spot sizes of $r_s = 8.1$ and 0.45 micrometres. With $D3 = 251$ mm we then have $X_a = 0.018$ and 0.327 micrometres respectively, both of which are small compared with a 100 micrometre condenser aperture. Thus we are justified in treating the illuminating aperture as incoherently filled in this case.

For a pointed tungsten filament with tip radius 1 micrometre, a similar argument gives $X_a = 0.26$ and 5.0 micrometres for the minimum and maximum C1 settings at 100 kV using eqn (4.15). This second value is no longer small compared with, say, a 40 micrometre illuminating aperture of the kind used in some minimum-exposure techniques for which eqn (4.16) would therefore underestimate the coherence width at the specimen. Note that if the illuminating aperture is coherently filled ($X_a > 2R_a$) the focused spot at the specimen is also perfectly coherent.

We can conclude that the effect of source size on image coherence will only become important for sources smaller than about 1 micrometre, such as a field-emission source, the finest pointed filament, or the smaller LaB_6 sources (see Table 7.1). Further, a comparison of the coherence conditions in two different experiments can only be made if both θ_s and θ_c are known. At higher voltages an even smaller source is required to produce partially coherent illumination of the illuminating aperture. An ingenious experimental technique for measuring the degree of coherence across the illuminating aperture is described by Dowell and Goodman (1973).

4.6 Coherence requirements in practice

Phase-contrast effects such as Fourier images, lattice images, and images of small molecules all show increased contrast as the coherence of the illumination is increased. The rapid change in image appearance with focus predicted by 'coherent' calculations of the type discussed in Chapters 5 and

6 will also only be seen if highly coherent illumination is used. In practice, this means a beam divergence of less than about 0.8 mrad. The fine structure of a high-resolution image, controlled by wave-optical interference, can be washed-out either through the use of large beam divergence or too large an energy spread in the incident beam (see Section 5.6). This attenuation of high spatial frequencies can be represented as a damping envelope which multiplies the coherent transfer function (see Section 3.3 and 4.2 and Frank (1973)). The result may be that important structural information about the specimen is lost. For periodic specimens, the safest procedure is to compute the image of interest and assess the effect of increasing beam divergence using eqn (4.1). Figure 4.9 shows the effect of increasing beam divergence on the computed image of a large unit-cell crystal. Much of this loss of fine detail is due to the phase shift introduced by spherical aberration, which becomes appreciable across the angular width of each diffraction spot, particularly those at large scattering angles where the transfer function oscillates rapidly. This degradation in image

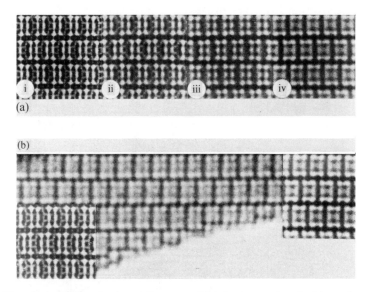

Fig. 4.9 The effect of increasing beam divergence (illuminating semi-angle θ_c) on the quality of high-resolution images. Here computer simulated images of $Nb_{12}O_{29}$ are shown for beam divergences of (i) 0.0 mrad, (ii) 0.6 mrad, (iii) 1.0 mrad, and (iv) 1.4 mrad. The loss of fine detail with increasing condenser aperture size is evident. This effect can be understood physically by extending the analysis of Section 5.2 to the three-beam case with near-axial illumination, or from Section 3.3. The simulated experimental conditions are $C_s = 1.8$ mm, 79 beams included within the objective aperture, $\Delta f = -60$ nm, 100 kV, crystal thickness $t = 5$ nm. These simulated images were produced by the method described by Billington and Kay (1974) (see Chapter 6), and provided by courtesy of M. O'Keefe and J. Sanders. (b) shows an experimental image of $Nb_{12}O_{29}$ taken by S. Iijima under the experimental conditions given above. The two inserts show computed images for $\theta_c = 0.0$ (left) and $\theta_c = 1.4$ mrad (right) as used experimentally. An excellent image match is obtained through the inclusion of the correct illumination angle.

quality with increasing beam divergence is borne out by experiment (Fig. 4.9). The severity of the effect depends on the particular specimen structure.

If the final image obtained at a beam divergence of, say 0.8 milliradians and high magnification (about 300 K) is insufficiently intense to allow accurate focusing, a reduced specimen height (shorter objective lens focal length) should be tried which will increase the objective pre-field focusing effect. This will increase the image current density and produce a more intense final image. In doing so, the illuminating spot size is decreased and the beam divergence is increased (the product of these is constant—see Section 2.2). A trade-off between coherence width X_c and specimen intensity j_0 is expected since $j_0 = \pi \beta \theta^2$ (eqn (7.1)) while $X_c = \lambda/2\pi\theta_c$. However, it appears that the reduction in spherical aberration accompanying this shorter focal length (Fig. 2.16) more than compensates for the loss of coherence, since the highest-quality images have generally been obtained at the shortest possible objective focal length, subject to the limitations of lens current and space within the pole-piece. The optimum choice of specimen position is further discussed in Appendix 3.

The relationship between coherence width and object current density has been studied experimentally by Harada, Gota, and Someya (1974). Figure 4.10 shows their results for a hair-pin filament and compares this with the

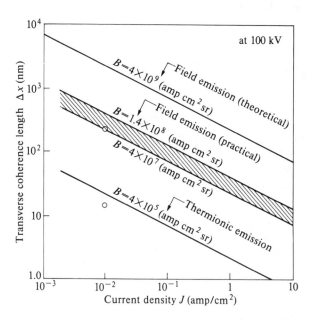

Fig. 4.10 Theoretical and experimental (circles) values for coherence width as a function of illumination current density at the specimen. The field-emission gun is compared with a hair-pin filament.

result obtained with a field-emission source. The ordinate in this figure is

$$X_c = \frac{\lambda}{2\theta_c}$$

which differs by a factor of π from that used by Born and Wolf (1975). This definition is common in the literature, the criterion for coherence being somewhat arbitrary. The theoretical curves of this figure assume an incoherently filled illuminating aperture.

The range of focal settings over which sharp 'Fourier images' (Cowley and Moodie 1960) can be seen has been analysed by Fujiwara (1974), showing explicitly the dependence of this range on the source coherence.

With non-periodic specimens, an impression of the coherence conditions can be obtained from a glance at the diffraction pattern. Figure 4.11 is a sketch of the diffraction pattern of an amorphous specimen formed with a large condenser aperture. Incoherent instabilities such as vibration and electronic fluctuations may limit the resolution to, say, 0.3 nm. This corresponds to the imposition of an 'aperture' (outer circle) of angular radius $\theta_1 = 0.82\lambda/0.3$, from the resolution limit of eqn (4.11). In the absence of a physical aperture, scattering outside this outer circle will contribute to the diffuse image background, resulting in a loss of image contrast. At the other extreme, eqn (4.17) gives the spacing of incoherently illuminated points as $X_i = \lambda/\theta_c$. Specimen spacings larger than this lie inside the inner continuous circle at $\theta_2 = \theta_c$ and are imaged under incoherent conditions. Their contrast cannot therefore be enhanced by the optimum-focus phase-contrast technique.

Notice that the coherence requirement becomes more severe as one attempts to form phase-contrast images of larger spacings. For some biological crystals with very large unit cells (e.g. catalase), a very large coherence width would be required for coherent imaging. Most published

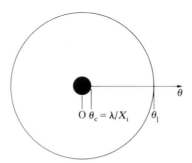

Fig. 4.11 The appearance of the diffraction pattern of an amorphous specimen. The central disc represents the 'unscattered' central beam, and the outer circle is drawn at the resolution limit of the microscope set by incoherent instabilities. Specimen spacings corresponding to scattering angles within the central disc cannot be enhanced by phase-contrast microscopy.

images of these specimens are formed under partially coherent imaging conditions.

When using pointed filaments, an LaB_6 source, or a field-emission source, the degrading effect of finite-beam divergence on image quality can usually be made negligible, since these sources generally provide adequate intensity for focusing with a small beam divergence (about 0.6 mrad). It is the increased brightness of these sources which makes them valuable, since this enables the image intensity obtainable with a hair-pin filament to be attained using a smaller beam divergence and therefore more coherent illumination. Hibi and Takahashi (1971) have published many images of biological specimens showing improved contrast and resolution when using a pointed filament which they claim gives more coherent illumination. Under particular experimental conditions this may be so (see Section 4.5) and we have seen that these include the focused spot size, the illuminating aperture size, the electron wavelength, and the extent of the objective lens pre-field. Certainly one would expect a change in experimental conditions which increases the number of Fresnel fringes observed to provide enhanced image contrast. A count of the number of fringes seen is probably the simplest qualitative guide to the coherence in the specimen plane (see Section 8.8 and Fig. 8.7) and can be used to compare sources if other experimental conditions are unchanged. An image recording must be taken, since the number of fringes seen through the binocular depends on the image intensity and can be misleading. The effect of increased source coherence on contrast is also demonstrated in Fig. 1.5.

For the laboratory not equipped for image simulation and analysis by computing methods the safest practical guide to coherence conditions is an analysis of optical diffractograms. Since the only focus setting which allows a simple image interpretation not requiring computer analysis is the Scherzer focus (eqns (6.16) and (4.12) with $n = 0$), a check on coherence conditions can be made by obtaining optical diffractograms from thin carbon-film micrographs, recorded at the Scherzer focus using several different condenser aperture sizes but otherwise identical conditions. The largest condenser aperture size is then selected for which the corresponding optical diffractogram shows good contrast across the entire inner ring out to the point-resolution limit of the instrument (eqn (6.16)). The theoretical basis for this procedure is described in Section 4.2 and Appendix 3.

4.7 Summary

The important ideas of coherence theory are outlined in this chapter. For the high-resolution microscopist the most important practical result of this chapter is that for thin specimens the smallest second condenser aperture should be used which allows the image to be focused accurately. This rule does not, however, apply if the focused electron beam spot at the specimen is less than a certain critical size (see Section 4.5), a situation only likely to arise when using a field-emission source or possibly the smallest LaB$_6$ or pointed filament sources. This chapter also refers the reader to experimental work showing the improved contrast possible using sources of high coherence and suggests methods for measuring coherence in the microscope (see also Section 8.8). The construction of annular apertures for conical illumination is also discussed. For the microscopist with some mathematical background, the most important section of this chapter is the discussion in Section 4.2 of the transfer function and damping envelopes due to partial temporal and spatial coherence. A clear understanding of these functions and their behaviour under changing conditions of focus, beam divergence, gun-bias, and specimen position in the microscope (see Appendix 3 and Section 2.7) is essential for the microscopist wishing to develop high-resolution techniques. This means that all the experimental parameters on which these functions depend must be measured—this can be done by the methods described in Chapter 8.

References

Barnett, M. E. (1974). Image formation in optical and electron transmission microscopy. *J. Microsc.* **102**, 1.

Born, M. and Wolf, E. (1975). *Principles of optics,* 5th edn. Pergamon, New York.

Burge, R. E., Dainty, J. C., and Thom, J. (1975). The spatial coherence of electron beams. In. *Proc. EMAG 1975,* Bristol (ed. J. Venables) p. 221. Academic Press, London.

Cowley, J. M. and Moodie, A. F. (1960). Fourier images IV: The phase grating. *Proc. Phys. Soc.* **76**, 378.

Dowell, W. C. T. and Goodman, P. (1973). Image formation and contrast from the convergent electron beam. *Phil. Mag.* **28**, 471.

Fejes, P. L. (1977). Approximations for the calculation of high-resolution electron microscope images of thin films. *Acta Crystallogr.* **A33**, 109.

Frank, J. (1973). The envelope of electron microscope transfer functions for partially coherent illumination. *Optik* **38**, 519.

Fujiwara, H. (1974). Effects of spatial coherence on Fourier imaging of a periodic object. *Opt. Acta* **21**, 861.

Gibson, J. M. and Howie, A. (1978/9). Investigation of local structure and composition in amorphous solids by high resolution electron microscopy. *Chemica Scripta* **14**, 109.

Harada, Y., Gota, T., and Someya, T. (1974). Coherence of field emission electron beam. In *Proc. 8th Int. Congr. Electron Microsc.,* Canberra, p. 110.

Hibi, T. and Takahashi, S. (1971). Relation between coherence of electron beam and contrast of electron image of biological substance. *J. Electron Microsc.* **20**, 17.

Holmes, S. M., Cockayne, D. J. H., and Ray, I. L. F. (1974). The influence of incident beam convergence upon images of lattice defects. In *Proc. 8th Int. Congr. Electron Microsc.,* Canberra, p. 290.

Hopkins, H. H. (1957). Applications of coherence theory in microscopy and interferometry. *J. Opt. Soc. Am.* **47**, 508.

Humphreys, C. J. and Spence J. C. H. (1981). Resolution and illumination coherence in electron microscopy. *Optik* **58**, 125.

Niehrs, H. (1973). Zur Formulierung der Bildintensitat bei ringformiger Objektbestrahlung in der Electronen–Mikroskopie. *Optik* **38**, 44.

Nagata, F. Matsuda, T., and Komoda, T. (1975). High resolution electron microscopy by an incoherent illumination method. *Japan J. Appl. Phys.* **14**, 1815.

O'Keefe, M. A. and Sanders, J. V. (1975). n-Beam lattice images. VI. Degradation of image resolution by a combination of incident beam divergence and spherical aberration. *Acta Crystallogr.* **A31**, 307.

Saxton, W. O., Jenkins, W. K., Freeman, L. A., and Smith, D. J. (1978). TEM observations using bright field hollow cone illumination. *Optik* **49**, 505.

Sellar, J. R. (1976). Partially coherent imaging of thick specimens in the scanning electron microscope. In *Proc. IITRI 1976* (ed. Om Johari), p. 369. IIT Research Institute, Chicago.

Thon, F. and Willasch, D. (1972). Imaging of heavy atoms in dark field electron microscopy using hollow cone illumination. *Optik* **36**, 55.

Wade, R. H. and Frank, J. (1977). Electron microscope transfer functions for partially coherent axial illumination. *Optik* **49**, 81.

Bibliography

A clear elementary account of partial coherence in imaging is given in the reference by Barnett, above. More advanced treatments can be found in:

Thompson, B. J. (1969). Image formation with partially coherent light. In *Progress in optics* (ed. E. Wolf) Vol. 7 North-Holland, Amsterdam.

Beran, M. J. and Parrent, G. B. (1964). *Theory of partial coherence.* Prentice-Hall, Englewood Cliffs, NJ.

Perina, J. (1971). *Coherence of light.* Van Nostrand, London.

5

HIGH-RESOLUTION IMAGES OF PERIODIC SPECIMENS

This chapter concerns the theoretical basis for the interpretation of HREM images of crystalline samples and their defects. A section on experimental methods is also included (see also Chapter 10). The applications of HREM imaging are reviewed in Chapter 12.

A long-term aim of electron microscopy has been the direct imaging of atoms in solids. To achieve this, a point-resolution of about 0.15 is required for most crystalline materials. This can now be obtained on the latest instruments by the more or less routine methods described in this chapter and in Chapter 12. At a resolution limit of 0.20 nm the image of a single atom changes little over a range of focus of, say, 5 nm, so that for specimens of this thickness or less, containing atoms which are well-separated laterally, we can at best hope to see a projection of the specimen structure in the direction of the incident beam. Optical sectioning, of the kind used with light microscopes, is not possible in transmission electron microscopy because of the small-angle scattering involved in HREM. In practice there are many other limitations on the interpretation of high-resolution lattice images and this chapter is devoted to the simplest theories necessary to understand them.

It is convenient to divide the imaging process into two parts—the interaction of the incident beam with the specimen and the subsequent electron optical imaging. The effect of electron optical parameters on simple two- and three-beam fringes is discussed first in Section 5.1. Then three simple theories giving analytic expressions for the diffracted beam amplitudes are summarized (Section 5.2) and their implications for the interpretation of lattice images are discussed. The second half of the chapter is devoted to recent work on many-beam lattice images, called 'structure images' (when they faithfully represent the crystal structure) and discusses some experimental aspects of high-resolution lattice imaging.

The experimental conditions which allow a simple interpretation of many-beam images are also discussed.

The chapter commences with some notes on three common imaging methods for simple lattice fringes. After a period of early optimism (for a review of the early pioneering work, see Menter (1958)) it became clear that only very limited information is given by these images formed from the interference between two or three beams. The most important use for these images now is for demonstrating the stability of electron microscopes. However a sound grasp of the theory of these few-beam lattice images is essential to an understanding of the many-beam structure images described later. The observation of two-dimensional many-beam images from large unit-cell crystals in 1972 (Cowley and Iijima (1972)), building on earlier work by J. Sanders, L. Hewatt, K. Yada, H. Hashimoto, T. Komoda and others (see Table 5.2 for references) marked an important turning point in the subject by showing that useful structural information could be obtained by the HREM method.

The topic of this chapter has been reviewed by Allpress and Sanders (1973). The review article by Cowley (1976a, 1985) also complements the material of this chapter.

5.1 The effect of lens aberrations on simple lattice fringes

In this section the effects of changes in lens parameters and illumination conditions on simple lattice fringes, formed from two or three diffracted beams, are investigated. Simple expressions are given which show the way in which lattice fringes are affected by focusing, spherical aberration, and electronic instabilities for the commonly used imaging methods. In a later section the effects of altering the specimen orientation (diffraction conditions) are discussed. For the purposes of this section the orientation of the specimen with respect to the incident beam is assumed fixed. The purpose of this section is to give the simplest theoretical expressions which give the microscopist an indication of the effects of common changes in experimental conditions.

The three common imaging methods used for two- and three-beam lattice images are shown in Fig. 5.1. If the complex amplitudes of the Bragg beams are $\Phi_{hk0} = \phi_{hk0}e^{i\varepsilon}$ (ϕ_{hk0} real), the image amplitude in general will be

$$\psi_i(\mathbf{r}_i, \Delta f, K_0) = \sum_{\mathbf{g}} \Phi_{\mathbf{g}} \exp(2\pi i \mathbf{g} \cdot \mathbf{r}) \exp\{i\chi(u_{\mathbf{g}})\} \qquad (5.1a)$$

For the simple case of an orthorhombic crystal with the beam parallel to the c-axis this becomes

$$\psi(x, y) = \sum \Phi_{hk0} \exp\{+2\pi i(hx/a + ky/b)\} \exp[i\chi(u_{h,k})] \qquad (5.1b)$$

where the sum is taken over all beams within the objective aperture and a and b are the orthogonal unit-cell dimensions normal to the electron beam

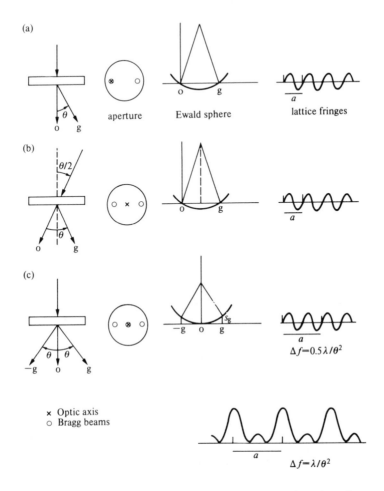

Fig. 5.1   Three imaging methods for simple lattice fringes. The first figure in each row suggests the directions of the incident and diffracted beams with respect to the crystal; the second shows the appearance of the objective aperture and the positions of the diffracted beams (○) and the optic axis (×); the third gives the corresponding Ewald sphere orientation (see Fig. 5.2), while the last figure shows the form of the lattice fringes produced. The cases of untilted illumination (a), tilted illumination (b), and three-beam fringes (c) are shown. In case (c) fringes are shown for two values of defocus Δf, to illustrate the occurrence of half-period fringes ($C_s = 0$). (d) Axial three-beam lattice imaging conditions summarized. Continuous lines are the locus of full-period images which coincide with the lattice. Broken lines show images of reversed contrast, while the dotted line shows one of the many half-period images. For an instrument with $1.4 < C_s < 2.6$ mm, images are only seen within the heavy parallelogram. Thus, if a high-contrast lattice image were recorded under these conditions the possible pairs of values of Δf and C_s would be greatly restricted. The depth of field Δz, the periodicity in focus Δf_t and the period $C_s(0)$ in spherical aberration are all shown. The figure is drawn for a lattice period $a = 0.313$ nm (silicon (111)) with $\theta_c = 0.0014$ rad at 100 kV and $\varepsilon = -\pi/2$.

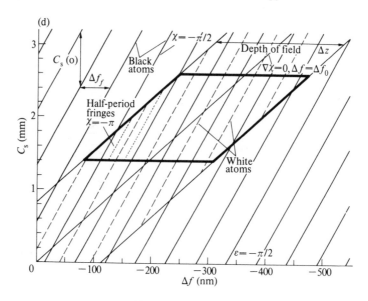

Fig. 5.1. (Continued)

direction. Here $\chi(u_{h,k})$ is the phase shift introduced by instrumental aberrations described in Section 3.3 and given, in the absence of astigmatism, by

$$\chi(u_{h,k}) = 2\pi/\lambda\{\tfrac{1}{2}\Delta f\,\lambda^2 u_{h,k}^2 + \tfrac{1}{4}C_s\lambda^4 u_{h,k}^4\} \tag{5.2}$$

where

$$u_{h,k}^2 = \frac{h^2}{a^2} + \frac{k^2}{b^2} = \mathbf{g}^2$$

The diffracted amplitudes Φ_{hk0} are functions of the specimen orientation and thickness. We now consider the one-dimensional cases shown in Figs 5.1(a), (b), and (c), and choose the Miller indices $\mathbf{g}$ of the reflection included within the aperture to be $(h, 0, 0)$. Then $u_h = h/a = \theta_0/\lambda$ where a is the lattice spacing and θ_0 is twice the Bragg angle. The image intensity in the kinematic case where the scattering phase $\varepsilon = -\pi/2$ is then, for the two beams shown in Fig. 5.1(a),

$$I(x) = \psi_i(x)\psi_i^*(x) \tag{5.3}$$

$$I(x) = \phi_0^2 + \phi_h^2 + 2\phi_0\phi_h \sin(2\pi h x/a - \chi(u_h)) \tag{5.4}$$

The form of this expression indicates that any change in $\chi(u_h)$ will produce a sideways movement of the fringe image. Fluctuations in objective lens current or high voltage cause Δf to vary (see Section 3.3) and so result in a small fluctuating fringe displacement during an exposure. The image resolution is therefore limited by these fluctuations. Lattice fringes formed under these conditions will be seen to move sideways (normal to their length) as the focus control is altered, often well beyond the edge of the specimen. Leaving aside the effect of spherical aberration, eqn (5.4) gives the displacement of the fringes due to defocus as

$$I(x) = \phi_0^2 + \phi_h^2 + 2\phi_0\phi_h \sin\{2\pi h(x - \Delta f\,\theta_0/2)/a\} \qquad (5.5)$$

In Fig. 5.1(b) the illumination has been tilted with respect to the optic axis resulting in an improvement in the incoherent resolution limit imposed by electronic instabilities (Komoda 1964; Dowell 1963). The image amplitude is then

$$\psi_i(x) = \Phi_0 \exp(i\chi(-u_h/2)) + \Phi_h \exp(-2\pi ihx/a + i\chi(u_h/2)) \qquad (5.6)$$

Since the function $\chi(u_h)$ is even, the same instrumental phase shift now applies to each beam. The image intensity is

$$I(x) = \phi_0^2 + \phi_h^2 + 2\phi_0\phi_h \cos((2\pi hx/a) - \varepsilon) \qquad (5.7)$$

which is independent of the focus condition (for perfectly coherent illumination). It is also therefore independent of fluctuations in the high voltage and lens currents within the accuracy of the illumination alignment. Chromatic aberration arising from the thermal energy spread of the incident electrons (Section 2.8.2) is similarly reduced by this technique. The highest 'resolution' two-beam lattice fringes are invariably formed using this imaging method, since it confers a limited immunity to these important image defects; however, it is extremely difficult to extract crystal structure information from these fringe images.

Figure 5.1(c) shows three-beam fringes formed with axial illumination, the method used to record the image shown in Fig. 8.3. The image amplitude is

$$\psi_i(x) = \Phi_0 + \Phi_h \exp(2\pi ihx/a + i\chi(u_h)) + \phi_{-h} \exp(-2\pi ihx/a + i\chi(-u_h)) \qquad (5.8)$$

and the image intensity becomes

$$I(x) = \phi_0^2 + \phi_h^2 + \phi_{-h}^2 + 2\phi_h\phi_{-h}\cos(4\pi hx/a) \\ + 4\phi_0\phi_h\cos(2\pi hx/a)\cos(\chi(u_h) + \varepsilon) \qquad (5.9)$$

In the simplified case where $\phi_h = \phi_{-h}$ and $\varepsilon = -\pi/2$ the image is seen to consist of constant terms plus cosine fringes with the lattice spacing, together with weaker cosine fringes of half the lattice spacing. These 'half-spacing' fringes are often seen on micrographs taken under these conditions. Frequently they cannot be seen through the viewing binocular and are only seen on close examination of the developed micrograph. The focus condition for observing the half-spacing if $\varepsilon = -\pi/2$ is obtained by setting $\chi(u_h) = n\pi$, giving

$$\Delta f = \frac{n\lambda}{\theta_0^2} - \frac{C_s\theta_0^2}{2}, \qquad n = 0, \pm1, \pm2, \ldots \qquad (5.10)$$

Some examples are sketched in Fig. 5.1(c). It has been suggested that since the half-period fringes are insensitive to focus, while the lattice-period fringe contrast may reverse with changes in focus, the effect of electronic instabilities (leading to a variation in $\chi(u_h)$) should be to allow the half-period fringes to accumulate during an exposure at the expense of the lattice-period fringes.

It is clear that any change of 2π in $\chi(u_h)$ occurring in eqn (5.8) leaves these axial three-beam fringes unaffected, while a change of π reverses their contrast. The change in focus needed to alter $\chi(u_h)$ by 2π is $\Delta f_f = 2n/(\lambda u_h^2)$. The change in C_s needed to change $\chi(u_h)$ by 2π is $C_s(0) = 4n/(\lambda^3 u_h^4)$ and the lattice fringes are thus periodic in both Δf and C_s. All these features of three-beam images are summarized in Fig. 5.1(d). The equation to the straight lines shown on the figure is obtained by setting $\chi(u_h) = (2n - \frac{1}{2})\pi$ and solving for C_s. From eqn (3.24) this gives

$$C_s = \frac{(4n - 1)}{(\lambda^3 u_h^4)} - \frac{2\Delta f}{(\lambda^2 u_h^2)}$$

as the condition for identical three-beam axial images. The phase angles $(2n - \frac{1}{2})\pi$ have been chosen since these result in an electron image which 'coincides' with the lattice in the sense that atom positions appear dark. This can be shown by using the kinematic amplitudes of eqn (5.15) in eqn (5.8) with $S_g = 0$. The 90° phase-shift ε due to scattering, represented by the factor $-i = \exp(-i\pi/2)$ in eqn (5.15), and the 90° lens phase-shift $\chi(u_h)$ (for $n = 0$, say) are then superimposed to give a total phase-shift of $-180°$. (This factor is responsible for the minus sign in equation 6.6, giving dark atom images). Diffracted beams whose phase differs from that of the central beam by an amount not equal to this 90° needed for simple image interpretation are common under dynamical scattering conditions, and may result in severely distorted images.

In practice fringes are not seen for all focus settings, only those near the 'stationary-phase' focus $\Delta f_0 = -C_s \lambda^2 u_h^2$ (see Section 5.19 and eqn (4.10)). This limited depth of field, resulting from the use of an extended electron source, is discussed in the next section and drawn in Fig. 5.1(d) as the envelope within which clear lattice fringes can be seen. The periodicity in focus in this case is actually a special simplified example of Fourier imaging discussed in Section 5.8. For a general one-dimensional many-beam image, a focus change of $\Delta f_f = 2a^2/\lambda$ changes the phase of all beams by 2π, and so leaves even many-beam images unaltered. For many crystals, this result can also be extended to cover two-dimensional lattice images (see Section 5.8). Then a focus change of $\Delta f_f/2$ results in an image shifted by half a unit cell along the cell diagonal (for cubic projections), which cannot usually be distinguished from the 'true' image through the viewing binocular. Half-period images can also be obtained at certain focus settings in two-dimensional images. Similar half-period fringes occur at certain specimen thicknesses (see Fig. 10.4(a)) and the spurious impression of high-resolution detail which they may give when used as an instrumental resolution test is discussed in Section 8.10.

In larger-unit-cell crystals, the Fourier image period Δf_f frequently becomes larger than the depth of field Δz, so that only a single lattice fringe image is seen. This may also occur for a certain choice of illumination semi-angle θ_c for any lattice spacing a. The general condition of restricting the depth of field to a single Fourier image period is that $a/\theta_c < 2a^2/\lambda$, or $\theta_c > \lambda/2a$. On combining this with the Bragg law, we see that this is just the condition that adjacent diffraction discs overlap.

5.2 The effect of beam divergence on depth of field for simple fringes

It is possible to estimate the range of focus Δz over which lattice fringes of a particular periodicity should be visible. A limit is set to this range by the finite angle θ_c of the illuminating cone of electrons. The depth of field obtained applies to lattice images formed with partially coherent illumination and should be distinguished from the depth of field given in Section 2.2, which applies to the incoherent imaging of amorphous specimens.

Taking each point within the effective source as an independent source of electrons (neighbouring points assumed incoherent), the total intensity can be obtained by summing the intensities of images formed by each source point. For the lattice fringes of Fig. 5.1(a) the methods of Chapter 3 give the image intensity when the illumination direction makes a small angle α with the optic axis as

$$
\begin{aligned}
I(x, \alpha) &= \Phi_0^2 + \Phi_h^2 + 2\Phi_0\Phi_h \cos\{\chi(-u_h - u') - \chi(u') + 2\pi u_h \chi\} \\
&= \Phi_0^2 + \Phi_h^2 + 2\Phi_0\Phi_h \cos\{2\pi h(x + \Delta f \alpha)/a + \pi \Delta f \theta_0^2/\lambda\}
\end{aligned}
\tag{5.11}
$$

with $u' = \alpha/\lambda$.

Physically, this means that fringes out of focus by an amount Δf suffer a translation normal to their length proportional both to the focus defect and to the illumination tilt. Assuming a uniform illumination disc, the total image intensity is then

$$
I_T(x) = \frac{1}{\theta_c} \int_{-\theta_c}^{\theta_c} I(x, \alpha) \, d\alpha
\tag{5.12}
$$

for an illuminating cone of semi-angle θ_c. Evaluating this shows that the fringe contrast $C' = (I_{max} - I_{min})/(I_{max} + I_{min})$ is proportional to

$$
C = \sin(\beta)/\beta
\tag{5.13}
$$

where $\beta = 2\pi \Delta f \theta_c/a$. The contrast falls to zero for $\beta = \pi$, so that the range of focus Δz over which fringes are expected is

$$
\Delta z = a/\theta_c
\tag{5.14}
$$

With $\theta_c = 1.5$ mrad and $a = 0.34$ nm, we have $\Delta z = 227$ nm. There would thus be little point in seeking lattice fringes under these conditions on an instrument whose smallest focal increment was much greater than about 200 nm.

It is instructive to derive a similar result using the damping envelope construction of Section 4.2. We now consider axial three-beam lattice fringes as shown in Fig. 5.1(c). It is shown in Section 5.19 (eqn (5.76)) that if the central beam is stronger than the diffracted beams, three-beam lattice fringes of highest contrast occur near the focus setting $\Delta f_0 = -C_s \lambda^2 u_h^2$ which makes the slope of $\chi(u_h)$ zero. (Neglecting the second bracketed term in eqn 4.10, this condition can also be obtained by setting $q = 0$ in eqn 4.10 and solving for Δf). If an arbitrary focus setting $\Delta f'$ is measured by its deviation from Δf_0, then the focus defect Δf measured from Gaussian focus is

$$\Delta f = \Delta f_0 + \Delta f'$$

The damping envelope due to beam divergence alone is (eqn (4.8))

$$A(u_h) = \exp(-\pi^2 u_c^2 q)$$

where

$$q = (C_s \lambda^3 u_h^3 + \Delta f \lambda u_h)^2 \quad \text{and} \quad \theta_c = \lambda u_c (\ln 2)^{1/2}$$

Combining these expressions gives, for the focus dependence of the fringe contrast,

$$A(u_h, \Delta f') = \exp(-B^2 \Delta f'^2) \quad \text{where} \quad B = \pi u_c \lambda u_h$$

This is a Gaussian of width $1/B$, so that the depth of field Δz is

$$\Delta z = \frac{1}{B} = \frac{(\ln 2)^{1/2}}{\pi} \frac{a}{\theta_c}$$

for a (100) type reflection ($u_h = 1/a$). This differs only by a numerical factor $(\ln 2)^{1/2}/\pi = 0.26$ from the result of eqn (5.14).

The contrast of these axial three-beam fringes will also be affected by the term $\exp(i\chi(u_h))$ of eqn (4.8), which has been neglected in the above discussion. In general then, these fringes will show a rapid variation of contrast with focus (due to the terms in $\cos \chi(u_h)$ of eqn (5.9)) modulated by the more slowly varying Gaussian envelope $A(u_h, \Delta f')$ described above and centred about the focus setting Δf_0. Note, also, that the term $\exp(i\chi(u_h))$ has the same value for all the 'Fourier image' focus settings (see Section 5.8) Δf_n which differ from the reference focus Δf_0 by integral multiples of $2n/(\lambda u_h^2)$ (n an integer). Thus, if a strong image occurs at the reference focus Δf_0 (this will be so, if, from eqns (5.9) and (5.76), $C_s = 2m/(\lambda^3 u_h^4)$ with m an integer) then similar strong images will occur at focus settings

$$\Delta f = \Delta f_0 \pm 2n/(\lambda u_h^2)$$

The contrast of these images will fall off smoothly as the focus is increased (or decreased) beyond Δf_0 due to the Gaussian damping of $A(u_h, \Delta f')$. This

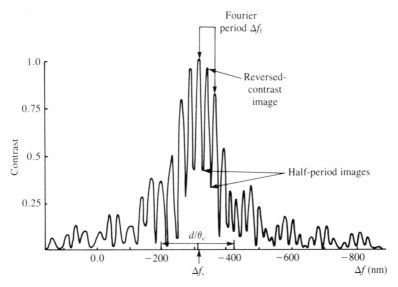

Fig. 5.2 Contrast of three-beam (111) lattice fringes in silicon as a function of focus. The depth of field Δz is limited by finite spatial coherence (beam divergence) and temporal coherence (chromatic effects). Contrast is a maximum at the stationary-phase focus Δf_s and high-contrast images would be seen in a through-focus series for a range $\Delta z \simeq d/\theta_c$ around Δf_s. (From Olsen and Spence (1981), reprinted by permission.)

behaviour is also seen in the calculated three-beam axial fringe contrast shown in Fig. 5.2 for silicon. This figure is the result of a dynamical calculation including both chromatic and spatial coherence effects, with $\theta_c = 1.4$ mrad, $C_s = 2.2$ mm, $\Delta = 5$ nm at 100 keV.

We may conclude that the finer the image detail, the smaller is the focal range over which it may be observed, for a fixed illumination aperture.

5.3 Approximations for the diffracted amplitudes

It is important to understand how the amplitudes of a set of Bragg reflections depend on the specimen structure, thickness, and orientation. In the previous section, we saw how these amplitudes form the Fourier coefficients in a series whose sum gives the image amplitude. The dimensionless diffracted amplitudes Φ_g express the total scattering by the crystal in a particular direction relative to a unit-amplitude incident wave. For all but the thinnest specimens, the possible directions are confined to a narrow range of angles around each reciprocal lattice point (see Fig. 5.3). This scattering is focused to a small spot by the objective lens in the lens back-focal plane. The intensities $|\Phi_g|^2$ are therefore proportional to the intensities of the diffraction spots seen when the microscope is focused for the specimen diffraction pattern.

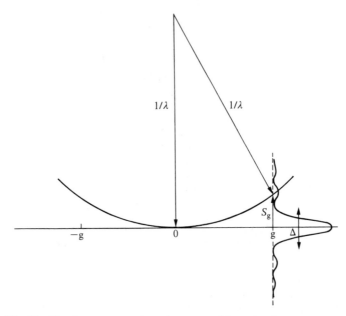

Fig. 5.3 The Ewald sphere construction. A vector of Length $1/\lambda$ is drawn in the crystal reciprocal lattice in the direction of the incident beam. In the zone-axis orientation as shown, this is normal to a plane of the reciprocal lattice. The kinematic amplitude diffracted into the beam **g** is proportional to the amplitude of the 'shape transform' $t \sin(\pi s_g t)/(\pi s_g t)$ at the point where the sphere crosses this function. The square of the function is suggested in the figure. The width Δ between first minima of this function is $2/t$, so that the angular range of incident beam directions which will produce scattering into the beam **g** increases as the specimen thickness decreases. A crystal of large transverse dimensions have been assumed. The excitation error s_g is drawn normal to the reciprocal lattice plane and is negative if the lattice point is outside the sphere. At 100 kV, $1/\lambda = 270$ nm^{-1} and with $|\mathbf{g}| = h/a \approx 2.5$ nm^{-1} the sphere is well approximated by a plane in the neighbourhood of the origin (i.e., some limited 'resolution') for a particular crystal thickness (which determines the width of the 'shape transform'). At higher voltage, the sphere becomes larger and the flat sphere approximation holds to larger crystal thickness—that is, with a narrower shape transform.

In the following sections, the kinematic and two-beam expressions for the diffracted amplitudes are given both for reference and for the information they supply about the effect of changes in experimental conditions on simple lattice fringes. In particular, we are interested in the best choice of specimen thickness and orientation for two-beam fringes.

5.3.1 The kinematic theory

For those ultra-thin, low-atomic-weight specimens for which the central beam suffers inappreciable scattering and is very much stronger than any diffracted beam, the kinematic or single-scattering theory of image formation may be applied. The results of this theory are easily understood and give a useful qualitative guide to the likely effects of changes in experimental conditions. The kinematic theory cannot be expected to give quantitative agreement with experiment except perhaps for low-atomic-number specimens a few nanometres thick (see Section 3.4). Measurable dynamical effects (multiple coherent scattering) leading to failure of this theory, have been observed in graphite flakes just 4 nm thick (P. Goodman, personal communitcation). Nevertheless, the simplicity of the kinematic model and the insight it gives into the physical scattering process have ensured its popularity in discussions on electron imaging.

The weak-phase object approximation (Section 3.4) can be extended to include the effect of Fresnel diffraction through the thickness of the specimen (and so take account of the curvature of the Ewald sphere) by integrating the contribution to scattering in a particular direction (Howie 1971). This gives the kinematic result for electron diffraction from a crystal with Fourier coefficients of potential v_g and thickness t as

$$\Phi_g = -i\sigma v_g t[\sin \pi s_g t/\pi s_g t] \exp(-i\pi s_g t) \qquad (5.15)$$

with

$$\Phi_0 = 1$$

and the sign convention for phases of Chapter 3. The orientation of the crystal with respect to the incident beam is specified by s_g, the excitation error, shown in Fig. 5.3. Practical examples of the way in which s_g can be measured from a diffraction pattern showing Kikuchi lines are given in Hirsch, Howie, Nicholson, Pashley, and Whelan (1965). Phase factors which depend on the lens focal length and object distance have been omitted. The exponential phase factor in eqn (5.15) depends on the choice of origin in the crystal—it can be eliminated by choosing an origin midway though the thickness of the specimen. Whatever origin is chosen, it is important that this same origin is used when reckoning the image focus defect. Equation (5.15) assumes that any focus defect is measured from the top surface of the specimen. Inserting this expression into eqn (5.6) gives the two-beam fringe intensity for tilted illumination as

$$I(x) = 1 - 2\sigma v_g t[\sin \pi s_g t/\pi s_g t] \sin(-\pi s_g t - 2\pi hx/a) \qquad (5.16)$$

This expression suggests the following:

1. Sinusoidal fringes are expected which, for a parallel-sided crystal, have the periodicity of the lattice. The positions of the fringe maxima will not in general coincide with the projected position of atoms in the crystal, but will depend on the specimen thickness and orientation.

2. For centrosymmetric crystal the v_g are real if absorption is negligible. Otherwise the v_g are complex and contain a phase factor which appears in the argument of the second sin function in eqn (5.16). The phases of the diffracted beams thus determine the relative displacements of the various fringes which make up an image. In this sense, image formation in the electron microscope 'solves' the phase problem of X-ray diffraction.

3. For fringes running parallel to the edge of a wedge-shaped specimen, we have a situation akin to phase modulation in electronic engineering. If the thickness varies with x (taken normal to the wedge edge) then the fringe spacing for a particular value of x will depend on the local variation of t with x. Only for a perfect wedge, with t proportional to x, would the spacing be constant. (See Section 5.4).

4. A similar variation in fringe spacing is expected for bent crystals, where s_g varies with x.

5. Fringes running perpendicular to the wedge edge are likely to be curved, since a different value of t applies to each chosen value of x (taken parallel to the wedge edge).

5.3.2   The two-beam dynamical theory

The highly restrictive assumption of an undepleted central beam in the kinematic theory is improved upon in the two-beam dynamical theory which provides a practical guide to the best specimen thickness for two-beam fringes. In this theory the real crystal is replaced by a fictitious crystal supporting only two diffracted beams. The likely accuracy of the theory can be judged from a glance at the diffraction pattern, which will indicate how much energy is scattered into reflections other than the central beam and the one other beam used for image formation. In this theory the sum of the intensities of the two beams is constant, and the contrast of two-beam lattice fringes at the exact Bragg angle is a maximum when the beams are of equal strength. This occurs when the specimen thickness is

$$t_g = \frac{\xi_g}{4} = \frac{\pi}{4\sigma v_g} \qquad (5.17)$$

where ξ_g is the extinction distance for the reflection **g**. Typical values at 100 kV are $t_{111} = 4$ nm (gold), $t_{111} = 14$ nm (aluminium), $t_{111} = 15$ nm (silicon), which give a very rough indication of the best specimen thickness for observing two-beam fringes. Values of t_g for other materials can be obtained from published scattering factors (e.g., Doyle and Turner (1968)). Scattering factor calculations are also discussed in Burge (1973), and Fourier coefficients for many elements can be found in Radi (1970). The most complete tabulation of high-energy electron scattering factors is that contained in the International Tables for X-ray Crystallography (1986), volume four.

If the scattering factors $f_j(\theta)$ are calculated from the first Born approximation, then

$$v_g = 0.04787 F_g / \Omega \qquad (5.18)$$

where Ω is the unit-cell volume in cubic nanometres and F_g is the kinematic structure factor obtained from the atomic scattering amplitudes according to

$$F_g = F_{h,k,l} = \sum_j f_j(\theta_g) \exp(2\pi i (hx_j/a + ky_j/b + lz_j/c)) \qquad (5.19)$$

where (x_j, y_j, z_j) are the coordinates of the atom j in the unit cell, and θ_g is the Bragg angle for the reflection with Miller indices h, k, l. Here v_g is in volts and F_g in Ångstroms, as usually tabulated (e.g., in the International Tables).

In practice the best specimen thickness for two-beam lattice fringes is generally found by trial and error; however, an indication of thickness is given by Pendellösung fringes. These broad dark and bright bands are seen parallel to the edge of wedge-shaped specimens at the Bragg angle if a small aperture is placed around the central beam. The thickness increment between successive fringe minima is approximately equal to the extinction distance ξ_g, where g refers to the reflection satisfied. The clearest lattice fringes are expected between the edge of the wedge and the first bright-field Pendollösung minimum.

The two-beam theory can also be used to predict the variation of lattice fringe contrast with orientation changes of the specimen. For reference, the complex amplitudes of the two diffracted beams are

$$\Phi_0 = \{\cos(\pi t(1 + w^2)^{1/2}/\xi_g)$$
$$- iw(1 + w^2)^{-1/2}\sin(\pi t(1 + w^2)^{1/2}/\xi_g)\}\exp(-i\pi s_g t) \quad (5.20)$$

and

$$\Phi_g = i(1 + w^2)^{-1/2}\sin(\pi t(1 + w^2)^{1/2}/\xi_g)\exp(-\pi i s_g t)$$

where $w = s_g \xi_g$. These expressions can be used with eqns (5.6) and (5.3) to give complicated expressions for the two-beam lattice fringe intensity, which are nevertheless easily plotted using a computer. The resulting expression gives a much better indication of the variation of lattice fringe contrast with specimen orientation and thickness, the only important approximation being the neglect of other diffracted beams. These neglected diffracted beams become increasingly important in crystals such as minerals with large unit cells. Whereas the width in reciprocal space of the central maximum of the kinematical 'shape transform' shown in Fig. 5.3 is $1/t$, the width of the corresponding two-beam dynamical function tends to $1/\xi_g$. The kinematic theory does however remain accurate to larger thickness for weak beams with large excitation errors or small values of v_g.

The theory of lattice imaging in the two-beam approximation is given in Cowley (1959). This and similar work (Hashimoto, Mannami, and Naiki 1961) using the two-beam expressions indicates that a reversal of fringe contrast is expected for every increase of half an extinction distance in thickness. In addition, the two-beam lattice image spacing is found to agree with the specimen lattice spacing only at the exact Bragg angle.

5.3.3   The thick-phase grating

We have seen that the commonly used phase object approximation (eqn (3.25)), while including multiple-scattering effects, is a projection approximation and does not take account of the curvature of the Ewald sphere, which is approximated by a plane. The development of the multi-slice theory of dynamical scattering suggests that this curvature corresponds physically to Fresnel diffraction within the specimen, to the extent that refractive index and propagation effects can be separated. Thus eqn (6.20), the condition for the neglect of Fresnel broadening of the electron wave, also represents the condition under which the Ewald sphere can be approximated by a plane (all s_g set to zero).

An approximation which takes account of multiple scattering and, in a limited way, the Ewald sphere curvature has been proposed by Cowley and Moodie (1962). This 'thick-phase grating' gives the object exit-face wave-function as

$$\psi(x, y) = \exp\left[-i\sigma t \sum_{h,k} v_{h,k,0}(\sin \pi s_{h,k,0}t / \pi s_{h,k,0}t) \exp(2\pi i(hx/a + ky/b))\right]$$

(5.21)

with a and b cubic unit-cell dimensions and the incident beam approximately in the direction of the c-axis. The Fourier coefficients of $\psi(x, y)$ then give the diffracted amplitudes Φ_g. We note that the first-order expansion of eqn (5.21) gives the kinematic result (eqn (5.15)) and that eqn (5.21) incorporates propagation effects if the $s_{h,k,0}$ are measured onto the Ewald sphere, as in Fig. 5.3. Since a plane in reciprocal space represents a projection in real space, the summed expression in eqn (5.21) above would represent simply a re-projection of the thin crystal potential if the $s_{h,k,0}$ were measured from the $(h, k, 0)$ zone onto a plane through the origin and normal to the electron beam. The resulting expression is then closely similar to the 'high-energy approximation' of Molière described in texts on the quantum theory of scattering. This expression, unlike the Born approximation, satisfies the optical theorem. The phase-grating or 'high-voltage limit' approximation of eqn (3.25) can also be derived by the method of partial wave analysis, or by a term-by-term comparison of the Born series (with $K_z = 0$) with the Fourier transform of the series expansion of eqn (3.25). The validity domain of eqn (5.21) has been investigated in unpublished work by A. F. Moodie.

5.3.4   *The projected charge density approximation*

Unlike the weak-phase object approximation, the projected charge density (PCD) approximation incorporates the effects of multiple scattering in a limited way, and is most useful for cases where the effects of spherical aberration are small.

To expose the principle of the PCD approximation, consider an image formed with no objective aperture and no spherical aberration. The exit-face wave amplitude for a phase object (including some multiple-scattering effects) is, from eqn (3.25),

$$\psi_e(x, y) = \exp(-i\sigma\phi_p(x, y))$$

(5.22)

with $\phi_p(x, y)$ the projected specimen potential in volt-nanometres. The back-focal plane amplitude is obtained by Fourier transformation as

$$\psi_d(u, v) = \mathscr{F}\{\exp(-i\sigma\phi_p(x, y))\} \exp(i\pi \, \Delta f \lambda(u^2 + v^2))$$
$$\approx \Phi(u, v)[1 + i\pi \, \Delta f \lambda(u^2 + v^2)] \tag{5.23}$$

if a small focusing error Δf is allowed and $\Phi(u, v)$ is the Fourier transform of $\exp(-i\sigma\phi_p(x, y))$. The image amplitude at unit magnification and without rotation is given by inverse transformation as

$$\psi_i(x, y) = \exp(-i\sigma\phi_p(x, y)) + i\pi \, \Delta f \lambda^{-1}(u^2 + v^2)\Phi(u, v)\}$$

A theorem from Fourier analysis can now be used (see, for example, Bracewell (1986)) which shows that if $f(x, y)$ and $\Phi(u, v)$ are a Fourier transform pair according to eqn (3.10), then

$$\mathscr{F}^{-1}((u^2 + v^2)\Phi(u, v)) = -\tfrac{1}{4}\pi^2 \, \nabla^2 f(x, y)$$

Using this result gives the image amplitude as

$$\psi_i(x, y) = \exp(-i\sigma\phi_p(x, y)) - (i \, \Delta f \lambda/4\pi) \, \nabla^2\{\exp(-i\sigma_p(x, y))\}$$
$$= \exp(-i\sigma\phi_p(x, y)) + (i \, \Delta f \lambda\sigma/4\pi) \exp(-i\sigma\phi_p(x, y)) \tag{5.24}$$
$$\times \{\sigma \, \nabla\phi_p(x, y) + i\nabla^2\phi_p(x, y)\}$$

so that the image intensity, to first order, is

$$I(x, y) = 1 - (\Delta f \lambda\sigma/2\pi) \, \nabla^2\phi_p(x, y) \tag{5.25}$$

Using Poisson's equation, $\nabla^2\phi_p(x, y) = -\rho_p(x, y)/\varepsilon_0\varepsilon$ we finally

$$I(x, y) \approx 1 + (\Delta f \lambda\sigma/2\pi\varepsilon_0\varepsilon)\rho_p(x, y) \tag{5.26}$$

A method for evaluating this by computer is given in eqn (5.71). Notice that a first-order (kinematic) expansion of the phase object expression has not been made so that, through the inclusion of multiple-scattering effects, this approximation can be expected to hold to greater thickness than the kinematic approximation. The PCD approximation does, however, assume a flat Ewald sphere. The quantity $\rho_p(x, y)$ is the projected total charge density including the nuclear contribution and not the electron charge density as measured in X-ray diffraction.

Using the sign convention for phases of Chapter 3, we see that the under-focused electron image (Δf negative) is deficient in regions of high specimen charge density. Here the developed micrograph is more transparent and a photographic print of the micrograph will appear dark in regions of high specimen charge density, such as around groups of heavy atoms. Also, for small defocus, the contrast is proportional to defocus. Experimental confirmation of eqn (5.26) is shown in Fig. 5.4.

The extension of this theory to include the effects of spherical aberration and a limiting objective aperture are discussed in the paper by Lynch, Moodie, and O'Keefe (1975). Their work shows that within a certain range

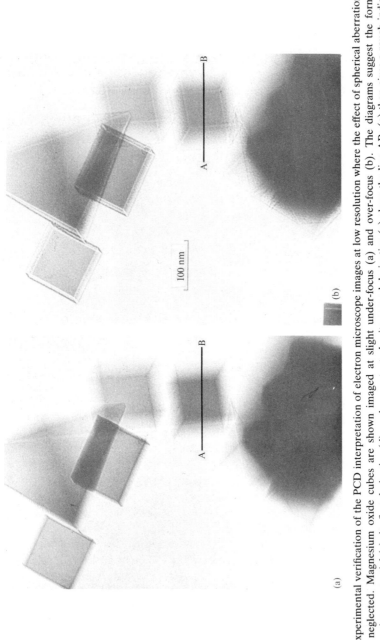

Fig. 5.4 Experimental verification of the PCD interpretation of electron microscope images at low resolution where the effect of spherical aberration can be completely neglected. Magnesium oxide cubes are shown imaged at slight under-focus (a) and over-focus (b). The diagrams suggest the form of the projected specimen potential (c), its first derivative (d), and a constant plus its second derivative (e) along the line AB; (e) then gives a rough indication of the contrast predicted by eqn (5.25). Part (f) shows the expected image with the sign of the focus reversed (Δf positive). That (e) and (f) correctly predict the contrast seen at the cube edges can be seen by comparison with (a) and (b). Notice that the cubes do not lie flat on the substrate and are balanced on a corner or edge. A clear plastic model cube or an ice block viewed in a direction close to one of its body diagonals will clarify the form of the projected potential in (c). The cubes appear as a cage of bright or dark wires.

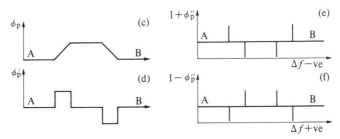

Fig. 5.4. (Continued)

of focus, specimen thickness, and resolution, eqn (5.25) can still be used as a basis for the interpretation of spherically aberrated images of limited resolution. That is, *within certain experimental conditions, the high-resolution image obtained represents the specimen projected total charge density as if seen through a spherically aberrated lens of limited resolution.* Note that PCD images are found in the neighbourhood of every Fourier image plane for periodic specimens (see Section 5.8). The experimental conditions necessary to allow such an interpretation (the aberrated PCD) are as follows.

1. At least 20 beams must be included within the objective aperture and contribute to the image. For a resolution limit of about 0.35 nm, this restricts the application of the method to specimens with a unit cell larger than about 0.7 nm. The method also holds for non-periodic specimens.

2. The crystal must be thin enough for the phase grating or 'high-voltage limit' approximation to hold (eqn (5.22)). This allows the use of specimens considerably thicker than the weak-phase object approximation (eqn (3.26)) used in Scherzer's theory. Detailed calculations of the amplitudes and phases of reflections have been made comparing the weak-phase object approximation with both the phase grating approximation and the correct coherent multiple-scattering solution (Lynch *et al.* 1975). These calculations suggest that the condition for the validity of the phase object approximation (eqn (6.20)) probably gives an overestimate of allowable thickness and that for specimens of medium atomic weight this approximation is likely to hold to a thickness of about 7 nm. The PCD approximation assumes a phase object for which no contrast is observed at Gaussian focus, other than the small effect of spherical aberration. At large specimen thickness, appreciable energy is scattered into beams outside the objective aperture, leading to aperture 'absorption' contrast. Thus an experimental check on specimen thickness can be made by observing the Gaussian focus ($\Delta f = 0$) image, which should show negligible contrast compared to the under-focus image if the simple PCD approximation is to be used. The largest aperture allowed by the resolution of the microscope should be used for this test. For some specimens it is possible that the aperture amplitude contrast will complement and enhance the phase contrast. O'Keefe and Sanders (1976) describe an image difference method for extracting the pure-phase contrast image component based on these observations. Note also that the PCD approximation holds to greater thickness for the systematics orientation (a single line of reflections excited), and with an increase in accelerating voltage.

3. The question of resolution, spherical aberration, and choice of focus must now be considered. These are related, since a requirement of the aberrated PCD approximation is that a first-order expansion of the transfer function (eqn (3.24)) is possible. This is so if

$$\pi \, \Delta f \lambda (u^2 + v^2) + \pi \lambda^3 C_s (u^2 + v^2)^2 / 2 \ll \pi/2 \tag{5.27}$$

where $(u^2 + v^2) = \theta^2/\lambda^2 = 1/d^2$ with θ the angle between the incident beam and a particular diffracted order. For the purposes of this chapter, d is taken as the image resolution if θ refers to the highest-order beam included within the aperture. A second condition limiting the application of the PCD approximation involves the approximation made between eqns (5.24) and (5.25). If C_s is known (see Section 8.2) and the resolution limit set by incoherent instabilities is also known, a maximum value for $(u^2 + v^2)$ can be found for the highest-order beam included within the aperture. Equation (5.27) can then be used to give the range of values of Δf for which the aberrated PCD image interpretation is possible. This range decreases with increasing resolution (aperture size). As an example, at 100 kV, images recorded with a central objective aperture which limits the resolution to 0.35 nm can be regarded as representing a spherically aberrated view of the specimen projected total charge density if recorded at an under-focus value between -30 and -80 nm ($C_s = 1.8$ mm). Notice that many combinations of C_s, Δf, and $(u^2 + v^2)_{\text{max}}$ do not allow any application of the PCD approximation—for example, there is no defocus value for which a 100 kV instrument with $C_s = 0.9$ mm can be used to give 0.28 nm resolution images interpretable on the aberrated PCD model.

To summarize, we see that the usefulness of the PCD approximation lies in its application to specimens of realistic thickness, since it includes the important effects of multiple scattering. Specimen thickness is best judged using Pendellösung fringes, as mentioned in Section 5.2.2. Often, sufficiently thin specimens can be made simply by crushing small particles of the material in an agate mortar under alcohol and picking these up onto holey carbon grids. A specimen is found near a hole in the grid and the carbon background is used to find a reference focus whose distance from Gaussian focus ($\Delta f = 0$) is known (see Sections 10.4 and 8.1).

The important limitation of the PCD approximation arises from the effect of spherical aberration, which, at the present stage of instrument development, prevents this approximation from being used at the level of atomic resolution. Under the experimental conditions used for single-atom imaging (see Chapter 6), eqn (5.27) is unlikely to be satisfied, so that the Scherzer condition and single-scattering approximation then become more useful guides to image interpretation.

The present state of the art in high-resolution transmission microscopy of oriented specimens thus allows a direct interpretation of images to about 0.20 nm resolution from the thinnest specimens without recourse to tedious image-matching methods. The extraction of finer detail from a micrograph requires detailed computer image-simulation, including all experimental parameters, and image-matching by trial and error. The improvement in resolution possible through the use of higher accelerating voltage is described in Section 6.3, where experimental images of the same crystal taken at 100 kV and 1 MeV accelerating voltage are compared.

5.4 Images of crystals with variable spacing—spinodal decomposition and modulated structures

A number of important image artifacts may arise in crystals in which either the lattice spacing or the chemical composition (or both) varies periodically. Before discussing the three most important artifacts, it is important to emphasize that, since the point-resolution of current HREM machines is about 0.15 nm, axial structure images of thin crystals of many of these materials may now be obtained which are free of electron optical 'artifacts'. Whether these thin crystals used for HREM specimens are representative of bulk material is another matter (see below). We first consider artifacts due to the lens alone, when using inclined illumination in order to obtain the highest 'resolution'. Following eqn (5.15) we saw that in wedge-shaped crystals the fringe spacing parallel to the edge will depend on the local variation of the phase ε of Φ_g with t. For the tilted illumination two-beam geometry of Fig. 5.1(b), the local periodicity observed in the image will be $1/u'_h(x)$, where

$$u'_h(x) = u_h + (1/2\pi)\,\mathrm{d}\varepsilon(x)/\mathrm{d}x \tag{5.28}$$

This result is obtained by analogy with the theory of phase modulation in electrical engineering. Equation (5.7) gives the image intensity variation. However, it is clear that any other spatially dependent contribution to the phase $\varepsilon(x)$ in eqn (5.7) will similarly influence the local image period. For crystals in which the lattice spacing varies, the objective lens phase factor $\chi(u)$ will also provide such a contribution, as follows.

Following Spence and Cowley (1979), we consider an idealized thin crystal containing a single atomic species for which the modulated crystal potential is proportional to

$$\phi(x) = \cos[2\pi u_0 x + A\sin(2\pi u_L x)]$$
$$= \cos[2\pi u_0 x + \phi'(x)]$$

Here A is the amplitude of the modulation and $1/u_L$ is its period. The true local period of this potential is $1/u'_x$ where

$$
\begin{aligned}
u'_x &= u_0 + (1/2\pi)\,\mathrm{d}\phi'/\mathrm{d}x \\
&= u_0 + u_L A\cos(2\pi u_L x)
\end{aligned}
\tag{5.29}
$$

We wish to compare this true period with that observed in the image. If $u_L \ll u_0$, the image intensity for tilted illumination is

$$
\begin{aligned}
I(x) &= |\Phi_0|^2 + 2\,|\Phi_0|\,|\Phi_u'|\cos[2\pi u_x' x + x(u_x) - x(u_0/2) + \varepsilon] \\
&= |\Phi_0|^2 + 2\,|\Phi_0|\,|\Phi_u'|\cos[2\pi u_x' x + \phi(x)]
\end{aligned}
\tag{5.30}
$$

where

$$
u_x = u_x' - u_0/2
\tag{5.31}
$$

We assume that the optic axis bisects the angle $2\theta_B = u_0\lambda$ between the direct 'beam' and that corresponding to the average period u_0. Here we have implicitly adopted the column approximation, (which may fail unless $u_L \ll u_0$), and we further assume that $\chi(u)$ varies slowly in the neighbourhood of u_0 (see Cockayne and Gronsky (1981) for a discussion of these points). The inverse local period observed in the image will be

$$
u_x'' = u_x' + (1/2\pi)\,d\chi/dx
\tag{5.32}
$$

The distorting effect of the lens may now be characterized by defining $F(x) = $ (local image spatial frequency—local object spatial frequency)/(local object spatial frequency). This gives a measure of the extent to which the lens 'amplifies' variations in object periodicity. Thus we define

$$
F(x) = (u_x'' - u_x')/u_x' = [d\chi/dx]/2\pi u_x'
$$

using eqn (5.32). Calculations of the maximum value of $F(x)$ (maximized over x) as a function of focus setting show (Spence and Cowley 1979) that F is minimized at the stationary phase focus $\Delta f = -C_s\lambda^2 u_0^2/4$. Under these conditions $\chi(u)$ varies least across the spectrum of satellite reflections around u_0 which result from the modulated structure.

This amplification effect may be as great as 400% under readily obtainable conditions, but may be only 90% at the optimum condition. It may thus explain observations of up to 33% variation in the lattice spacing of Ni–23%Au alloy films showing spinodal decomposition. However, other more subtle effects may be equally important. Cockayne and Gronsky (1981) have studied the effect of periodic segregation of the atomic species on these images, using the artificial superlattice method (see Section 5.11). They then find image spacings (again for tilted beam illumination) differing from the object spacing by as much as 50%. In addition, they find that the number of lattice planes in the spinodal period $1/u_L$ may be represented incorrectly in the image. However, if differences in atomic number can be ignored, these workers find agreement between object and image spacings within 10% for $C_s < 1$ mm at 100 kV (for $u_0/u_L = 18$ and small values of A).

The effects of elastic relaxation must also be considered in these very thin films. Shear stresses have been shown (Gibson and Treacy 1984) to distort the local unit cell dimensions appreciably. This effect is present for all imaging conditions, including axial structure imaging at atomic resolution. In particular, the elastic response of a thin modulated film will cause a bending of the lattice planes about the Y axis (normal to the beam and x). Calculations for this important effect (Gibson and Treacy 1984) show that it may lead to differences of up to 33% between the bulk lattice spacing and that present in very thin films. A comprehensive analysis of TEM images of spinodally decomposed $In_xGa_{1-x}As_yP_{1-y}$ films may be found in Treacy, Gibson, and Howie (1985), who also discuss in detail the elastic relaxation expected in these very thin films.

We may conclude that the determination of local atomic composition from measurements of lattice image spacings in conjunction with Vegard's law is fraught with very great difficulties if the tilted illumination geometry is used. Useful results have, however, been obtained in a few cases after extensive and careful analysis—for example Hall, Self, and Stobbs (1983) were able to relate lattice spacings to the carbon content of retained austenite in a dual-phase steel. In the axial geometry structure, images may, however, be obtained, for which a straightforward interpretation is possible provided that sufficiently thin samples are used and that the point-resolution exceeds the lateral interatomic spacing. No information, however, is obtainable from these images on composition variations on the beam direction.

5.5   Are the atom images black or white? A simple symmetry argument

Since a change of focus of $\Delta f_f/2$ (see Sections 5.1 and eqn (5.63b)) reverses the contrast of lattice images, the question arises as to how one can tell whether dark or light regions in an image coincide with the atomic column positions. If a diffractogram is obtainable, and the thickness is known, a determination can be made by image-matching techniques (see Section 5.11). It is sometimes possible, however, to use the point group symmetry of a defect of known structure to resolve this question (Olsen and Spence 1981). As an example, Fig. 5.5 shows the structure of an intrinsic stacking fault in silicon viewed along $[1\bar{1}0]$. At limited resolution, the image appears as shown in Fig. 5.5(b). The stacking fault contains two twofold symmetry axes (in projection) as shown at A and B. Since both are known *a priori* to fall on a layer of atoms (rather than tunnels), they may be used to determine whether a given image shows white tunnels or white atoms. (A 'layer' is a line of dots running across the figure). This is not possible in perfect crystal. Figure 5.6 shows a series of dynamical computed images which demonstrate this idea. The contrast along a horizontal line across these images will be found to be that of the atoms, not the tunnels. This powerful argument holds for any thickness of sample and focus setting if the image is correctly aligned and stigmated. A similar argument can be used to identify the position of a sub-nanometer diameter electron probe in coherent microdiffraction studies (see Section 11.4).

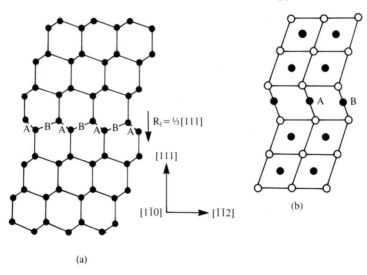

(a)

(b)

Fig. 5.5 (a) The structure of an intrinsic stacking fault projected along [1\bar{1}0] in the diamond structure. (b) The simplified projection seen at limited resolution. The open circles are tunnel sites, filled circles are pairs of atomic columns. The screw axis at A falls on a filled circle, B is a twofold axis. Arrows indicate the layers which change position with focus. (From Olsen and Spence (1981), reprinted by permission.)

5.6   The multislice method and the polynomial solution

The multislice theory of dynamical electron diffraction was developed before computers became available commercially (Cowley and Moodie 1959). It was not, therefore, originally intended as a numerical algorithm, but as an attempt to obtain a closed-form expression for the dynamical Bragg beam amplitudes Φ_g. The result is (Moodie 1972)

$$\Phi_g = \sum_{n=1}^{\infty} E_n(g)Z_n(g) \tag{5.33}$$

where the terms E_n depend only on structure factors, while the functions $Z_n(g)$ involve polynomials which are functions of the geometric factors only (excitation errors, wavelength, lattice constant). This polynomial result has subsequently been obtained by other authors using different methods, and can be shown to reduce to the phase-grating approximation, to a power series in thickness, or to the two-beam approximation under suitable approximations (see Cowley (1975) for references). Equation (5.33) has been used to explain the occurrence of dynamically forbidden reflections (see Section 5.10). With the advent of electronic computers, a form of the multislice theory more convenient for computational work was applied in calculations first by Goodman and others (see Goodman and Moodie (1974)). This recursion relation, which has become the basis of all modern multislice computations, is

$$\Phi_g(n) = \sum_h \Phi_h(n-1)P_{\Delta Z}(\mathbf{h})Q(\mathbf{g}-\mathbf{h}) \tag{5.34}$$

where $P_{\Delta Z}(\mathbf{h})$ is given by eqn (3.16) and $Q(\mathbf{g})$, the 'phase grating amplitudes' are given by the Fourier coefficients of eqn (5.22). Here the crystal has been divided into slices normal to the beam (not necessarily equal in thickness to the period of the lattice in that direction) and $\Phi_g(n)$ is the set of Bragg diffracted beams emerging from slice n. Thus $\phi_p(x, y)$ is the specimen potential projected through a single slice. Computational aspects of eqn (5.34) are discussed in Section 5.11. $\Phi_g(1)$ are taken to be the Fourier coefficients of the incident plane-wave. The derivation of the time-independent Schrödinger equation from eqn (5.34) is outlined in Goodman and Moodie (1974), where the relationship of this approach to other dynamical theories is fully discussed (see also Section 5.11). Research workers wishing to write a computer program based on the multislice algorithm are referred to Section 5.11, and, in particular, to the articles by Self, O'Keefe, Buseck, and Spargo (1983) and Ishizuka (1985). The propagation function $P_{\Delta Z}(\mathbf{g})$ can be written

$$P_{\Delta Z}(\mathbf{g}) = \exp(-2\pi i S_g \,\Delta Z) \tag{5.35}$$

where S_g, the excitation error for beam $\mathbf{g}$, is taken negative for reflections outside the Ewald sphere (see Section 5.12). A review of 'real space' multislice methods may be found in van Dyke and Coene (1984).

The effects of steeply inclined boundary conditions on the multislice formulation have been discussed by Ishizuka (1982a). Anstis (1977) has shown that a unity sum of beam intensities is not a sufficient condition to ensure that the calculation includes enough beams. The calculation will be normalized for any number of beams in the limit of sufficiently small slice thickness. The method may be extended to include three-dimensional diffraction effects by choosing the slice thickness to be much less than the periodicity in the beam direction. An early example of such a calculation, used to analyse the occurrence of 'forbidden' termination reflections (see Section 12.8) can be found in Lynch (1971).

5.7 Bloch wave methods, bound states and 'symmetry reduction' of the dispersion matrix

A solution to the N-beam dynamical problem of kilovolt electron diffraction was first given by H. Bethe (Bethe, 1928) for both the Bragg and Laue geometries. Pedagogically sound treatments of this theory can be found in several texts. Amongst the best reviews and texts are: Hirsch et al. (1977), Metherell (1975), Dederichs (1972), Reimer (1984), Humphreys (1979), and Howie (1971). We consider here only those aspects of dynamical theory which are relevant to lattice imaging, such as symmetry reduction of the

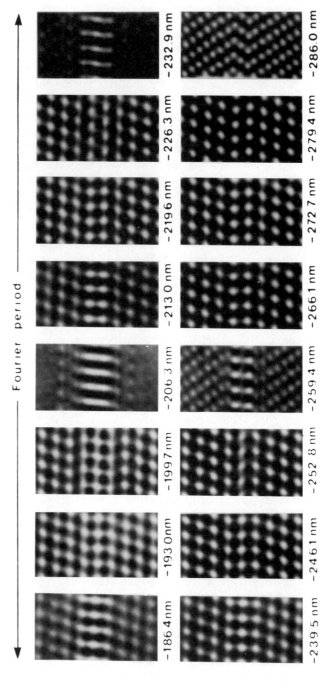

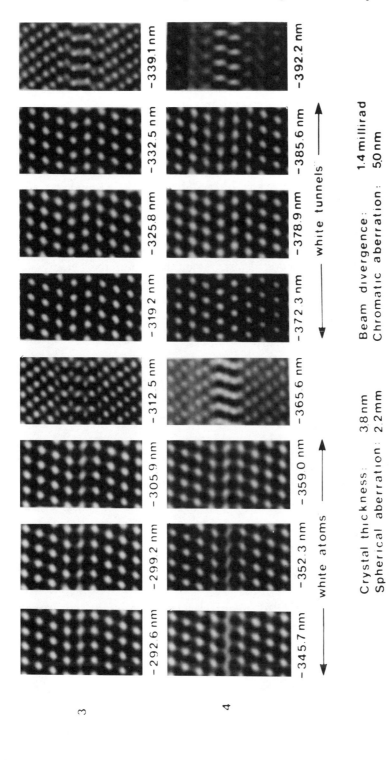

Fig. 5.6 Computed images of an intrinsic stacking fault in silicon. The focus increment between images is $\Delta f_f/8$. The stationary-phase image occurs near $\Delta f_s = -305.9$ nm. Images differing from this by Δf_f have similar black/white contrast (e.g. $\Delta f = -252.8$ nm) but different dot positions. Half-period images are also seen (e.g., Δf, $\Delta f = -286.0$ nm). Since the Fourier imaging is exact in the [111] direction only, image dot displacements in the [$\bar{1}\bar{1}2$] directions can be used to estimate focus.

dispersion matrix, the theory of dynamically forbidden reflections, and the direct observation of separate Bloch waves in real space. It is important to point out that the 'quantum-mechanical' sign convention used in this section is the opposite of the 'crystallographic convention' adopted elsewhere throughout this book (see Section 5.12). All the above reviews use the following quantum-mechanical sign convention.

It is conventional to define quantities

$$U_g = \frac{\sigma v_g}{\pi \lambda} = \frac{1}{\lambda \xi_g} = \frac{2m \, |e| \, v_g}{\hbar^2} \tag{5.36}$$

where v_g are the Fourier coefficients of crystal potential (in volts) (see also eqn (5.18)). Note that σ here involves the relativistically corrected electron mass m and wavelength λ (see eqn (2.4) and following). It has been shown by K. Fujiwara and A. Howie (see Humphreys (1979) and Gevers and David (1982) for reviews) that electron spin effects are negligible in transmission electron diffraction. Thus, rather than solve the Dirac equation, we may derive a relativistically corrected Schrödinger-like equation as follows. The relativistic energy–momentum conservation equation for an electron accelerated to a potential $e(V_0 + \phi)$ is

$$p^2 c^2 + m_0^2 c^4 = [|e| \, (V_0 + \phi) + m_0 c^2]^2 = W^2 \tag{5.37a}$$

Here V_0 is the accelerating voltage and $\phi = \phi(\mathbf{r})$ is the crystal potential, whose Fourier coefficients are given by eqn (5.18). The total relativistic energy is

$$W = mc^2 = |e| \, (V_0 + \phi) + m_0 c^2 = \gamma m_0 c^2$$

with γ given by eqn (6.18).

Neglecting a term $|e|^2 \phi^2$, these equations give;

$$p^2 c^2 - 2mc^2 \, |e| \, \phi = |e| \, V_0 (2m_0 c^2 + |e| \, V_0)$$

Replacing p by the operator $-i\hbar\nabla$, and operating on Ψ gives the Shrödinger-like equation:

$$(\hbar^2 / 2m_0) \, \nabla^2 \Psi(\mathbf{r}) + |e| \, \gamma \phi(\mathbf{r}) \Psi(\mathbf{r}) + |e| \, V_r \Psi(\mathbf{r}) = 0 \tag{5.37b}$$

where $\gamma = m/m_0$ and V_r, the relativistically corrected accelerating voltage, is given following eqn (2.4).

The solution to eqn (5.37a) for a free electron, with $p = hk_i$ and $\phi = 0$, gives the vacuum electron wavevector as

$$|\mathbf{k}_i| = \frac{W\beta}{hc} = \frac{1}{\lambda} = \frac{\gamma \beta m_0 c^2}{12.4}$$

Here if $m_0 c^2 = 511$ (kV) then $|\mathbf{k}_i|$ is given in (Ångstroms)$^{-1}$. This result is consistent with eqn (2.4).

The solution to eqn (5.37b) is taken to be a linear sum of Bloch waves of the form

$$\Psi(\mathbf{r}) = \sum_j \alpha_j b(\mathbf{k}^{(j)}, \mathbf{r}) \tag{5.37c}$$

where

$$b(\mathbf{k}^{(j)}, \mathbf{r}) = \sum_\mathbf{g} C_\mathbf{g}^{(j)} \exp((2\pi i \mathbf{k}^{(j)} + \mathbf{g}) \cdot \mathbf{r}) \tag{5.37d}$$

which describes the total electron wavefunction inside the crystal. Inside the crystal the beam electron wavevector corrected for the effect of the mean inner potential is

$$\mathbf{k}_0^2 = (1/\lambda)^2 + U_0 = \mathbf{k}_i^2 + U_0 \tag{5.37e}$$

where $U_\mathbf{g}$ has dimensions (length)$^{-2}$ and λ is given by eqn (2.5). One Bloch wave is associated with each branch j of the dispersion surface (see above references), and values of $C_\mathbf{g}^{(j)}$ and $\mathbf{k}^{(j)}$, the Bloch wave coefficients and wavevectors, are obtained by numerical solution of the eigenvalue equation (5.38a) which results from substituting eqn (5.37d) into the Schrödinger equation (5.37b). The Fourier coefficients of $\phi(\mathbf{r})$ are $v_\mathbf{g}$. If backscattered waves are neglected and the small-angle approximation

$$\mathbf{k}_0^2 - (\mathbf{k}^{(j)} + \mathbf{g})^2 \approx 2\mathbf{k}_0(S_\mathbf{g} - \gamma^{(j)}) \tag{5.37f}$$

is made, where $\gamma^{(j)} = k_z^{(j)} - |\mathbf{k}_0|$, and we define $U_{\mathbf{g}-\mathbf{g}} = 2k_0 S_\mathbf{g}$, we then obtain

$$\sum_\mathbf{h} U_{\mathbf{g}-\mathbf{h}} C_\mathbf{h}^{(j)} = 2 |\mathbf{k}_0| \gamma^{(j)} C_\mathbf{g}^{(j)} \tag{5.38a}$$

which can be written

$$\begin{bmatrix} 2|\mathbf{k}_0| S_\mathbf{h} & \cdots & U_{\mathbf{g}-\mathbf{h}} & \cdots & U_{\mathbf{l}-\mathbf{h}} \\ \vdots & & \vdots & & \vdots \\ U_{\mathbf{h}-\mathbf{g}} & \cdots & 2|k_0| S_g & \cdots & U_{\mathbf{l}-\mathbf{g}} \\ \vdots & & \vdots & & \vdots \\ U_{\mathbf{h}-\mathbf{l}} & \cdots & U_{\mathbf{g}-\mathbf{l}} & \cdots & 2|k_0| S_\mathbf{l} \end{bmatrix} \begin{bmatrix} C_\mathbf{h}^{(j)} \\ \vdots \\ C_\mathbf{g}^{(j)} \\ \vdots \\ C_\mathbf{l}^{(j)} \end{bmatrix} = 2 |\mathbf{k}_0| \gamma^{(j)} \begin{bmatrix} C_\mathbf{h}^{(j)} \\ \vdots \\ C_\mathbf{g}^{(j)} \\ \vdots \\ C_\mathbf{l}^{(j)} \end{bmatrix} \tag{5.38b}$$

or

$$\mathbf{AC} = 2 |\mathbf{k}_0| \gamma^{(j)} \mathbf{C} \tag{5.39}$$

Here the eigenvalues $\gamma^{(j)}$ measure the deviation of the dynamical dispersion surfaces from spheres of radius $|\mathbf{k}_0|$ centred on each reciprocal lattice point. We take $\mathbf{A}$ to be an $n \times n$ matrix. The spheres correspond to the dispersion surfaces of plane waves in the 'empty lattice' approximation, in which all $U_g = 0$. Here $\mathbf{k}^{(j)2} = k_z^2 + k_t^2$. The excitation errors $S_\mathbf{g}$ (see Fig. 5.3) are related to the two-dimensional vector $\mathbf{K} = -\mathbf{k}_t^{(j)}$ drawn from the origin of reciprocal space to the centre of the Laue circle by

$$2 |\mathbf{k}_0| S_\mathbf{g} = 2\mathbf{K} \cdot \mathbf{g} - \mathbf{g}^2 \tag{5.40}$$

For most HREM work in axial orientations $K = 0$ so that

$$S_g = -\lambda g^2/2 \qquad (5.41a)$$

in accordance with the sign convention of Fig. 5.3. In deriving eqn (5.38) the Laue (transmission) geometry has been assumed for a parallel-sided slab of crystal; backscattering has been neglected, and the projection approximation assumed. Thus, only reflections in the zero-order Laue zone are considered important.

The effects of steeply inclined boundary conditions on lattice images are discussed in Ishizuka (1982) from the multislice viewpoint. A discussion of this problem in the Bloch wave forumulation can be found in Metherall (1975) and Gjønnes and Gjønnes (1985). Three-dimensional diffraction effects in lattice imaging are discussed in Section 12.8. For this case, where $g_z \neq 0$, it is necessary to renormalize the eigenvectors $\mathbf{C}$ (see Kambe and Moliere (1970), Buxton (1976), and Lewis, Villagrana and Metherall (1978)), and eqn (5.38) cannot be used. For the most general case of a parallel-sided slab of crystal traversed by an electron beam inclined to the surface normal $\hat{n}$ the renormalized equation to be solved is (Zuo, Spence and O'Keeffe (1988))

$$\frac{B_g(2\mathbf{k}^{(j)} \cdot \mathbf{g} + \mathbf{g}^2)}{1 + g_n/k_n} - \sum_h \frac{B_h U_{g-h}}{(1 + g_n/k_n)^{1/2}(1 + h_n/k_n)^{1/2}} = -2k_n\gamma B_g \qquad (5.41b)$$

Here $g_n = \mathbf{g} \cdot \hat{n}$, $k_n = \mathbf{k}_0 \cdot \hat{n}$ and the result incudes the effects of multiple scattering in the beam direction (three-dimensional dynamical diffraction).

In the absence of absorption effects, the dispersion or structure matrix $\mathbf{A}$ is Hermitian (therefore $\gamma^{(j)}$ are real) and, for centrosymmetric crystals, it is real and symmetric ($U_g = U_{-g}$). It reflects the point symmetry about the centre of the Laue circle in the Brillouin zone. The following relations can be also shown to hold amongst the eigenvectors and eigenvalues:

$$C_g^{(j)}(\mathbf{K} + \mathbf{h}) = C_{g-h}^{(j)}(\mathbf{K}) \qquad (5.42)$$

$$\gamma^{(j)}(\mathbf{K} + \mathbf{h}) = \gamma^{(j)}(\mathbf{K}) + S_h \qquad (5.43)$$

$$\mu^{(j)}(\mathbf{K} + \mathbf{h}) = \mu^{(j)}(\mathbf{K}) \qquad (5.44)$$

$$\gamma^{(j)}(\mathbf{K}) = \gamma^{(j)}(-\mathbf{K}) \qquad (5.45)$$

$$C_g^{(j)}(\mathbf{K}) = C_{-g}^{(j)*}(-\mathbf{K}) \qquad (5.46a)$$

$$\sum_g C_g^{(i)} C_g^{(j)*} = \delta_{ij} \qquad (5.46b)$$

$$\sum_j C_g^{(j)} C_h^{(j)*} = \delta_{gh} \qquad (5.46c)$$

Thus, the eigenvectors form a complete orthogonal and normalized set. The matrix $\mathbf{C}$ whose columns are the complex eigenvectors $\mathbf{C}^{(j)}$ is unitary. Here $\mathbf{K}$, the vector to the centre of the Laue circle, is assumed to be 'reduced', to lie within the first Brillioun zone, and $\mu^{(j)}$ is the absorption parameter for the jth Bloch wave. In the axial orientation, the point group of the dispersion matrix is equal to that of the projected crystal structure at the point of highest symmetry. The symmetries of the Bloch wave coefficients can be found from group character tables (see below). From eqns (5.42–5.44) we see that it is only necessary to solve for $C_{\mathbf{g}}^{(j)}$, $\gamma^{(j)}$, and $\mu^{(j)}$ within the two-dimensional reciprocal-space unit cell in order to obtain solutions for all incident beam directions. Note that while the Bloch waves given by eqn (5.37d) are periodic in reciprocal space, the total wavefunction given by eqn (5.37c) (which incorporates the effects of boundary conditions) is not. Bloch waves are numbered in order of decreasing $k_z^{(j)}$, so that the wave with largest $k_z^{(j)}$ has index 1. All Bloch waves correspond to the same total energy. Whereas the dispersion matrix expresses the point group symmetry of the crystal, the translational symmetry is expressed by the total wave function (eqn (5.37c)), which depends on boundary conditions. The excitation amplitudes α_j are obtained by matching the incident wavefunction and its gradient to the crystal wavefunction at the top surface of the crystal. We then find that $\alpha^{(j)} = C_0^{(j)*}$. The result for the total HREM image wavefunction, including the effects of lens aberrations, is

$$I(\mathbf{r}) = \sum_{\mathbf{g},\mathbf{h}} \sum_{i,j} C_0^{(i)} C_0^{(j)} C_{\mathbf{g}}^{(i)} C_{\mathbf{h}}^{(j)} \exp(-2\pi(\gamma_{\mathrm{Im}}^{(i)} + \gamma_{\mathrm{Im}}^{(j)})t)$$

$$\times \exp\{i[2\pi(\gamma_{\mathrm{R}}^{(i)} - \gamma_{\mathrm{R}}^{(j)})t + 2\pi(\mathbf{g} - \mathbf{h}) \cdot \mathbf{r} - \chi(\mathbf{g}, \Delta f) + \chi(\mathbf{h}, \Delta f)]\}$$

$$\tag{5.47}$$

$$= |\psi_{\mathrm{i}}(\mathbf{r}, \Delta f, \mathbf{K}_0)|^2 \tag{5.48}$$

The image is seen to contain, in general, all possible 'false' periodicities $(\mathbf{g} - \mathbf{h})$, including the half-spacings with $\mathbf{h} = -\mathbf{g}$. Only under linear imaging conditions (small t) is this intensity simply related to the crystal potential (see Pirouz (1981)). Here $\chi(\mathbf{h}, \Delta f)$ is defined by eqn (3.24) (Δf positive for overfocus), and $C_{\mathbf{g}}^{(i)} = C_{\mathbf{g}}^{(i)}(K)$, etc. The quantities $\gamma_{\mathrm{Im}}^{(i)}$ are the imaginary parts of the eigenvectors $\gamma^{(i)} = \gamma_{\mathrm{R}}^{(i)} + i\gamma_{\mathrm{Im}}^{(i)}$, introduced to account for absorption. This is described by the use of an optical potential $U(\mathbf{r}) = U_{\mathrm{R}}(\mathbf{r}) + iU_{\mathrm{Im}}(\mathbf{r})$ so that the Fourier coefficients $v_{\mathbf{g}}$ are now replaced by complex coefficients $(v_{\mathbf{g}} + iv_{\mathbf{g}}')$. Values of $\gamma_{\mathrm{Im}}^{(i)}$ may be found (for given $U_{\mathbf{g}}'$) by matrix diagonalization of the complex matrix $\mathbf{A}$. A good approximation, however, based on perturbation theory, expresses $\gamma_{\mathrm{Im}}^{(i)}$ in terms of $U_{\mathbf{g}}'$ and $C_{\mathbf{g}}^{(j)}$ (see Humphreys (1979) for a review). Some difficulties in accurately accounting for absorption effects in HREM are further discussed in Section 5.9.

For thin crystals showing a centre of symmetry in projection, we may take $U_{\mathbf{g}}' = \gamma_{\mathrm{Im}}^{(i)} = 0$; eqn (5.47) then becomes the fundamental equation for lattice imaging in the Bloch wave formulation. Partial coherence effects are discussed in the next section.

For real coefficients $U_{\mathbf{g}}$ the solution of the eigenvalue problem of eqns (5.38) may be accomplished by standard computational methods. Programs are also available for complex $U_{\mathbf{g}}$, although slower. A perturbation expression for $\gamma_{\mathrm{Im}}^{(j)}$ has also been given (Hirsch et al. 1977) which avoids the need to work with a complex dispersion matrix in centrosymmetric projections.

Further treatments of lattice imaging from the Bloch wave point of view can be found in the work of Desseaux, Renault, and Bourret (1977), Pirouz (1981), Marks (1984), Fujimoto (1978), and Kambe (1982). Pirouz (1981) has related the projected charge density approximation (see Section 5.3.4) and the projected potential approximation (see Section 3.4) to the Bloch wave formulation through a second-order expansion of the term $\exp(i\gamma^{(j)}t)$ (see eqn (5.47)). In the theory of electron channelling and channelling radiation (see Section 11.1), it is customary to use a modified Bloch wave theory in which the total wavefunction is written as a product of a z-dependent plane wave of very high kinetic energy (a 'free' state), and a transverse wavefunction. The eigenvalue problem for the transverse states is solved, and these may be either free or bound, depending on whether the corresponding transverse energy $h^2 k_0 \gamma^{(j)}/m$ is either larger or smaller than the maxima in the interatomic potential (see Kambe, Lehmpfuhl and Fujimoto (1974) for more detail). This approach has been applied to the problem of lattice imaging by several authors. Fujimoto (1978) has used it to analyse the apparent periodicity in thickness of some lattice images (see Section 5.14), while Marks (1984) has combined this approach with the 'k.p' method of band structure calculation to assess the effects of small misalignments on lattice images (including instrumental aberration effects). Kambe (1982) has shown that, for Ge [110] images, a similar three-Bloch-wave analysis indicates that at certain thicknesses two of these interfere destructively, so that one may obtain directly by HREM real-space images of individual Bloch waves at certain thicknesses. (This analysis does not make the 'independent Bloch wave' approximation.) It is a feature of these transverse-wavefunction treatments that the role of the excitation errors is disguised. They have the advantage, however, of exposing the rather small number of bound states normally excited, even in cases where the number of Bragg beams is very large (Vergasov and Chuklovskii 1985).

The computation of HREM images in axial orientations by direct application of eqn (5.39) involves a large amount of redundant calculation if the crystal structure possesses some symmetry. There have been two (related) approaches to the problem of reducing the dimension of the dispersion matrix in axial orientations of high symmetry, and these are now discussed. In non-axial orientations some symmetry may also be preserved if, for example, $\mathbf{K}$ lies on a symmetry element. These methods have variously been known as 'beam reduction', 'Niehrs reduction' or 'symmetry reduction'.

General group-theoretical approaches to this problem have been given by Tinnappel (1975) and Kogiso and Takahashi (1977). Here the known point group symmetry of the crystal projection is used to find a matrix $\mathbf{T}$ which, by a similarity transformation, gives $\mathbf{A}$ in block-diagonal form. Eigenvalues and eigenvectors can then be found separately for each of these blocks which are of lower dimension than that of $\mathbf{A}$. The columns of $\mathbf{T}$ belong to the irreducible representation of the group of the incident wavevector, and may be obtained from group character tables (Cracknell 1968).

A second approach relies on more physical arguments, and inspection of the possible symmetries of the Bloch waves which are consistent with the known symmetry of the crystal projection. Simple worked examples can be found in Hirsch *et al.* (1977). The most extensive applications of this method have been made by Fukuhara (1966), Taftø (1979), and Howie (1971). Other examples can be found in Serneels and Gevers (1969) for the systematics case and Desseaux *et al.* (1977) for [011] projections of the diamond structure. In this manner cases involving as many as twelve beams (related by symmetry) can be reduced to a 2×2 matrix eigenvalue problem. We now proceed to describe a simple method for setting up reduced dispersion-matrices for evaluation by computer. To illustrate the method, we will also take as an example relevant to HREM work the soluble axial seven-beam case (Blume (1965)) for a crystal projection with sixfold axis of symmetry parallel to the incident beam. Figure 5.7 (from Taftø (1979)) provides a representation for the symmetries of Bloch waves for the 10 two-dimensional crystallographic point groups. For each of these are shown the allowed symmetries of the Bloch waves, including some of the reversal point groups (Cochran 1952).

Figure 5.8 shows the relevant reciprocal lattice vectors for a seven-beam example. The essential point of symmetry reduction is that, if all the Fourier coefficients U_{g-h} which are equivalent by crystal symmetry are re-labelled, say, V_n with a single index, it will be found that many of the equations (5.38a) become identical (for one Bloch wave j). The number of equations will be equal to the number of lattice points in the 'asymmetric sector' shown in Fig. 5.8. Thus, in simple cases (such as Fig. 5.8), the number of Bloch waves N excited in axial orientations is equal to one more than the number of distinct excitation errors ($N = 2$ in this case). In addition, by collecting terms it will be found that the number of terms in each equation is also reduced. Now the eigenvector elements C_g can be written

$$C_g = c_g \exp(\alpha_g) \qquad (5.49)$$

where values of α_g are obtainable from Fig. 5.7. Values of c_g are equal for all symmetry-related points, and are further constrained by the orthogonality relations (5.46b) and (5.46c).

In order to obtain a general procedure for beam reduction, we represent each $\mathbf{g}$-vector by two indicies, and write eqn (5.38a) as

$$\sum_{k=1}^{N} \sum_{l=1}^{h_k} U_{i,l-k,l} \exp(\alpha_{kl}) c_{k,l}^{(j)} = 2 |\mathbf{k}_0| \gamma^{(j)} C_{i,l}^{(j)} \qquad (5.50)$$

for Bloch wave j. Here the sum on k extends over the N non-equivalent reciprocal lattice points within the 'asymmetric sector' shown in Fig. 5.8, and the sum on l extends over all symmetry related points in other sectors. Unlike the n equations of (5.38b), however, we now have just N equations

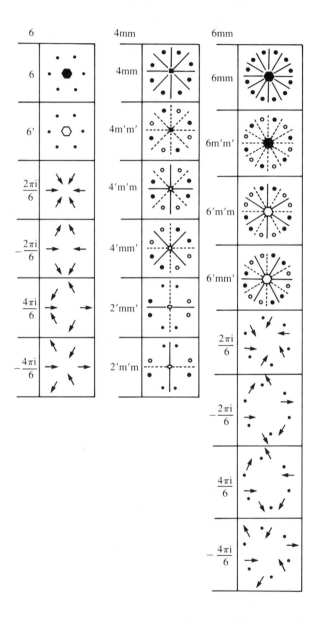

Fig. 5.7 The 10 two-dimensional point groups according to which a projected crystal structure may be classified. The arrows indicate the phases α_g of the complex coefficients of the allowed Bloch waves, which are consistent with a given crystal symmetry. Open and filled circles indicate reversed signs.

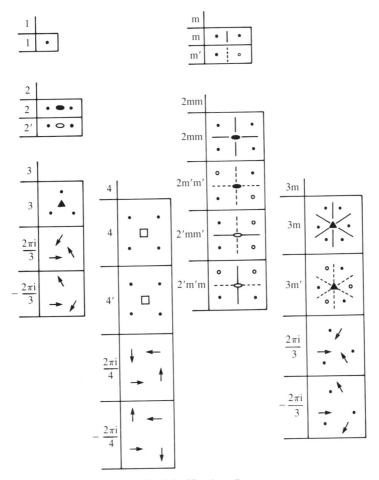

Fig. 5.7. (Continued)

(one for each i). Values of k and l are indicated on the example. It will transpire that the dimension of the reduced matrix is equal to N (adjusted for 'multiplicities'). For our seven-beam example it is convenient to define

$$V_1 = U_{1,1-2,1} = U_{1,1-2,2} = U_{1,1-2,3} = U_{1,1-2,4} = U_{1,1-2,5} = U_{1,1-2,6}$$

For the first Bloch wave ($j = 1$) we find from Fig. 5.7 (point group 6) that $\alpha = 0$. Equation (5.50) then becomes, for $i = 1$,

$$U_{1,1-1,1}c_{1,1} + (U_{1,1-2,1} + U_{1,1-2,2} + U_{1,1-2,3} + U_{1,1-2,4}$$
$$+ U_{1,1-2,5} + U_{1,1-2,6})c_{2,1} = 2\,|\mathbf{k}_0|\,\gamma c_{1,1}$$

In terms of V_i and $U_{1,1-1,1} = 0$ we have

$$6V_1 c_{2,1} = 2\,|\mathbf{k}_0|\,\gamma c_{1,1} \tag{5.51}$$

In a similar way the second of eqns (5.50) ($i = 2$) becomes (still for the

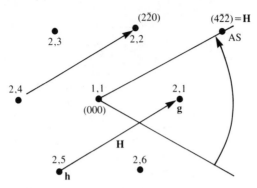

Fig. 5.8 Example of symmetry reduction. The seven-beam case for a [111] zone-axis pattern in cubic crystals. The asymmetric sector (AS) is indicated, and Miller indices (e.g. (220)) and values of (k, l) are shown. Two vectors $(\mathbf{g} - \mathbf{h}) = \mathbf{H}$ are shown for which the corresponding values of $U_{\mathbf{H}}$ would be equal. All $\{220\}$ excitation errors are equal.

first Bloch wave $j = 1$)

$$V_1 c_{11} + (2\,|\mathbf{k}_0|\,S_\mathbf{g} + 2V_1 + 2V_2 + V_3)C_{21} = 2\,|\mathbf{k}_0|\,\gamma c_{2,1} \qquad (5.52)$$

where $U_{2,1-2,1} = 2\,|\mathbf{k}_0|\,S_\mathbf{g}$ and $V_2 = U_{2,1-2,3}$, etc., and $V_3 = U_{2,1-2,4}$, etc. We also define

$$X = (2V_1 + 2V_2 + V_3 + 2\,|\mathbf{k}_0|\,S_\mathbf{g})/2V_1$$

Equations 5.51 and 5.52 can then be written

$$\begin{bmatrix} -2\,|\mathbf{k}_0|\,\gamma & 6V_1 \\ V_1 & 2\,|\mathbf{k}_0|\,(S_\mathbf{g} - \gamma) + 2V_1 + 2V_2 + V_3 \end{bmatrix} \begin{bmatrix} C_{11}^{(1)} \\ C_{21}^{(1)} \end{bmatrix} = 0 \qquad (5.53)$$

which can be solved for two values of $\gamma^{(j)}$. If eqn (5.46c) and Fig. 5.7 are also used, both $C_{11}^{(1)}$ and $C_{21}^{(1)}$ and the remaining five values of $C_{k1}^{(1)}$ can be found. The procedure continues with $\alpha = \pi$ for the second Bloch wave (Fig. 5.7). The result for all cases may be expressed as a matrix whose columns (j) contain the eigenvector elements $C_\mathbf{g}^{(j)}$ for each Bloch wave (j). We find*

$$C = \begin{bmatrix} X_+ & 1/\sqrt{6} & 0 & 1/\sqrt{3} & 0 & 1/\sqrt{3} & X_- \\ X_+ & -1/\sqrt{6} & \frac{1}{2} & \dfrac{1}{2\sqrt{3}} & \frac{1}{2} & -\dfrac{1}{2\sqrt{3}} & X_- \\ X_+ & 1/\sqrt{6} & \frac{1}{2} & -\dfrac{1}{2\sqrt{3}} & -\frac{1}{2} & -\dfrac{1}{2\sqrt{3}} & X_- \\ X_+ & -1/\sqrt{6} & 0 & -1/\sqrt{3} & 0 & 1/\sqrt{3} & X_- \\ X_+ & 1/\sqrt{6} & -\frac{1}{2} & -\dfrac{1}{2\sqrt{3}} & \frac{1}{2} & -\dfrac{1}{2\sqrt{3}} & X_- \\ X_+ & -1/\sqrt{6} & -\frac{1}{2} & \dfrac{1}{2\sqrt{3}} & -\frac{1}{2} & -\dfrac{1}{2\sqrt{3}} & X_- \\ \sqrt{6}X_- & 0 & 0 & 0 & 0 & 0 & -\sqrt{6}X_+ \end{bmatrix} \qquad (5.54)$$

* See note on p. 195.

where

$$X_{+,-} = (\tfrac{1}{12}[1 \pm X/\sqrt{6+X^2}])^{1/2} \qquad (5.55)$$

Here $C_{11}^{(1)} = C_{21}^{(1)} = X_+$, etc. The eigenvectors in eqn (5.54) have been converted to real form for convenience. Two of the eigenvalues are found to be

$$\gamma^{(1)} = [V_1(X + (6+X^2)^{1/2})]/2\,|\mathbf{k}_0|$$
$$\gamma^{(7)} = [V_1(X - (6+X^2)^{1/2})]/2\,|\mathbf{k}_0| \qquad (5.56)$$

Note that since $\alpha^{(j)} = C_0^{(j)*}$, the excitation amplitudes for all the Bloch waves except two (the first and last column in C) are zero. Only waves 1 and 7 are excited.

Returning now to the general problem for computation, we wish to replace the matrix $\mathbf{A}$ by a matrix containing elements only for those N points within the asymmetric sector. The elements of the reduced dispersion matrix are therefore, from eqn (5.50),

$$h_{ik} = \sum_{l=1}^{h_k} U_{i1-kl}\exp(\alpha_{kl}) \qquad (5.57)$$

with the sum l over symmetry-related points. For such a matrix $h_{ik} \neq h_{ki}^*$, and it is more convenient for computational purposes to work with an equivalent reduced matrix whose elements are given by (Taftø 1979)

$$a_{ik} = \left(\frac{n_i}{n_k}\right)^{1/2} \sum_{l=1}^{n_k} U_{i1-kl}\exp(\alpha_{kl}) \qquad (5.58)$$

Here n_k is the number of lattice points linked by symmetry and n_i is the multiplicity of the set of points. This is equal to m for an m-fold symmetry axis, to 2 for points on mirror lines, 1 at the origin and 0 for points on anti-mirror lines. Thus for the examples given in Fig. 5.8, $n_i = 6$.

When account is taken of the incident boundary conditions for a parallel-sided slab of crystal, the complete solution may then be written

$$\Phi(\mathbf{g}) = \mathbf{S}\Phi(0) \qquad (5.59)$$

with

$$\mathbf{S} = \mathbf{C}\{\mathbf{D}\}\mathbf{C}^{-1} = \exp(2\pi i \mathbf{A} t) \qquad (5.60)$$

Here $\Phi(\mathbf{g})$ is a column vector containing the complex amplitudes of the diffracted beams, $\Phi(0)$ is a similar vector describing the incident beam and $\mathbf{D}$ is a diagonal matrix whose diagonal elements are $\exp(2\pi i \gamma^{(i)} t)$, with t the crystal thickness. For the seven-beam case, it will be found that, because of the zeros in eqn (5.54), the solution $\Phi(\mathbf{g})$ depends only on the two eigenvalues of eqn (5.56). Thus, this axial seven-beam case is said to be reducible to two beams. Since $\mathbf{C}$ is unitary, its inverse is equal to the transposed conjugate form of $\mathbf{C}$.

The matrix $\mathbf{S}$ is known as the scattering matrix, and was first introduced by Sturkey (1950). Matrix $\mathbf{A}$ is known as the structure matrix, or the dispersion matrix if the quantities $2\,|\mathbf{k}_0|\,\gamma^{(j)}$ are included on the diagonal.

Since **S** is unitary, we see that dynamical electron diffraction in the Laue geometry consists of a unitary transformation (Sturkey 1962).

It is clear that the development of a completely general computer program which will reduce and diagonalize the dispersion matrix of any crystal of given symmetry is a large undertaking, and this fact probably accounts in part for the popularity of the multislice method for HREM computations. Symmetry reduction is also possible in that method. The existence of specialised computer hardware for fast Fourier transforms can make the evaluation of the convolution in eqn 5.34 very rapid. We have yet to see the development of specialised hardware for matrix diagonalisation. An important advantage of the Bloch-wave method is that a single diagonalisation gives results for all thicknesses.

5.8   Partial coherence effects in dynamical computations—beyond the product representation. Fourier images

All partial temporal and spatial coherence effects may be included exactly using eqns (4.1) and (5.47) (for the Bloch wave method) or, for the multislice method, using eqn (4.1) and the results of the recursion relation (5.34) combined with eqn (5.1a). The results of such a calculation are shown in Fig. 4.9. In this way, variations in diffraction conditions (i.e., rotation of the Ewald sphere) within the angular range θ_c of the beam divergence are correctly incorporated, and no approximations are made to the aberration function $\chi(u_g)$.

Even using modern computers, calculations of this type are extremely time-consuming and, for the small values of θ_c and thin samples normally used for HREM work, unnecessarily accurate. If variations in diffraction conditions over the angular range θ_c are neglected (a good approximation for thin crystals—see Fig. 5.3 and Section 5.3.2) and a first-order Taylor expansion is made for $\chi(u_g)$ about u_g over the angular range θ_c, then the expression for a single Fourier coefficient of the image *intensity* becomes (O'Keefe and Saxton 1983; Fejes 1977),

$$I_g = \sum_h \Phi_h \exp(i\chi(\mathbf{h}))\gamma\{(\nabla\chi(\mathbf{h}) - \nabla\chi(\mathbf{h} - \mathbf{g}))/2\pi\}\Phi^*_{\mathbf{h}-\mathbf{g}} \qquad (5.61)$$

$$\times \exp\{-i\chi(\mathbf{h} - \mathbf{g})\}E\{\tfrac{1}{2}(\mathbf{h}^2 - |\mathbf{h} - \mathbf{g}|^2)^2\}$$

Here $\gamma(\mathbf{r})$, the transform of the angular distribution of source intensity, has been defined in eqn (4.5), while

$$E(q) = \exp(-2\pi^2 \Delta^2 \lambda^2 q^2) \qquad (5.62)$$

is the Fourier transform of $B(\Delta f)$, the distribution of focus settings (or, equivalently, incident beam energies) given in eqn (4.1). The linear imaging approximation of eqns (4.8) and (A3.1) involves the term $E(\mathbf{K}^2/2)$. The quantity $\Delta = C_c Q$ describes the variances in the statistically independent fluctuations in accelerating voltage, lens current, and source energy spread (see eqn (4.9)).

Equation (5.61) should be used for the computer simulation of dynamical HREM images where θ_c is small under the 'crystallographic' sign convention (see Section 5.12). It assumes that the effective illumination aperture is perfectly incoherently filled, and so is appropriate for LaB_6 and tungsten hair-pin filaments under normal conditions (see Section 4.5 for discussion). For the other sign convention, the sign in front of $\chi(\mathbf{g})$ should be reversed everywhere. Since the longitudinal (or temporal) coherence time $\Delta\tau = L/V = h/\Delta E$ is much shorter than typical film exposure times, eqn (5.61) performs an incoherent addition of intensities for electrons of different energies. (Here L is the coherence length, V is the electron velocity, and ΔE is the source energy spread.)

For field-emission sources, the illumination aperture may be coherently filled if $X_a > 2R_a$ (see Section 4.5). Transfer functions for linear imaging in this case are derived in Humphreys and Spence (1981). The expression corresponding to eqn (5.61) (not restricted to linear imaging) then becomes (O'Keefe and Saxton 1983)

$$I_{\mathbf{g}} = \sum_{\mathbf{h}} \Phi_{\mathbf{h}} \exp(i\chi(\mathbf{h}))\gamma\{(\nabla\chi(\mathbf{h})/2\pi)\}\Phi_{\mathbf{h-g}} \tag{5.63a}$$
$$\times \exp\{-i\chi(\mathbf{h}-\mathbf{g})\}\gamma\{\nabla\chi(\mathbf{h}-\mathbf{g})/2\pi\}E\{\tfrac{1}{2}(\mathbf{h}^2 - |\mathbf{h}-\mathbf{g}|^2)^2\}$$

Here again wavefunction components of different energies have been added incoherently. Equations (5.63a) and (5.61) predict identical images for a disc-shaped source in the weak-phase object approximation (Humphreys and Spence 1981).

For $\theta_c = 0$ and $\Delta = 0$, the image intensities given by eqns (5.61) and (5.63) are periodic functions of both focus setting Δf and spherical aberration constant C_s, as first pointed out by Cowley and Moodie (1960) (see also the discussion of particular cases in Sections 5.1 and 5.15). Here we consider the general n-beam case for a projection which allows orthogonal unit-cell axes (possibly non-primitive) to be chosen in directions normal to the electron beam. Let these cell dimensions be a and b. Firstly we consider the case $a = b$, so that $g^2 = (h^2 + k^2)/a^2$ where h and k are integers. Then it is easily shown, from eqn (4.8b), that

$$\chi(\mathbf{g}, \Delta f + 2na^2/\lambda) = \chi(\mathbf{g}, \Delta f) + 2n\pi(h^2 + k^2)$$

where n is an integer. For ideal monochromatic illumination $\gamma(\mathbf{r}_0) = E(q) = 1$ and eqns (5.61) and (5.63a) involve only $\exp(i\chi(\mathbf{G}))$, where $\mathbf{G}$ is any reciprocal lattice vector. Then $\exp(i\chi(G, \Delta f)) = \exp(i\chi(G, \Delta f + 2na^2/\lambda))$ for all $\mathbf{G}$, since n, h and k are all integers. Hence, changes in focus by

$$\Delta f'_f = 2na^2/\lambda \tag{5.63b}$$

leave the two-dimensional image intensity distribution unchanged. This result holds for any number of beams. In practice the effects of finite coherence mean that Fourier images can be seen only over a limited range of focus. As pointed out in Section 5.1, the general condition for restricting the depth of field to a single Fourier image half-period is that

$$\theta_c > \lambda/(2a)$$

This is just the condition that adjacent diffraction discs overlap.

Equation (5.63b) also limits the number of dynamical images which need to be computed in image simulations—these need only cover the range Δf_f. By a similar argument it can be shown that images are periodic in C_s (see Section 5.1).

For $a \neq b$ it will be found that the images are also periodic if $a^2/b^2 = p$, where p is an integer. This condition holds for many projections of simple structures. For non-orthogonal axes with angle γ, periodic images occur if $a^2/b^2 = p$ and $\cos \gamma = (b/a)m$, where m is an integer. For smaller changes in focus, such as $\Delta f_f/2$, an image of reversed contrast is formed. (See Fig. 5.2 for the three-beam case.) It is instructive to plot out the transfer function $\sin \chi(K)$ for two focus settings differing by Δf_f in order to confirm that these functions differ everywhere except at reciprocal lattice points.

5.9 Absorption effects

'Absorption' of the elastic electron wavefunction in HREM may refer either to the exclusion of elastic scattering from the image by the objective aperture, as discussed in Section 6.1, or to the depletion of the elastic wavefield by inelastic scattering events in the specimen. Only the second effect is considered here. For most medium voltage HREM work with sample thicknesses of less than about 20 nm this effect is generally negligible. However, there has been a striking failure amongst HREM research workers to obtain agreement between computer-simulated images and experimental images for the thicker crystals for which the effects of absorption are not negligible. This problem therefore stands as an important challenge for the future. Magnesium oxide smoke forms in perfect cubes of sub-micron dimensions, providing [110]-oriented wedges of accurately known thickness, and there have been several measurements of the electron absorption coefficients for this material. It would therefore appear to provide an excellent opportunity for testing HREM theory in the presence of absorption. (See Appendix 5). Equation (5.47) gives the lattice image in terms of $\gamma_R^{(j)}$ and $\gamma_{Im}^{(j)}$ which can be determined if $U_g + iU_g'$ is known. In the multislice approach (using the crystallographic sign convention), the Fourier coefficients v_g are replaced by complex coefficients $v_g - iv_g'$ (see eqn (5.36)). Here v_g are the Fourier coefficients of the 'real' potential $\phi_R(x, y)$, while v_g' are the coefficients of the 'imaginary' potential $\phi_i(x, y)$, given in eqn (6.2). Values of v_g' (or, equivalently, U_g') are spread throughout the literature (see Reimer (1984) for a summary), and measurements exist for silicon, magnesium oxide, and germanium and a few other materials. However, there are some important difficulties with the direct use of these coefficients for HREM image simulation, which we now discuss.

Microdiffraction patterns from regions of thicker crystal used to form high-resolution images show a considerable amount of scattering in non-Bragg directions, even from perfect crystals. In the absence of defects, this scattering arises from inelastic scattering of the fast electrons by the various crystal elementary excitations, chiefly plasmons, phonons, inner-shell excitations, and valence-band 'single-electron' excitations (see Section 6.6). The question arises of what contribution, if any, these scattered electrons make to a high-resolution image.

The usual assumption has been that electrons that lose energy ΔE in traversing the specimen will, if the objective lens is correctly focused for the elastic or 'zero loss' electrons, be out of focus by an amount $C_c(\Delta E/E_0)$ due to the chromatic aberration of the lens, and so contribute only a slowly varying blurred background to the HREM image, as indicated by eqn (2.34). However, it is important to appreciate that there are two independent effects which influence this background, namely the size of the objective aperture used and the strength of the inelastic scattering. In addition to contributing a low-resolution background to HREM images, a second important effect of inelastic scattering is to deplete the elastic wavelength, and so modify the amplitudes and phases of the beams used for image formation. A full account of this process for high-resolution imaging has not been given, however, several historically distinct bodies of literature have a bearing on the problem—these include the theory of the Debye–Waller factor, the justification for the use of an 'absorption' or 'optical' potential in electron diffraction, and work on the effects of absorption on diffraction contrast images.

The Debye–Waller factor expresses the attenuation of the elastic Bragg beams due to the excitation of phonons in a crystal according to the kinematic theory. When used in the dynamical theory, it does not affect the normalization of intensities or cause 'absorption'. The use of a complex optical potential to describe depletion of the elastic wavefield by inelastic processes in the dynamical theory (absorption) was justified by Yoshioka (1957). It is not inconsistent to include both a Debye–Waller factor and a contribution to V'_g from phonon excitations in dynamical calculations for the elastic wavefield (Ohtsuki 1967). In the quantum picture, the Debye–Waller factor describes the modification to the real parts V_g due to virtual inelastic scattering which results from the creation and annihilation of phonons in a single interaction (at the same time). The imaginary part of the potential V'_g however, describes the redistribution of elastic scattering which results from 'real' inelastic phonon scattering through non-Bragg angles. In the classical picture, the Debye–Waller factor describes the time-averaged periodic crystal potential responsible for the purely elastic Bragg scattering in Bragg directions. For a review of the electronic contribution to the absorption potential, see Ritchie and Howie (1977).

For isotropic materials, the Debye–Waller factor is incorporated by replacing the F_g of eqn (5.19) by

$$F'_g = F_g \exp(-2\pi^2 Bg^2) \qquad (5.64)$$

where the constant B describes the temperature-dependent lattice vibration amplitude. Tabulated values of B can be found in the International Tables for X-ray Crystallography.

In principle, the 'classical' method of treating phonon absorption effects in HREM is straightforward. The interaction time of a fast electron traversing a thin crystal is much shorter than that of any of the inelastic crystal excitations. Therefore a time-average is required of the intensity of the dynamical, many-beam, aperture-limited image for every instantaneous configuration of the crystal potential. This would require a complete description of all the atomic displacements due to the thermal motion of the atoms. The time-averaging must be performed on the dynamical image, rather than on diffracted beams or the crystal potential. A second aspect of such a calculation would be the estimation of the distribution of diffuse inelastic scattering due to electronic excitations, from which the background due to those inelastic electrons which pass through the objective aperture could be estimated. The contribution of these electronic processes to the imaginary part of the optical potential must also be determined. The result would give the high-resolution image observed without the use of an energy filter, that is, the image due to both elastically and inelastically scattered electrons.

While such a calculation is possible in principle, it would be impossibly laborious in practice, and a variety of approximations must be made for realistic calculations. In addition, this 'classical' approach ignores the quantum nature of phonons, which absorb or give up a quantum of energy to the fast electron during their creation or annihilation.

Since, due to the presence of chromatic aberration, only electrons within a small range of energies around the accelerating potential will contribute to the in-focus HREM image, we may think of lattice imaging as a kind of energy filtering for elastically scattered electrons. Then two effects must be considered—the modification to the amplitudes and phases of the elastic beams due to inelastic scattering, and the contribution to the low-resolution background from inelastically scattered electrons which pass through the objective aperture. The existence of virtual inelastic scattering and exchange effects between the fast electron and crystal electrons also makes a small contribution to V_g (the elastic potential); however, Rez (1977) has shown that these effects are small. We note in passing that these exchange effects between the fast electron and crystal electrons are fundamentally different from the exchange effects included in band-structure calculations, which take account of exchange amongst crystal electrons. Thus, the potential measured by electron diffraction, while sensitive to bonding effects, is not the same as that used for band-structure calculations. The bulk of the work in the literature on imaging, however, is concerned with estimating the effect of inelastic absorption on a particular elastic Bragg beam, which may then be related to a diffraction contrast image formed from this beam alone (using a small objective aperture) through the column approximation. This approach is not directly applicable to HREM many-beam images, since these are formed using larger apertures at higher resolution. The straightforward application of coefficients intended for use in diffraction contrast image calculations to HREM image calculations is thus unlikely to be correct for several reasons.

1. Careful recent experimental measurements of V'_g for silicon (Voss, Lehmpfuhl, and Smith 1980) indicate that earlier theoretical estimates (Radi 1970) are about three times too large. Here the measured values of V'_g are found to be about one-hundredth of V_g.
2. Inelastic processes which preserve diffraction contrast (such as plasmon excitation and small-energy-loss single-electron excitation) may not be included in an estimate of V'_g, or may be included in a way which makes V'_g a function of the objective aperture size used. This aperture will generally be much smaller than that used in HREM.
3. The case of phonon scattering raises special difficulties (see below).
4. The use of the independent Bloch wave model in some of these calculations must be questioned for the thin specimens used in HREM work.

In summary, it may be said that the few reliable values of V'_g useful for HREM calculations are those which have been measured experimentally by convergent-beam or similar diffraction methods (see, for example Voss, Lehmpfuhl, and Smith (1980)). Empirical relationships, such as

$$V'_g = V_g(A \, |\mathbf{g}| + Bg^2) \tag{5.65}$$

have also been used, where A and B are fitting parameters, and $\mathbf{g}$ is a reciprocal lattice vector.

The second step in correcting HREM image calculations for the effects of inelastic scattering is to estimate the low-resolution background contribution to the images from inelastically scattered electrons which pass through the objective aperture. This problem has received little attention, no doubt owing to the difficulty of determining the detailed angular redistribution of multiple inelastic scattering within the objective aperture. In practice a simpler approach would be to include as background an appropriate fraction of the energy lost from the elastic wavefield as a consequence of the use of an absorption potential.

The situation is less clear for phonon scattering. Transmitted electrons which lose energy owing to the creation of phonons in a crystalline sample give up very small amounts of energy (less than 1 eV even for multiple losses) and so remain 'in focus', and, in addition, the angular distribution of phonon-loss electrons shows a broad maximum for $0.5 < (q) < 1 \, \text{Å}^{-1}$, in addition to peaks at the Bragg positions. Much of this broad maximum falls within the resolution limit of modern HREM instruments. The question therefore arises of what contribution this scattering makes to HREM images. Recently it has been shown (Cowley 1988) that phonon-loss electrons will make a high-resolution contribution to the many-beam image, and that images with atomic resolution may thus be obtained from inelastically scattered electrons.

Detailed calculations for elastic HREM images of Au_4Mn have shown that these are drastically altered (for thicknesses of several hundred Ångstroms) by the inclusion of absorption effects (Van Dyck 1985, personal communication). Theoretical calculations showing the strong effects of absorption on lattice images in 'thick' crystals have also been published by Pirouz (1979).

5.10 Dynamical forbidden reflections

The observation of intense Bragg reflections whose kinematic structure factors are zero will be familiar to most electron microscopists. For example, in the diamond structure the (200) reflection, for which $F_{200} = 0$ (see eqn (5.19)), is commonly observed as a result of double scattering, involving the $(11\bar{1})$ and $(1\bar{1}1)$ reflections. All these reflections lie in the [011] zone axis pattern, and we note that $(11\bar{1}) + (1\bar{1}1) = (200)$.

It has been shown, however, that for certain incident beam directions, certain reflections remain absent, despite the effects of multiple scattering, *for all thicknesses and accelerating voltages* (Cowley and Moodie 1959; Gjønnes and Moodie 1965). The reflections which remain absent are those whose absence results from the existence of screw or glide translational symmetry elements, and are not 'accidental' absences such as those due to atoms which lie on special positions in the lattice. The incident beam directions which cause the continued absence of these space-group-forbidden reflections are given below, and include the case, important for lattice imaging, in which the beam runs normal to the screw axis or glide plane. This powerful theoretical prediction was first confirmed experimentally by Goodman and Lehmpfuhl (1964) and Fujime, Watanabe, and Ogawa (1964). It forms the basis of a general method of crystal space-group determination by convergent-beam electron diffraction (see Section 11.4). In this brief summary, we consider only the results in the projection approximation for which only those reflections in the zero-order Laue zone (ZOLZ) are considered. For the general three-dimensional result, the reader is referred to the classic paper of Gjønnes and Moodie (1965).

The importance of this result for HREM work lies in the fact that only if these space-group-forbidden reflections do not contribute to the image will the projected potential seen in the HREM image reveal the true crystal structure. Yet the angular range of incident beam directions for which these reflections remain absent is rather small (typically about a tenth of a degree, depending on crystal thickness and structure), so that small errors of alignment may result in misleading images of the crystal structure.

In the projection approximation, the only plane groups corresponding to a twofold screw axis or glide plane normal to the beam are *pg*, *pmg*, or *pgg* in the rectangular system or $p4g$ in the square system. Screw axes other than twofold may be thought of as giving rise to accidental forbidden reflections in projection. To expose the principle of the effect, we consider the simple case of a crystal with projected symmetry pmg, of which CdS forms an example if the electron beam is normal to the c-axis. CdS is hexagonal, with a 6_3 screw axis along $\mathbf{c}$. In projection, this screw axis becomes a glide line. Because of this translational symmetry element, the structure factors have the signs shown in Fig. 5.9(a), with

$$F_{hk} = (-1)^k F_{\bar{h}k} \qquad (5.66)$$

and

$$F_{0k} = 0 \qquad \text{for} \qquad k = 2n + 1 \qquad (5.67)$$

(see Stout and Jensen (1968)). Here the index k is parallel to $\mathbf{c}$, and the beam direction is normal to the page. When these conditions are imposed on eqn (5.33), it will be found that the resulting series of terms can be arranged in pairs with products of F_{hk} of equal magnitude and opposite sign. This occurs only if the excitation errors also preserve certain symmetry, in particular we require that K (the projection of the incident electron wavevector in the ZOLZ) lie on the bold cross shown in Fig. 5.9(a). Here, either the (010) reflection is at the Bragg condition or the beam lies in the plane containing the $(0k0)$ systematics and the zone axis. Physically, the extinction arises because of the exact cancellation (by destructive interference) of all pairs of multiple-scattering paths (such as x and y in Fig. 5.9(a)) which are related by crystal symmetry. Thus, near the axial orientation (and within the projection approximation) the experimental convergent beam discs F_{0k} appear as shown in Fig. 5.9(b), for a twofold screw axis normal to the beam (Goodman and Lehmpful 1964).

For a glide plane normal to the beam, again in the projection approximation, the period of the projected potential will be halved, leading to the extinction of alternate reflections in the ZOLZ plane.

For a glide plane which contains the beam direction, the situation (after projection of the crystal structure in the beam direction) is the same as that for the twofold screw axis.

The effects of dynamically forbidden reflections on lattice images have been studied by several workers. Nagakura and Nakamura (1983), for example, observed image artifacts in cementite and studied their dependance on beam misalignment. Figures 5.10(a) and (b) show the intensity of the first-order dynamically forbidden reflection in this material as a function of incident beam misalignment in the directions K_x and K_y in Fig. 5.9(b). For this structure, the incident beam must be aligned to within less than 1 mrad of the zone axis in order to produce a faithful representation of the projected crystal structure in the HREM image. A review of the effects of dynamically forbidden reflections in several materials, including TiO_2, SiC, Mo_5O_{14}, CuAgSe, and SnO_2 can be found in Smith, Bursill, and Wood (1985). Figure 5.11 (from that work) shows clearly the artifactual 4.6 Å fringes which appear in thicker regions of this sample unless the alignment is

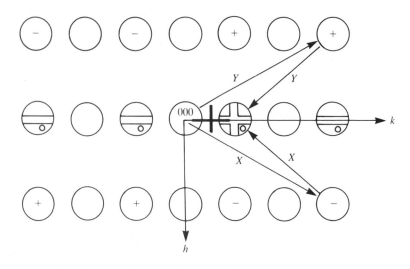

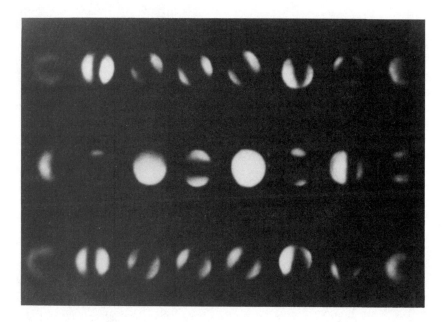

Fig. 5.9 (a) The principle of dynamical extinction. For a screw axis along k normal to the beam the structure factors have the signs indicated. Those containing O are forbidden. Unmarked discs have equal structure factors. Double scattering along X into the (010) reflection is exactly cancelled by that along Y. The bold cross indicates the locus of K along which extinction is expected in the projection approximation. (b) Experimental CBED pattern from CdS showing dynamical extinctions. The beam is approximately normal to the screw axis, (which runs across the page) along which every second reflection remains absent, despite multiple scattering.

within about 0.2 mrad of the exact zone axis orientation. Note that it is important to distinguish the linear image fringes due to the (0, 2) interaction (see Fig. 5.9(a), with (0, 1) dynamically forbidden) from the (0, 1) (0, $\bar{1}$) interaction (producing fringes of the same period) in thicker crystals due to non-linear imaging effects.

The more subtle effects of translational symmetry elements whose translational component lies parallel to the beam direction are described in Ishizuka (1982b). A discussion of diffraction symmetries and reciprocity in terms of the interaction Hamiltonian can be found in Portier and Gratias (1981). Many experimental examples, including three-dimensional effects, can be found in the excellent compilation of CBED (Convergent-beam electron diffraction) patterns produced by Tanaka and Terauchi (1985). Figure 5.12 shows the experimental confirmation of the remarkable extinction which occurs only for a small disc in the case of a glide plane normal to the beam with three-dimensional scattering (Tanaka, 1987, personal communication).

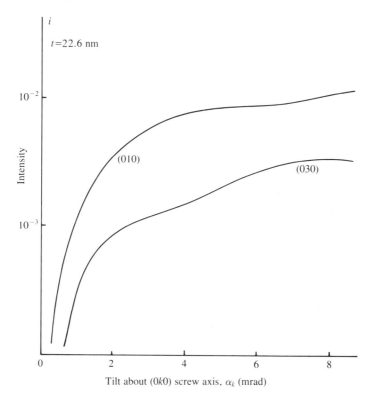

Fig. 5.10 Computed Bragg beam intensities for two dynamically forbidden reflections due to a screw axis along [0k0] in cementite at 200 kV as a function of α_k (in mrad) at a thickness of 22.6 nm (Fig. 5.10(a)) and as a function of α_h (Fig. 5.10(b)). Here α_k is the angular misalignment from the exact zone axis orientation and describes a rotation about the k axis in Fig. 5.9(a) (similarly for α_h). The angular width of the extinction band decreases with increasing crystal thickness. (From Nagakura and Nakamura (1983).)

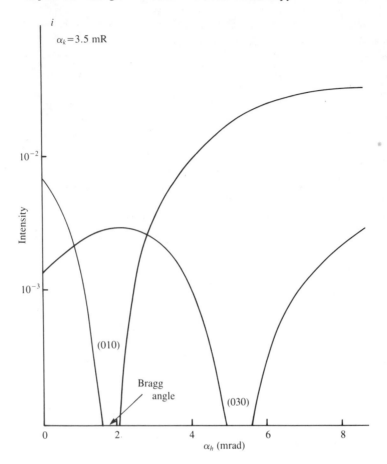

Fig. 5.10. (Continued)

5.11 Computational algorithms and the relationship between them. Supercells and image patching

Computational algorithms for solving the dynamical electron diffraction problem have mainly been based on the Bloch wave method (see Section 5.7) or the multislice method (Section 5.6). Other methods include the Feynman path-integral formulation (van Dyck 1978; Jap and Glaeser 1978) and numerical evaluation of the polynomial solution (eqn (5.33)). The method of Sturkey (1962) is based on the expansion (from eqn (5.60))

$$S = \exp(2\pi i \mathbf{A}t) = [\exp(2\pi i \mathbf{A}t/n)]^n = [\exp(2\pi i \mathbf{A}\,\Delta z)]^n \qquad (5.68)$$

where n is chosen sufficiently large to ensure that the series expansion of $\exp(i\mathbf{A}\,\Delta z)$ converges rapidly. The differential equation method of Howie and Whelan (see Hirsch *et al.* (1977)) can be obtained by differentiating the

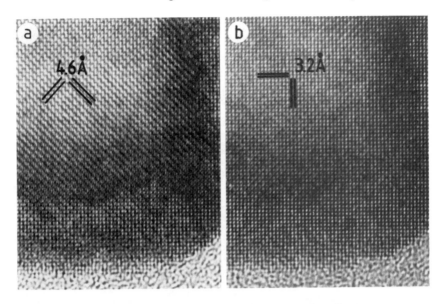

Fig. 5.11   A lattice image of a wedge-shaped thin crystal of rutile recorded at 500 kV in the [001] projection. Artifactual 4.6 Å fringes can be seen in thicker crystal, resulting from a slight error in alignment. The contribution of these dynamically forbidden reflections has been eliminated in (b), following an alignment correction of less than 0.12 mrad.

fundamental equation given above, while the multislice iterative technique is obtained from this equation by writing

$$\mathbf{A} = \mathbf{V} + \mathbf{T} \tag{5.69}$$

where $\mathbf{T}$ is a diagonal matrix containing only the excitation errors. Since $\mathbf{V}$ and $\mathbf{T}$ do not commute, a theorem due to Zassenhaus must be used to obtain the multislice iterative formula from this expansion as described by Goodman and Moodie (1974) in more detail. The formal relationship between all these algorithms is summarized in Fig. 5.13.

It is often instructive to consider two limiting cases of the fundamental equation (5.60) when attempting to understand dynamical scattering effects. For $\mathbf{T} = 0$ (all excitation errors zero) a term-by-term comparison of the series expansion of $\mathbf{S}$ with that of the Fourier transform of the phase-grating expression (eqn (3.25)) shows that the observed column of $\mathbf{S}$ contains just the Fourier coefficients of $\psi_e(x, y)$. Thus, the simple phase-grating approximation is recovered from the fundamental equation as the accelerating voltage tends to infinity ($S_g \rightarrow 0$), in which limit $\sigma \rightarrow 2\pi/(hc)$, the Compton interaction parameter (c is the speed of light, h is Planck's constant). A term-by-term comparison of this same series with the Born series for electron scattering also confirms that the phase-grating approximation (known elsewhere as the Moliere high-energy approximation) provides an exact summation for the series in the limiting case where the component of

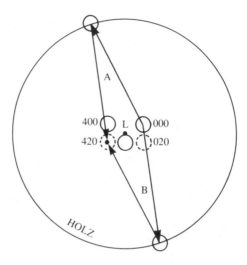

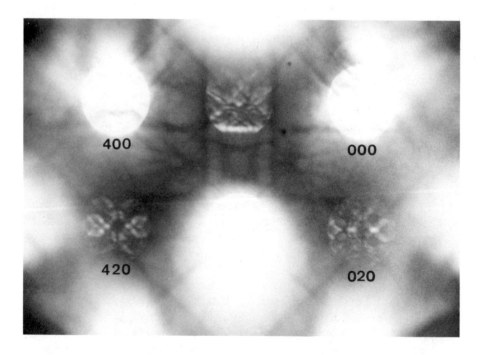

Fig. 5.12 Convergent beam pattern of spinel, showing three-dimensional dynamical extinction at a single point. (Courtesy of M. Tanaka.) The dynamic extinction is due to a glide plane which is set perpendicular to the incident beam, and occurs at the exact Bragg position of a kinematically forbidden reflection. The extinction occurs owing to the interference of two 'Umweg-reflections via HOLZ'. These paths are schematically shown in (a) and are symmetric with respect to the position of the projection of the Laue point (L). The micrograph (b) was taken from a (001) spinel ($MgAl_2O_4$) at an accelerating voltage of 50.5 kV. The (420) reflection shows the extinction (dark spot) due to a d-glide plane parallel to the (001) plane.

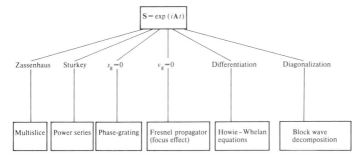

Fig. 5.13 Algorithms and approximations in dynamical theory.

the scattering vector in the beam direction is zero. The phase-grating approximation also gives an approximation for Pendellösung effects, since the Fourier coefficients of $\psi_e(x, y)$ are periodic in t, unlike the kinematic expression (eqn (5.16)), which predicts intensities proportional to t^2 for $S_g = 0$. By varying the direction of projection in eqn (3.25) the phase-grating approximation can also be seen to give an estimate of the angular dependence of diffracted beams (their 'rocking curve').

The second limiting case of the fundamental equation is obtained when $V \rightarrow 0$, in which case, since A is diagonal, the observed column of S contains entries

$$\phi_g = \exp(2\pi i S_g t) = \exp(-2\pi i \lambda u_g^2 t / 2) \tag{5.70}$$

since $S_g = -\lambda u_g^2 / 2$ in the zone axis orientation. Comparison with eqn (3.24) shows that this phase shift corresponds to free-space Fresnel propagation over a distance equal to the crystal thickness, as expected in the 'empty lattice' approximation of zero crystal potential.

Attempting to match computed images with experimental images is a tedious business, reminiscent of the early trial-and-error methods used in X-ray structure analysis. The parameters of specimen thickness, focus, and specimen structure must be varied in turn until an acceptable match is obtained. The aperture size needed for the calculated images to obtain agreement may not coincide with that used experimentally, since the resolution of the instrument is rarely known with precision and beams may be included in the experimental image which contribute only a diffuse image background and do not increase the image resolution.

For perfect crystals, it should be noted that the images will be periodic functions of the focus setting (see Section 5.8). Therefore, it is only necessary to compute images through a single 'Fourier image' period in order to obtain all possible images for a particular thickness. In addition it can be shown that dynamical solutions are functions only of the product $\lambda t = X$ at non-relativistic energies ($V_0 < 100$ kV). All ((non-relativistic) diffraction patterns for the same value of X are identical for a given structure, so that halving the specimen thickness is equivalent to halving the electron wavelength. Specimens thus 'look' thinner to the fast electron as its energy is increased.

For periodic images obtained under the conditions for which the PCD approximation holds, the calculation of simulated images can be greatly speeded up by the direct application of eqn (5.25). The projected charge density is obtained from

$$\nabla^2 \phi_p(x, y) = -16\pi^2 t \sum_{h,k} S^2(h, k, 0) v_{h,k,0} \exp(2\pi i(hx/a + ky/b)) \quad (5.71)$$

where $s = \sin \theta / h$ with θ the Bragg angle for the reflection $(h, k, 0)$ and the sum is extended over all reflections within the objective aperture.

The articles by Goodman and Moodie (1974) and Self et al. (1983), and the books by Head, Humble, Clarebrough, Morton, and Forwood (1973), Cowley (1975), and Hirsch et al. (1977) contain ample information to enable workers with some Fortran programming experience to write a dynamical program which will compute the complex Bragg diffracted amplitudes from a crystal of given structure and orientation. Computing methods based on the multislice technique are most efficient for the simulation of large unit-cell crystals and thin specimens. For the simulation of diffraction contrast (single-beam) images of defects based on the column approximation, the Runge–Kutta (or a similar) technique for the solution of coupled differential equations has traditionally been used. Matrix methods, which depend on the diagonalization of the structure matrix $\mathbf{A}$, are most efficient for thick, perfect specimens of the small-unit-cell crystals of interest to metallurgists. All these numerical methods provide solutions to the same fundamental equation, and have close similarities to the numerical techniques used in solid-state electron band-structure calculations. Several workers have demonstrated the equivalence of the various computing techniques. In the multislice technique, most of the computing time is occupied performing a numerical convolution. Since this operation can be replaced by equivalent numerical fast Fourier 'transforms', calculations can be speeded up through the use of array processor hardware.

For the multislice method the computing time per slice for n beams is proportional to $n \log_2 n$, whereas for the Bloch-wave method, matrix diagonalization times are proportional to n^2. Thus, the multislice method is faster for $n > 16$ (Self et al. 1983). Computing space increases roughly as n^2 for the diagonalization method, and as n for the multislice method. If the convolution in eqn (5.34) is replaced by Fourier transforms executed on an array processor and a simple product, an even faster algorithm results. A recent study of the use of inexpensive array processors with small computers for multislice calculations (Rez 1985) has shown that this algorithm requires less than 0.1 second per slice for 4096 beams for two-dimensional calculations. This does not include the time needed to set up the arrays. A typical slice thickness is about 0.3 nm. The crucial requirement for high speed is that all three arrays (phase-grating, beams, propagator) are stored in memory with direct access to the array processor.

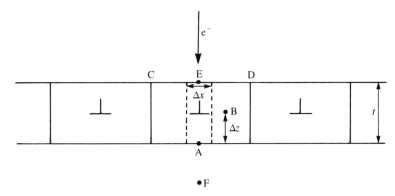

Fig. 5.14 The principle of periodic continuation. Each identical 'supercell' contains the defect of interest. Because the wavefunction at A is almost independent of the potential at B, dynamical images can be computed in 'patches'.

The importance of this is that it is now possible to build efficient single-user systems, based on small computers and array processors, which allow *interactive* dynamical structure-image calculations. Thus, using an image intensifier and video recording system (see Section 11.3), an image-matching search through model structures, thickness, and focus parameters can be completed rapidly.

An important question which arises in image simulations concerns the lateral spread of the electron wavefunction. For the simulation of images of defects it is common to use the method of periodic continuation (see Section 5.11), as shown in Fig. 5.14. We wish to know how large a cell is needed for a given defect. Thus we wish to know the extent to which the dynamical wavefunction at A in Fig. 5.14 (immediately below a defect) depends on the crystal potential at B. Within the phase-object approximation (eqn (3.26)) there is clearly no spreading; however, in thicker crystals this may not be so. This problem has been studied theoretically in two main ways. Firstly, dynamical calculations for a STEM probe wavefunction incident at E have been made for various thickness. If the width of this wave packet at E is small, then its width at A gives a measure of the lateral spreading of the wavefunction. The results of these calculations (Spence 1978; Humphreys and Spence 1979; Marks 1985) show that, in the axial orientation for a specimen whose thickness is less than $t = 50$ nm, this broadening is very small and given very approximately by $\Delta x = 2t \tan \theta_c$. A more realistic estimate for imaging calculations (based on the 'Takagi triangle' construction—see Hirsch *et al.* (1977)) might be $\Delta x = 2t \tan \theta$, where θ is twice the largest Bragg angle which contributes to the dynamical diffraction process. Alternatively, we might argue that the contribution at A from a spherical wave source at B will be negligible if the lateral distance between A and B exceeds the width of the first Fresnel fringe formed at 'defocus' Δz. (This gives an estimate of the effects of free-space propagation alone.)

From eqn (8.8) we have, approximately

$$\Delta x = (\lambda \, \Delta z)^{1/2} \tag{5.72}$$

for the propagation effect. In practice this equation is found to overestimate Δx. Finally, we see that in the phase-grating approximation (see Fig. 2.2 and eqn (3.25)), in which the wavefunction is computed along the single optical path EA in Fig. 5.14, there is no lateral spreading of the wavefunction. A more straightforward method of analysing the spreading of the wavefunction is to perform dynamical image simulations for the same defect within 'superlattices' of varying dimension CD. The dependence of the wavefunction at the point A can then be studied as function of the super-cell size CD and the thickness t. The results of extensive computational trials of this type (Spence, O'Keefe, and Iijima 1978; Matsuhata, Van Dyck, Landuyt, and Amelinckx 1984) show that this broadening is typically less than 0.4 nm for thicknesses of less than 50 nm. However, it will depend, weakly, on the structure and atomic number of the crystal, on the accelerating voltage, and on the order of the image Fourier coefficients considered.

In summary, this important qualitative result indicates that the axial, dynamical image wavefunction on the exit-face of a crystal is locally determined by the crystal potential within a small cylinder whose axis forms the beam direction and whose diameter is always less than a few Ångstroms for typical HREM conditions. This is essentially a consequence of the forward-scattering nature of high-energy electron diffraction. Thus, for the purposes of finding a few particular atomic column positions, a rather small artificial super-cell can be used, with CD typically about 2.5 nm if $t < 20$ nm. This argument only applies to the axial orientation.

The importance of this result is that for the purposes of the image simulation of defects at interior points of a super-cell such as A, there is no need to obtain perfectly smooth periodic continuation of the crystal potential at the boundaries C and D. An abrupt discontinuity of the potential at C will only influence the dynamical image within a few Ångstroms (laterally) of this point. *It follows that images of large defects can be simulated in 'patches' using separate computations.* The image patches may subsequently be joined together, if sufficient allowance has been made at the borders for discontinuities in the potential. For example, figure 8 of Olsen and Spence (1981) actually consists of three panels, resulting from separate computations, which have been joined at overlapping regions.

It is important to emphasize that these techniques of patching and lack of smooth periodic continuation cannot be used for computation of dynamical electron diffraction patterns.

At large focus defects, the most important broadening effect will occur beyond the exit-face of the crystal, between A and F in Fig. 5.14. It has therefore been suggested (Matsuhata et al. 1984), that, since most of the computer time is devoted to the multislice, it may be possible to use a small super-cell for the multislice calculation within the crystal, followed by an image-synthesis calculation referred to a much larger super-cell to account for propagation between A and F. A knowledge of the crystal structure surrounding the defect is assumed.

5.12 Sign conventions

Since high-energy electron diffraction is described by the Dirac equation (or the relativistically corrected Schrödinger equation), all the signs in the scattering and image formation process are fixed by this equation if the time dependence is included. Most modern quantum mechanics texts write the time-dependent Schrödinger equation as

$$(\hbar^2/2m)\,\nabla^2\psi = -i\hbar\,\frac{d\psi}{dt} \tag{5.73}$$

with the time-dependent solution

$$\psi_t(\mathbf{r}) = A\,\exp(+2\pi i\mathbf{k}\cdot\mathbf{r} - i\omega t) \tag{5.74}$$

Note that $\psi_t^*(\mathbf{r})$ is *not* a solution of eqn (5.73), since this would require an imaginary wavevector. If the minus sign on the right-hand side of eqn (5.73) is replaced by a plus sign, $\psi_t^*(\mathbf{r})$ becomes a solution. This form is used in some texts. For the time-independent Schrödinger equation, either

$$\psi_s(\mathbf{r}) = A\,\exp(2\pi i\mathbf{k}\cdot\mathbf{r}) \tag{5.75}$$

or $\psi_s^*(\mathbf{r})$ are suitable solutions. While most textbooks give the Dirac equation and the Schrödinger equation in forms which require $\psi_t(\mathbf{r})$ as the solution, others (e.g., Jauch and Rohrlich (1955)) adopt forms for these equations which require $\psi_t^*(\mathbf{r})$ for the positive-energy, electron plane-wave states. In this book (with the exception of Section 5.7 on Bloch wave methods), ψ_s^* is used throughout. Section 5.7 uses $\psi_s(\mathbf{r})$, for conformity with most texts on Quantum mechanics. The theory of inelastic electron scattering (for which a time-dependent Schrödinger equation may be used) is not discussed in detail in this book (see Section 6.6).

The situation is complicated by the need for conformity with the established conventions of X-ray crystallography, as codified in the International Tables for X-ray crystallography. This is desirable since most modern HREM work involves crystallographic problems and one often wishes to compare the results of X-ray and electron work, to use existing X-ray computer software, or to make use of the extensive tabulations of structure factors given in the International Tables.

The X-ray tradition has been based like this book, on the use of $\psi_s^*(\mathbf{r})$ (von Laue 1960; International Tables for X-ray Crystallography, Vol. 1, p. 353), and indeed this form appears to have been the most common in the early days of electron and X-ray diffraction, out of which the new quantum mechanics grew. Either sign convention, if used consistently will produce

Table 5.1 Sign conventions used in HREM

	Quantum-Mechanical Convention	Crystallographic Convention (used in this book)
Free-space wave	$\exp\{+2\pi i(\mathbf{k}\cdot\mathbf{r}-\omega t)\}$	$\exp\{-2\pi i(\mathbf{k}\cdot\mathbf{r}-\omega t)\}$
Transmission function	$\exp\{+i\phi_p(x)\cdot\Delta z\}$	$\exp\{-i\sigma\phi_p(x)\cdot\Delta z\}$
Phenomenological absorption	$\phi_p(x)\rightarrow\phi_R(x)+i\phi_i(x)$	$\phi_p(x)\rightarrow\phi_R(x)-i\phi_i(x)$
Propagation function	$\exp\{2\pi i S_g(u)\cdot\Delta z\}$	$\exp\{-2\pi i S_g(u)\cdot\Delta z\}$
Wave aberration function (modification of diffracted wave by objective lens)	$\psi'(u)=\psi(u)\exp\{-\pi\lambda u^2 \times(\Delta f+\frac{1}{2}\lambda^2 C_s u^2)\}$	$\psi'(u)=\psi(u)\exp\{+i\pi\lambda u^2 \times(\Delta f+\frac{1}{2}\lambda^2 C_s u^2)\}$
Object to diffraction space	$\int\psi(\mathbf{r})\exp\{-2\pi(\mathbf{u}\cdot\mathbf{r})\}\,dr$	$\int\psi(r)\exp\{+2\pi i(\mathbf{u}\cdot\mathbf{r})\}\,dr$
Diffraction to object space/reciprocal to real	$\int\psi(\mathbf{u})\exp\{+2\pi(\mathbf{u}\cdot\mathbf{r})\}\,du$	$\int\psi(\mathbf{u})\exp\{-2\pi(\mathbf{u}\cdot\mathbf{r})\}\,du$
Structure factors	$\sum_j f_j\exp\{-2\pi i(\mathbf{u}\cdot\mathbf{r}_j)\}$	$\sum_j f_j\exp\{+2\pi i(\mathbf{u}\cdot\mathbf{r}_j)\}$

σ	Electron interaction constant $=2\pi me\lambda/h^2$; $m=$ (relativistic) electron mass, $\lambda=$ electron wavelength, $e=$ (magnitude of) electron charge, $h=$ Planck's constant
ϕ_p	Crystal potential averaged along beam direction (positive)
Δz	Slice thickness
ϕ_i	Absorption potential (positive)
S_g	Excitation error (negative for reflections lying outside the Ewald sphere)
Δf	Defocus (negative for underfocus)
C_s	Spherical aberration coefficient
f_j	Electron scattering factor for jth atom
$\mathbf{r}_j$	Position of jth atom

identical images. These sign conventions are summarized in Table 5.1 (from Saxton, O'Keefe, Cockayne and Wilkens (1983), with corrections), and it is therefore proposed that these be known as the 'quantum-mechanical convention' (for $\psi_s(\mathbf{r})$) and the 'crystallographic convention' (for $\psi_s^*(\mathbf{r})$). Note that in *both* conventions in this book it is also assumed that a weakened lens (underfocus, first Fresnel fringe bright, toward Scherzer focus) is represented by a *negative* value of Δf. (The opposite convention has been used in some papers.)

An instructive case concerns the comparison of stacking faults imaged by X-ray topography and by electron microscopy. The various sign conventions used in both fields have been reviewed by Lang (1983). Here the distinction between extrinsic and intrinsic faults depends on the contrast of the first thickness fringe, which may be reversed by a sign error in calculations.

5.13   Testing image-simulation programs. The accuracy of atom position determinations

Most image-simulation software programs consist of three sections: a program to compute the Fourier coefficients of crystal potential, a dynamical scattering program, and an image-simulation program. The first of these (based on eqn (5.18)) may be checked by hand (note the sign convention required in Table 5.1 and eqn (5.19)). The second program must reproduce experimentally observed thickness fringes and convergent beam patterns (see Section 11.4). An approximate check can be made by running the program for a small-unit-cell crystal under 'two-beam' conditions (Bragg condition satisfied for beam g). The approximately sinusoidal variation of the intensity of beam g with thickness should have a 'period' within about 15% of that given by ξ_g in eqn (5.17). If no resolution limits are imposed, the imaging program should give images in very thin crystal ($t < 20$ Å) which agree exactly (apart from a uniform background) with a map of the projected crystal potential as given by the first program. It is worth checking that the phase change of 90° introduced by the lens (for moderate spatial frequencies) at Scherzer focus has the *same* sign (i.e., adds to) that which arises from scattering (the factor i in eqn (3.2b)). Thus a total phase shift of 180° (*not* 0°) is required for phase contrast. Finally, the images at all thicknesses should have the same symmetry as the projected potential, unless a resolution limit has been imposed, in which case the image symmetry may be higher.

The ultimate test of an image-simulation procedure is its ability to reproduce experimental results. The most convincing image simulations are those in which computed and experimental images are matched as a function of some parameter, usually thickness or defocus. It is desirable to determine these parameters independently. The focus setting may be determined by the methods described in Section 8.1. Specimen thickness may be determined by using the image width of an inclined planar fault known to lie on a particular crystallographic plane, or from the external crystal morphology. Perhaps the simplest case for testing a program is that of MgO smoke crystals, which form in perfect cubes. Images taken with the beam parallel to [110] therefore allow an exact thickness calibration, so that the 'turning points' (at which the fringe contrast reverses) can be compared with the results of calculation over a range of thickness (O'Keefe, Spence, Hutchinson, and Waddington 1985; Hashimoto 1985). (See Appendix 5.) These may depend sensitively on the absorption potential used in thicker crystal. In this way, in principle, the degree of ionicity for a crystal may be 'measured' (Anstis, Lynch, Moodie, and O'Keefe 1973).

As shown in Fig. 5.20, the image blobs may not coincide with true atom positions. An analysis of the silicon [110] case has been given by Krivanek and Rez (1980). For sufficiently thin crystals, we have seen (eqns (5.26) and (3.29)) that peaks of crystal potential must coincide with corresponding peaks in the image intensity if no resolution limits are imposed. For dynamical images there is, in general, no one-to-one correspondence between the crystal exit face wavefunction and the projected crystal potential (see the discussion of resolution in Section 8.10). A detailed analysis of this problem (Saxton and Smith 1985) shows that even under kinematic scattering conditions, lateral shifts between the position of a maximum in the projected crystal potential and that of the corresponding image 'peak' of about 0.03 nm are common at some focus settings. These results are somewhat model-sensitive; however, it is also found that the choice of focus which maximizes image contrast is not that which minimizes the lateral image shift. Dynamical calculations for model structures give broadly similar results, and we may summarize these by saying that image peak separations may vary by up to 20% as a function of thickness and focus. These calculations were all performed for the case in which the instrumental point-resolution (eqn (6.17)) is comparable to the interatomic spacing. For spacings much larger than the point-resolution in very thin crystals, these image shifts will become negligible.

We note that the errors found in this study are a small fraction (about 10%) of the point-resolution of the instrument (about 0.2 nm). Just as the position of the maximum in a sine wave can be located to within a small fraction of its wavelength, so it is common in X-ray diffraction to locate atom positions to within an accuracy far smaller than the d-spacing of the highest-order reflection used in the refinement. Similarly, if it is known *a priori* that an image feature consists of a *single* atomic column, then the centre of this image 'peak' may be located to within a small fraction of the point-resolution. This is possible because of the interferometric nature of electron images, for which a difference in optical path length of one electron wavelength between adjacent image points may cause a contrast reversal. The problem of *distinguishing* two adjacent or overlapping columns in a crystal of unknown structure is discussed in Section 8.10. These ideas have been applied to a determination of the structure of $K_{8-x}Nb_{16-x}W_{12+x}O_{80}$ by Hovmoller, Sjorgen, Farrants, Sunberg, and Marinder (1984). An accuracy in atomic coordinate determination of 0.01 nm is claimed, due mainly to the noise reduction which results from averaging over many 'identical' unit cells. The assumption of constant crystal thickness over the field used is critical to the success of this work, and dynamical image simulation is essential.

A procedure for reducing the 'errors' in atom position determination through the use of a through-focal series has also been proposed (Saxton and Smith 1985).

5.14   Image interpretation in germanium—a case study

In this section a summary is given of the results of a detailed study of the imaging conditions in crystalline germanium. Germanium and silicon are particularly convenient specimens for electron microscopy—they are not contaminated rapidly, they are sufficiently good conductors to prevent accumulation of specimen charge, both are brittle and so allow stable images of immobile defects to be obtained, and finally, when viewed in the [110] direction the diamond structure which they share exposes tunnels whose size lies within the resolution limit of most modern microscopes.

Specimen preparation equipment for silicon and germanium is now available commercially, based on the successful early methods of Booker and Stickler (1962) (see also Goodhew (1972)). A wedge-shaped region of specimen is sought and this is oriented to give the diffraction pattern shown in Fig. 5.15. If a through-focus series of images of such a specimen is recorded using the objective aperture A shown at 100 kV, images similar to those shown in Figs 10.4(a) and (b) will be obtained, with the following characteristics.

1. The images are approximately periodic with thickness. In particular, identical images resembling the specimen structure are found at thickness increments of about 12 nm, the first occurring at 7.5 nm thickness. All these

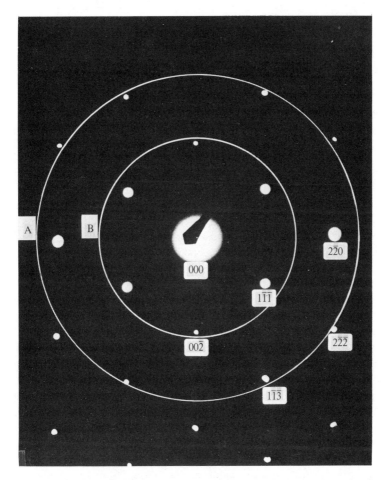

Fig. 5.15 The diffraction pattern from a germanium crystal oriented with the [110] direction parallel to the electron beam. An aperture placed at A, from which a 13-beam image can be formed, contains the smallest number of beams necessary to resolve every atom in the structure.

Fig. 5.16 A model of the diamond structure viewed approximately in the [110] direction. The white 'atoms' form a line in the [110] direction which runs down the exposed six-sided tunnels. The positions of the columns of atoms seen in this projection coincide with those shown in the computed micrograph in Fig. 5.17.

images are of 'reversed' contrast in which the atoms appear white in a positive print. Figure 5.16 shows the diamond structure, while Fig. 5.17 shows three identical computed images expected at three different thicknesses, each resembling the crystal structure. In Fig. 5.17(d) is shown a curious image obtainable at certain thicknesses, in which the atoms appear to have a hole in their centre. These annular single-atom images have also been observed by H. Hashimoto in crystals of gold, as shown in Fig. 5.18. The image appearance is a consequence of multiple scattering and does not reveal the internal structure of the atom! Images of conventional contrast from germanium crystals in which the atoms appear dark in a positive print are only possible at thicknesses of less than 3 nm at 100 kV accelerating voltage.

The occurrence of images which repeat with increasing thickness has been observed experimentally in other materials (Fejes, Iijima, and Cowley 1973) and is found to occur only in certain structures illuminated by electrons of a particular wavelength. The effect indicates that in these special cases all beams oscillate with the same periodicity in depth through the crystal. The special 'tuned' accelerating voltages which produce this effect are analysed in Glaisher and Spargo (1985). An explanation has also been given using the methods of electron channelling theory (Fujimoto 1978). In the simple two-beam case (and cases reducible to it), we see that all beams have the same Pendellösung depth period. Thus, we can expect periodic imaging (in thickness) for the many simple crystal structures for which only two Bloch waves are strongly excited, or bound within the potential well (see Section

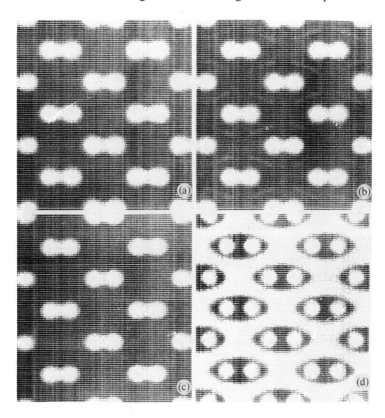

Fig. 5.17 Similar images predicted for three different thicknesses of germanium. All are 13-beam images computed for $C_s = 0.7$ mm and $\Delta f = -60$ nm at 100 kV accelerating voltage. The thicknesses are (a), $t = 7.6$ nm, (b) $t = 18$ nm, and (c) $t = 30.8$ nm; (d) shows the annular type of image found at certain thicknesses ($t = 3.6$ nm in this case). The annular atom image shape is a consequence of multiple scattering in the specimen. A similar effect is seen in the experimental image shown in Fig. 5.18.

5.7). A full understanding of this effect for large numbers of beams may have important implications for structure analysis by electron diffraction, since it implies a return to kinematic scattering conditions at large thickness.

2. The images are periodic with focus. That is, as predicted by the theory of Fourier images (Section 5.8), identical images are found at focus settings $\Delta f \pm 2nd^2/\lambda$, where n is an integer, Δf is an arbitrary focus setting, and d is the large unit-cell dimension normal to the electron beam. For germanium, $d = 0.5658$ so that at 100 kV identical images of perfect crystal recur at focus increments of 173 nm on either side of the Scherzer (or any other) focus (see Section 6.2). This periodicity with focus greatly reduces the labour of image calculation, since a computed series of images covering a range of 173 nm therefore contains all possible images. A further simplification arises since, over this range, only four distinct images of high contrast appear.

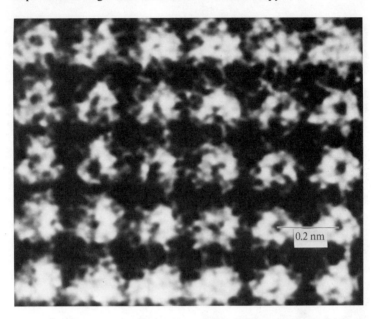

Fig. 5.18 Images of gold atoms (courtesy of Professor H. Hashimoto). The incident beam direction is [100], the specimen thickness about 19 nm and an accelerating voltage of 100 kV was used. The annular appearance of these atoms is an artifact resulting from multiple scattering.

These would be likely to be selected by a microscopist in practice and are shown in Fig. 5.19. Two of these resemble the crystal structure; one of them is displaced by a half a unit cell from the true atom positions. These two images recur several times within the Fourier image period—in all, there are seven different focus settings within the range $\Delta f = +26.5$ nm to $\Delta f = -146.5$ nm which give useful images of the crystal structure similar to either Fig. 5.19(a) or 5.19(b).

The theory of Fourier images has been worked out in detail by Cowley and Moodie (1960) and we can expect these useful image 'artifacts' to become increasingly important as more images are obtained from small-unit-cell crystals. Fourier images have not often been observed in the complex oxides since, with d large, the Fourier image defocus period exceeds the range of focus over which partial coherence effects allow sharp imaging (see Section 5.2). In practice, Fourier images are familiar to any microscopist who has noticed that simple axial three-beam lattice fringes fade and recur as the focus setting is adjusted—several focus settings appear to give equally sharp fringes. It is important to emphasize that defects may not be imaged sharply in Fourier images (see Section 5.18).

The general result for predicting the occurrence of Fourier images is that identical images of a periodic object are formed at successive focal increments Δf_0 satisfying

$$\Delta f_0 = 2na^2/\lambda \qquad \text{and} \qquad \Delta f_0 = 2mb^2/\lambda$$

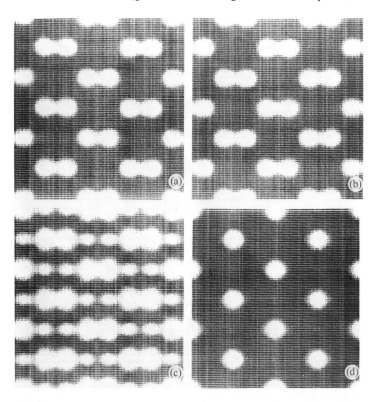

Fig. 5.19 The four important image types seen in any through-focus series of crystalline germanium taken under the following conditions: $C_s = 0.7$ mm, $V_0 = 100$ kV, 13 beams included in the objective aperture. The specimen thickness is close to any of the thicknesses shown in Figs. 5.17 (a), (b), and (c). Only (a) and (b) are a good representation of the crystal structure. The focus values are (a) -60 nm, (b) $+26.5$ nm, (c) -16.8 nm, and (d) -68.0 nm.

where a and b are the orthogonal two-dimensional unit cell dimensions and m and n are integers. In many crystallographic projections with non-orthogonal primitive unit cell vectors, a large orthogonal cell can be found satisfying $a^2/b^2 = m$ and therefore giving periodic images. For the diamond structure viewed in the [110] direction, these equations simplify, since $a = b/\sqrt{2}$, so that the first equation is automatically satisfied if the second is satisfied.

A fuller discussion of imaging conditions in the diamond structure can be found in Spence, O'Keefe, and Kolar (1977) and Nishida (1980) using multislice calculations, and in Desseaux *et al.* (1977) for the Bloch wave methods. Figure 5.20 shows an image of crystalline silicon in which every atom appears to be resolved; however, this Fourier image in fact shows a false structure. The image has been photo-processed to reduce noise by superimposing photographic images displaced by one unit cell. This processing method cannot be used for the more challenging problem of defect imaging.

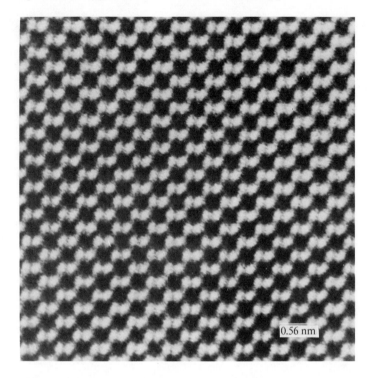

Fig. 5.20 Electron micrograph of silicon viewed in the [110] direction (courtesy of K. Izui, *J. Electron Microsc.* **26,** 129 (1977)). This image was recorded using a pole-piece with $C_s = 0.7$ mm fitted to the JEOL 100C electron microscope. The objective aperture used included about 100 beams; however, many of these do not contribute to the image owing to the effects of electronic instability (see Section 3.3). Since the interatomic spacing (0.14 nm) is smaller than the point-resolution (0.28 nm), this is not a true structure image, and gives incorrect atom positions, as described in Section 5.13.

The point-resolution of present-day 100 kV machines exceeds the primitive unit-cell dimensions of all the simple projections of silicon, germanium, and diamond. Therefore, the special considerations outlined at the end of Section 5.16 apply to these crystals, and the lattice images of highest contrast will occur at rather large under-focus values and not at the Scherzer focus, as for large unit-cell crystals. This is a most important practical consideration. In particular, these high-contrast fine-lattice images (such as that shown in Fig. 5.20) do not reveal simply the structure of any defects present.

5.15 Images of defects in crystalline solids

The prospect of determining the detailed atomic structure of defects in solids has provided perhaps the strongest motivation for the development of the techniques of high-resolution electron microscopy. Many of the mechanical, electrical, chemical, and thermal properties of solids depend on the presence of point, line, and planar defects whose atomic structure is at present largely unknown. As an example, the strength of metals depends in part on the resistance which a crystal structure imposes on the motion of a dislocation (the Peierls force). This force depends crucially on the arrangement of atoms at the dislocation core, whose structure is at present unknown. The atomic structure of point defects is also largely a matter of informed speculation at present. Finally the atomic rearrangements, and corresponding charges in total crystal energy, which occur during phase transitions are of the greatest interest to solid-state physicists.

In this book we are not concerned with the methods of diffraction contrast which have been used very successfully over the past three decades to obtain images showing the strain-field surrounding a defect. The practical application of these methods is described in the books by Loretto and Smallman (1975) and Edington (1976). These single-beam images are generally limited in resolution to perhaps 1.5 nm and image interpretation has traditionally been based on the 'column approximation', which assumes a slowly varying strain field and so does not apply at atomic resolution. More information on methods for predicting the form of diffraction contrast images of defects can be found in the two volumes by Hirsch et al. (1977) and Head et al. (1973).

Defects in many-beam lattice images have, however, been observed from the very early days of electron microscopy (Menter 1958). The interpretation of these images has only recently become possible in certain favourable cases and this field remains a field of active research where, unfortunately, simple procedures and rules for image interpretation cannot always be given. Nevertheless, the methods of high-resolution electron micrsocopy have been outstandingly successful in revealing the variety of microphases which exist in minearals. This fine polycrystalline structure can give rise to completely misleading results from X-ray structure analysis. Unit-cell resolution in the electron microscope is often sufficient to reveal the structure and size of the various phases present. Point defects in complex oxides have even been identified by these methods (Iijima, Kimura, and Goto 1974). Some examples of the complications which may arise in lattice images of dislocations are described in the paper by Cockayne, Parson, and Hoelke (1971), who were able to show that the two-beam lattice image of a single inclined dislocation may show one, three, or more terminating fringes, depending on the diffraction conditions. In addition, the terminating lattice fringe image does not necessarily occur at the site of the corresponding terminating lattice plane in the specimen. Nevertheless, experimental conditions can be found which are favourable to a simple image interpretation—for example an edge dislocation viewed end-on. The electrons scattered from a crystal containing a defect travel in directions other than the Bragg directions, and it is just this 'diffuse elastic' scattering which carries the high-resolution image detail. The most important complication for image interpretation in these specimens arises from multiple scattering from the elastic diffuse scattering into new directions which differ from the old by twice the Bragg angle. Multiple scattering amongst the diffuse scattering itself can often be neglected as a first approximation. Given a sufficiently large computer it is now possible to simulate exactly the high-resolution electron image from a crystalline defect of known structure. We are thus in the position of having to work backwards, as in X-ray diffraction, for defect-structure analysis. An intelligent guess for the defect structure (a trial structure) is used as the basis for computer-simulated images which can subsequently be compared with experimental micrographs. Details of the 'artificial superlattice' methods of image simulation used in these studies can be found in the articles by Fields and Cowley (1978), Humphreys, Diamond, Hart-Davis, and Butler (1977), MacLagan, Bursill, and Spargo (1977), and Spence (1977) (see also Section 5.11). The confusing effects of multiple scattering can be minimized by working at higher accelerating voltage with the thinnest possible specimens. The following considerations are important for this and similar studies.

1. At 100 kV accelerating voltage for all but the lowest-atomic-number specimens, multiple scattering is likely to be important in any specimen sufficiently thick to preserve the bulk structure of a defect of interest. 'Image forces' arising from the presence of the two bonding surfaces will have an important influence on the form of defects seen at high resolution. The high-resolution image obtained from a thin specimen under favourable experimental conditions shows a projection of the crystal structure. A useful structure image can therefore only be obtained if the crystal is periodic in the direction of electron illumination, that is, if the atoms are aligned in columns parallel to the electron beam. In the neighbourhood of a defect, these columns may be quite disordered in directions normal to the electron beam. The hope is that for some defects for which the strain field is independent of distance in the direction of the electron beam a favourable projection may be obtained through sufficient thickness of crystal to provide an image of high contrast. Any completely disordered material on the top and bottom surfaces of the crystal can then be expected to give only a background of 'Fresnel noise' in the final image, leading to a slight reduction in contrast.

2. For perfect crystals, an image which faithfully reproduces the crystal structure may only be obtainable from a narrow range of specimen thickness.

3. The field of view seen through the electron microscope binocular is likely to include a large region of perfect crystal in addition to the small defect of interest. This perfect crystal region will produce images which are periodic with focus (Fourier images—see Section 5.8). For simple metals and elements, these Fourier images recur for every few tens of nanometres change in focus, so that the practical problem arises of selecting the true 'structure image' (which may correctly expose the defect structure) from the many periodic Fourier images (in which the defect will be almost invisible). Fourier images from a crystal containing a small isolated defect will closely resemble images of the perfect crystal, as is demonstrated in the optical simulation shown in Fig. 5.21. Thus, the presence of small defects may be masked in Fourier images, whose periodicity is artificially inhanced. For reliable image interpretation, then, images must be recorded at a known focus setting (see Section 8.1) and compared with computed images.

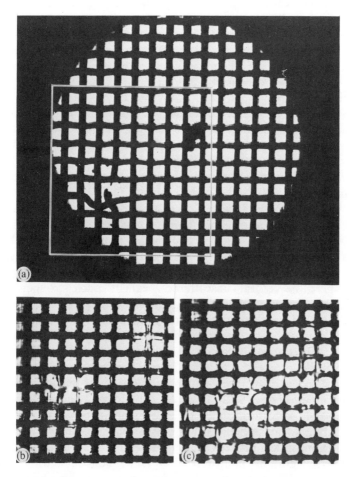

Fig. 5.21 Optical images of a piece of wire gauze illuminated with laser light recorded at three different focus settings. A crystal 'defect' has been simulated by piercing the gauze with a pin. Figure 5.21(a) shows the exact focus image, while Fig. 5.21(b) shows the first-order Fourier image from the enclosed region of (a) at defocus $\Delta f_0 = 2d^2/\lambda$. The spacing of the gauze is $d = 0.19$ mm, and using $\lambda = 632.8$ nm gives $\Delta f_0 = 114$ mm. The second-order Fourier image recorded at $2\,\Delta f_0$ is shown in (c). Note that the Fourier images (b) and (c) appear periodic, so that the effect of the defect is masked. An intermediate focus setting can be found at which the contrast of the defect alone is a maximum. Similar effects are seen in electron microscope images of defects in crystals.

4. The possibility that multiple scattering effects can be used to advantage has also been considered. There is some evidence that in crystals in which the perfect crystal produces images periodic in thickness, defects may also produce interpretable images at fixed thickness increments. This problem is discussed for the particular case of dislocation core lattice images in Spence, O'Keefe, and Iijima (1976).

5. The prospects for imaging point defects in metals have been investigated in some detail (Fields and Cowley 1978). This work outlines the conditions under which clear images of split interstitial defects can be obtained using a 1 MeV microscope, despite strong multiple scattering in crystals of gold and aluminium. The problem of containing these mobile defects in thin foils, perhaps by using a low-temperature stage, remains to be investigated. 'Knock-on' damage caused by the electron beam (see Section 6.6) must also be considered.

Three studies in particular illustrate the power of defect analysis by high-resolution electron microscopy in materials science, solid-state chemistry, and mineralogy respectively, and these are discussed in turn below. Further examples can be found in Chapter 12.

The first is a determination of the atomic structure of grain boundaries in germanium by Krivanek, Isoda, and Kobayashi (1977). Germanium has the diamond structure, shown in Fig. 5.16. Calculations similar to those shown in Fig. 10.4(a) for silicon establish that for crystals of germanium in [110] orientation less than 8 nm thick at 500 kV (or 6 nm thick at 100 kV), the imaging is kinematic (see Section 5.3.1); that is, in practical terms, one is examining regions of crysal to the left of the first dark Pendellösung band in Fig. 10.3 where the central 'unscattered' beam is much stronger than any other. For this crystal thickness in the symmetrical [110] orientation, the images may therefore be interpreted in a straightforward way in terms of the central structure as 'structure images'. These authors found small evaporated films (normally amorphous) less than 10 nm thick. These contained a wide variety of crystalline faults. The core structure of a 39° tilt boundary was selected for investigation from images obtained on the Kyoto 500 kV machine. Figure 5.22 shows the boundary selected for analysis. Since the crystals were surrounded by amorphous material, the micrograph used for analysis (a magnified portion only of which is shown in Fig. 5.22) also provided an optical diffractogram (see Section 8.7). This diffractogram was used to determine the point-resolution limit, the focus setting (an image satisfying eqn (6.16) was selected from a through-focus series), and to ensure accurate astigmatism correction. Confident of having satisfied the conditions of focus, specimen thickness, and orientation which allow a simple image interpretation at 500 kV, these workers proceeded to construct a plausible atomic model for the grain boundary based on chemical knowledge of likely bounding arrangements in tetrahedrally coordinated crystals. Each of the simple image-interpretation theories discussed earlier (Sections 3.4 and 5.4) predicts an image dark in regions

Fig. 5.22 Grain boundary and (111) twinning planes (shown by arrows) in germanium. This structure image was recorded at 500 kV accelerating voltage from a specimen less than 10 nm thick in the [110] symmetrical orientation. The inner seven Bragg beams in the [110] zone fall within the resolution limit of the instrument and contribute to the perfect crystal images. Faulted regions of crystal are imaged by elastic diffuse scattering, which an optical diffractogram shows is well transmitted out to a resolution limit of about 0.28 nm. No objective aperture was used while recording this image, resolution is thus limited by the 'virtual aperture' described in Section 3.3 and Fig. 5.30.

where a large number of atoms are seen in projection, forming columns in the beam direction. The white areas of Fig. 5.22 are therefore interpreted as the six-sided tunnels seen in the [110] projection of germanium (see Fig. 5.16). At the grain boundary a line of slightly enlarged blobs is seen; these are interpreted as seven-membered rings. To confirm this interpretation, a three-dimensional model of the grain boundary was constructed from tubular metal and photographed in the [110] projection, as shown in Fig. 5.23. The out-of-focus optical image shows good agreement with the experimental electron image. To obtain this agreement, leaving no bonds 'dangling', it was necessary to postulate a boundary containing interspersed five- and seven-membered rings together with many twinning planes indicated by the broken 'bonds' in Fig. 5.24. Studies such as these allow theories of grain boundary formation and energetics to be tested while also suggesting mechanisms of crystal growth. The recent improvement in point-resolution to about 0.16 nm has since opened up to the methods of structure imaging the much larger range of problems in materials science and metallurgy which require the elucidation of the structure of defects in elemental, small-unit-cell crystals. Workers interested in defect analysis in small-unit-cell crystals are referred to Section 10.3 for more details. In particular, the cautionary comment at the end of Section 5.15 should be

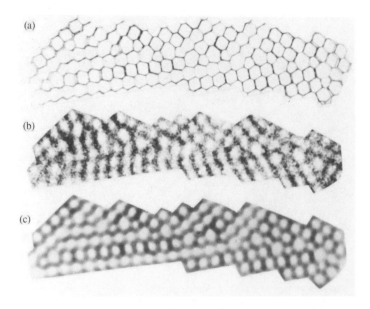

Fig. 5.23 (a) An optical image of a three-dimensional model constructed to agree with the grain boundary image shown in Fig. 5.22. The out-of-focus image of this model shown in (c) is in substantial agreement with the experimental electron image shown in (b). Here (b) is an enlarged portion of the boundary shown in Fig. 5.22. A clear idea of this projection of the diamond structure can be obtained from Fig. 5.16.

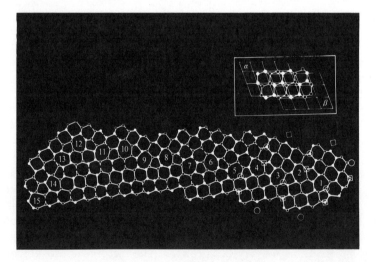

Fig. 5.24 The grain boundary model proposed to account for the image shown in Fig. 5.23 drawn out more clearly. Empty and full circles denote atoms in alternate (022) planes. Twinning planes are indicated by broken bonds and the anomalous seven-sided rings needed to accommodate the change in crystal orientation across the boundary are shown numbered. Five-membered rings can also be seen. The larger circles and squares mark two interpenetrating coincidence site lattices which can be used to characterize a grain boundary. (Figures 5.22, 5.23, 5.24 by permission of the *Philosophical Magazine*).

noted. Here it is shown that the focus setting which produces lattice fringes of highest contrast (and is therefore likely to be selected by the microscopist) is not, on 100 kV machines with $C_s > 0.5$ mm, that which allows a simple interpretation of defect structure in small-unit-cell crystals.

Perhaps the most dramatic impact of high-resolution structure imaging methods has been felt in mineralogy. Here, resolution even at the unit-cell level is often sufficient to allow deductions about the geological history of specimens to be made. While X-ray diffraction techniques produce statistically averaged information from 'large' regions of specimen, structure-imaging methods are capable of exposing the crystalline microstructure, revealing a variety of polytypes, twins, anti-phase domains, shear planes, and other faults, which may account for departures from stoichiometry at the atomic level. A good example of this work can be found in Buseck and Iijima (1975), who studied terrestrial and meteoritic samples of inter-grown orthorhombic and monoclinic enstatite, taking these to reveal the incomplete (unequilibrated) solid-state reaction between the two phases. Likely transformation mechanicms such as shearing, high-temperature quenching from protoenstatite, or slow static transformation can be distinguished by examining the structure images. Studies such as these can clarify the thermal, chemical, and mechanical history of minerals.

Finally, we consider the work of Skarnulis, Iijima, and Cowley (1976) as representative of applications in solid-state chemistry. This work is an investigation of the atomic structural basis for the known non-stoichiometry of the transition-metal oxide whose approximate chemical composition is $GeNb_9O_{25}$. It is an important study because it represents a first attempt to obtain quantitative information on point-defect densities from high-resolution micrographs. An idealized model of the structure is shown in Fig. 5.25 and consists of octahedral blocks in two layers similar to that shown in Fig. 5.28(b); however, the tunnels between blocks here form a two-by-two pattern. The purpose of the study was to determine the concentration of niobium and germanium atoms which fill the shaded tunnels shown. The structure provides both tetrahedral and octahedral coordination sites within these columns. Since the images show only a projection of the crystal structure, a unique atomic structure cannot be determined; however, the number of metal atoms at octahedral sites within the shaded column needed to account for the observed black contrast can be determined. The occupancy of these sites was varied in the computed images until the image match shown in Fig. 5.26 was obtained. The analysis was made possible by the rapid change in contrast (from black to white) seen in computed images caused by the introduction of just a few 'defect' slices, making the image very sensitive to the number of these faulty crystal layers. An important point is that the ability to match images over the large focus range shown gives added confidence in the defect-structure model deduced (see Section 5.13). Changing the focus setting alters the weight of the various image Fourier components (see Section 6.2) and so acts as a filter—a structure accurately matched at many focus settings has therefore been tested over

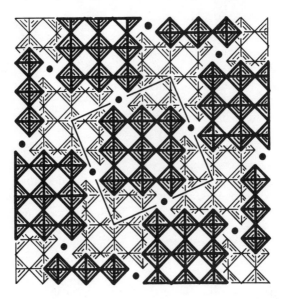

Fig. 5.25 Idealized model for the structure of $GeNb_9O_{25}$. The darker (upper) and lighter (lower) square are octahedra of ReO_3 type structure. The unit cell is outlined and the filled circles represent tetrahedrally coordinated Nb or Ge atoms. The unit-cell dimensions are $a = 1.56$ nm, $c = 0.384$ nm. (c into page).

the full range of spatial frequencies within the microscope's passband. In addition, the ability to match images over a range of focus gives an assurance that the instrumental parameters (C_s, Δf, θ_c, Δ, C_c—see Section 3.3) are accurately known.

More examples of the applications of HREM can be found in Chapter 12.

5.16   Experimental aspects of many-beam lattice images

The precautions necessary when recording high-resolution images are discussed generally in Chapters 9 and 1. There are, however, some special problems with many-beam lattice images and these are discussed below.

1. A short-focal-length objective lens ($f < 2$ mm) must be used with a specimen holder whose orientation is continuously variable. It has only become possible to obtain useful information from many-beam structure images since the development of stable high-resolution goniometer stages. Decreasing the lens focal length to the minimum allowed by the lens current supplied, results in a more intense final image (owing to the objective pre-field focusing effect (Section 2.9)), an increase in magnification and a possible reduction in spherical aberration (see Fig. 2.16). On some machines a variable height stage is available which is useful for this work—on other machines it may be necessary to place washers between the

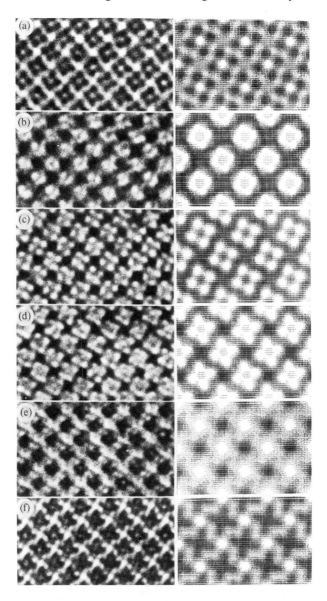

Fig. 5.26 Computed (right) and experimental (left) structure images of GeNb$_9$O$_{25}$ matched for various focus settings. The computed images are based on the defect structure refinement described in the text. The focus values for the experimental and theoretical images in nanometers are: (a) -10, -10; (b) -45, -42.5; (c) -70, -70; (d) -95, -97.5; (e) -120, -117.5; and (f) -150, -150. Image (c) is the optimum focus image which can be directly interpreted in terms of the crystal structure. Focal step calibration was obtained by the method of Section 8.1. Beam-divergence effects have not been included in the calculations, which assumed an objective aperture cut-off at 2.63 nm^{-1} and $C_s = 1.8$ mm. Small differences between computed and experimental focus values may be due to vibration from the plate transport mechanism following each exposure, which may alter the specimen height (see Section 10.5).

specimen and the top-entry cartridge in order to lower the specimen in the pole-piece. A lower limit to focal length is set by the lower cold-finger (which must not touch the specimen holder for any specimen orientation), by saturation of the lens pole-piece, or possibly by the maximum lens current available. The determination of the optimum specimen position is discussed in Section 1.2. A tilting top-entry specimen holder designed for use with the JEOL ultra-high-resolution pole-piece ($C_s = 0.7$ mm) is available form the JEOL Corporation of Japan.†

2. The top-entry stages preferred for high-resolution work—particularly for long-exposure dark-field recordings—are not generally eucentric. Small changes in specimen orientation may results in large lateral displacements of the image, so that a small crystal of interest may be lost from the field of view with a very small change in orientation. Foot controls should be used with an electric servo-controlled goniometer to control the specimen orientation so that corrections can be made with the specimen translation controls while tilting a small specimen. An accuracy of approximately 1 mrad in setting the orientation of the crystal is required; however, this will depend on the beam divergence used, the resolution sought, and the spherical aberration constant. Dynamical image calculations reveal the sensitivity of the image to small changes in orientation, and these become increasingly important for small beam divergence and high resolution. An important experimental problem in setting the orientation of very thin crystals is the lack of visible Kikuchi—one must rely on an orientation setting obtained from a slightly thicker portion of the crystal and judge any bending of the crystal from changes in the image symmetry. The method of orienting specimens is given in Section 10.4.

3. For many large unit-cell specimens, computed images have shown that the image of highest contrast in a through-focal series is also the image which can be simply interpreted in terms of the specimen structure. This image is easily selected by the microscopist, so that a through-focal series may not be necessary for such a specimen.

4. The choice of illuminating aperture (second condenser aperture) is important. Image resolution will steadily improve for thin specimens (phase objects) as θ_c is reduced, so that the smallest possible aperture should be used which provides a sufficiently intense image for focusing at high magnification. The loss of resolution accompanying a large beam divergence is roughly equivalent to the use of a smaller objective aperture. If the angle subtended by the highest-order beam included within the objective aperture is θ, then a resolution $d = \lambda/\theta$ can only be expected from an instrument with $C_s = 1.8$ mm if the beam divergence is less than 0.6 mrad (O'Keefe and Sanders 1975). The degradation of image quality with increasing beam divergence is further discussed in Sections 4.6, 3.3, and 4.2. The advantage

---

† Groups interested in constructing a similar holder should address correspondence concerning patent rights to Dr. A. Spargo, Department of Physics, Melbourne University, Parkville, Victoria, Australia.

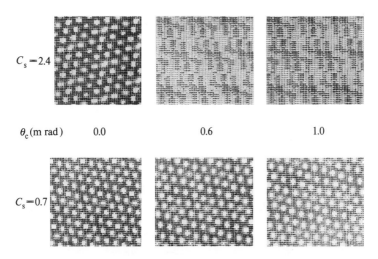

$C_s = 2.4$

θ_c (m rad)   0.0   0.6   1.0

$C_s = 0.7$

Fig. 5.27 The variation in image resolution with beam divergence θ_c for different objective lens spherical aberration constants. The crystal structure used for these computed images is 4Hb–TaS$_2$. The detailed way in which resolution is lost with increasing beam divergence depends on the structure considered; however, it is likely that for most structures, the use of a lens with a small spherical aberration constant will allow a larger illuminating aperture to be used (for the same resolution), resulting in a more intense final image.

of a small spherical aberration constant from the point of view of the illumination conditions is indicated in Fig. 5.27. Image resolution in these computed images is seen to fall off less rapidly with increasing beam divergence when using a lens with a low spherical aberration constant than for a lens of greater spherical aberration constant. Thus, a low-spherical-aberration objective lens allows a larger illuminating aperture to be used (for the same resolution), resulting in a more intense final image for focusing at high magnification.

   5. The effect of variations in the microscope gun-bias setting (see Section 7.3) on the contrast and resolution of lattice images can be understood using the damping envelope concept described in Section 4.2 (see Fig. 4.3). Appendix 3 assesses the relative importance of the choice of gun-bias as a function of the accelerating voltage. For practical purposes, when 200 kV machines are used to image large-unit-cell crystals at the Scherzer focus (eqn (6.16)), it is only necessary that the resolution limit imposed by chromatic aberration (eqn (4.11)) exceed that due to spherical aberration (eqn (6.16)). Using the figures of Appendix 3, we see that this is always true for modern instruments for all gun-bias settings. However, for lattice imaging of small-unit-cell crystals and image-processing studies in which one is interested in information beyond the point-resolution limit of eqn (6.16), there is clearly an advantage in operating the microscope at the largest gun-bias setting (smallest beam current), since, as shown in Appendix 3, this determines the absolute 'information limit' on present-day 200 kV instruments. This conclusion does not always hold for high-voltage machines, for which the electronic stability of the objective lens supply is likely to control the information resolution limit.

An example of a many-beam structure image is shown in Fig. 5.28. The specimen is an oxide of niobium, $Nb_{22}O_{54}$. Figure 5.28(b) shows the idealized structure seen in the projection used to obtain the image of Fig. 5.28(a). The structure can be thought of as made up of edge- and corner-sharing octahedra with a niobium atom at the centre of each octahedron and oxygen atoms at each corner. The centre of each crossed square is the apex of an octahedron. The unit cell contains two layers in the projection direction: the upper and lower layers are indicated by thick and thin lines respectively. Blocks of 3×4 octahedra can be seen in the upper layer, the lower layer contains blocks of 3×3 octahedra. There are additional niobium atoms shown by the black dots. The unit-cell boundary is shown dotted. The image is interpreted as follows. Since Fig. 5.28(a) is a positive print of an under-focus image (Δf negative in eqn (3.20)), we expect regions from which atoms are absent to appear white. From Fig. 5.28(b), we see that a block in the upper layer contains a regular pattern of 3×2 tunnels ('empty' spaces) between the octahedra, while the lower layer contains a 2×2 grid of tunnels. This pattern of 3×2 and 2×2 white blobs is clearly seen in Fig. 5.28(a). The additional niobium atoms are surrounded by four niobium atoms in octahedra, so that here the crystal projected charge density is large, with the image correspondingly dark at this point. The contrast is seen to degrade with increasing specimen thickness. An amorphous carbon support film can just be seen extending beyond the specimen edge. It is the defects which can occur in these structures, and their movements, which are of considerable interest to crystal chemists (see, for example, Iijima et al. (1974)).

A cautionary comment must, however, be made in connection with the

---

Fig. 5.28 (a) Electron micrograph of a specimen of niobium oxide, $Nb_{22}O_{54}$. The image interpretation is described in the text. The experimental conditions used to obtain this picture were: focus defect $\Delta f = -60$ nm, $\theta_c = 0.5$ mrad, $C_s = 1.8$ mm, $V_0 = 100$ kV, electron–optical magnification 400 000, optical magnification about 10, about 60 beams included in the objective aperture, and an exposure time of 10 s. The importance of contrast due to the exclusion of beams by the objective aperture (rather than phase contrast) can be estimated by observing the contrast at Gaussian focus with the aperture in place. (b) The idealized structure of the material. Each crossed square is an end-on view of an octahedron along its body diagonal. There are two layers, and oxygen atoms attached at the corners of the octahedra with niobium atoms in the centre. The upper layer in (b) is bounded by a heavy line. Additional niobium atoms are shown as black dots. The squares not crossed with diagonals are tunnels and correspond to white dots in the image. These and similar structures are best understood with the use of plastic or cardboard models, but it should be borne in mind that the tetrahedra are a convenient idealization which may not give exact atomic positions. The unit cell is shown dotted in (b) and is outlined on the experimental image. It has dimensions $a = 2.12$ nm, $b = 1.56$ nm, and $c = 0.38$ nm. (Image courtesy of Dr. L. A. Bursill.)

(a)

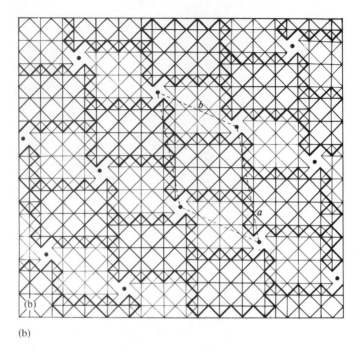

(b)

Fig. 5.28. (Continued)

PCD interpretation of images. We have seen that for current values of C_s, this approximation holds only for image detail larger than about 0.4 nm, and then only under certain experimental conditions. On the latest generation of microscopes, it is not difficult to record images showing detail on a scale finer than this by using a suitably chosen objective aperture. The complications in image interpretation which then arise when eqn (5.27) is not satisfied have been suggested for bright-field images in Fig. 6.3. 'Diffraction ripple', arising from the truncated Fourier series used to form the image, may be present and the overlapping subsidiary minima of the Airy's disc function can easily lead to a false impression of structure. In order to exclude these image artifacts it may be necessary to limit the resolution of a modern microscope to less than its fullest capability through the use of a suitable objective aperture in order to use the PCD model for image interpretation.

An example of an image which cannot be interpreted on the PCD model is shown in Fig. 5.29, taken under conditions not satisfying eqn (5.21). The specimen is hollandite. The image bears no obvious resemblance to the structure and can only be interpreted by detailed matching with computer-simulated images. The image shows two features characteristic of many-beam structure images.

    1. A bright Fresnel fringe is seen at the specimen edge, indicating that the image was recorded slightly under-focus (objective lens too weak).

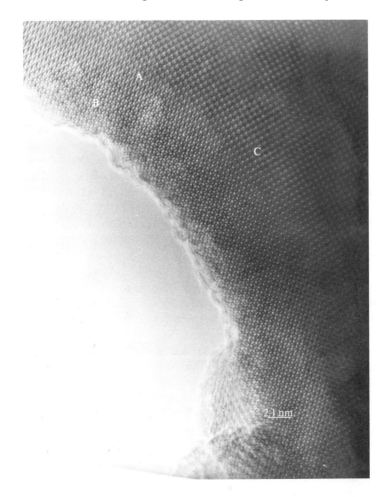

Fig. 5.29   A high-resolution image of hollandite. The structure of this material is described in Dryden and Wadsley (1958). This image was obtained under similar conditions to those given for Fig. 5.28 but with all beams included in the aperture and $\theta_c = 0.3$ mrad. Although the image bears no simple resemblance to the specimen structure, good agreement has been obtained between computed and experimental images down to a point resolution of 0.3 nm (see Bursill and Wilson 1977). The effect of specimen bending on the form of the image can be seen at B, while a comparison of areas A and C illustrated the effect of increasing specimen thickness.

Sometimes this fringe can be seen extending inside the specimen itself, as predicted by eqn (3.1) for a phase edge.

2. The low beam divergence used makes the image sensitive to small orientation changes. A region of altered image contrast is seen at A, where the specimen bends out of the zone axis ('cross grating') orientation. At B the image has the same symmetry as the observed diffraction pattern. In the thicker region C, multiple scattering and

Fresnel diffraction produce a complicated pattern of image contrast. If taken under controlled conditions, images such as these can be used to determine whether the tunnels of the hollandite structure are filled by metal atoms, and so test theories about the crystal's stoichiometry and degree of ordering (see Bursill and Wilson (1977)).

The advantages of using higher accelerating voltage are dramatically illustrated in Fig. 5.30, where images of the same crystal structure ($Nb_{12}O_{29}$) are shown as recorded on 100 kV and 1 MeV machines in (a) and (b) respectively. This crystal has a similar structure to that shown in Fig. 5.28(b), except that the octahedra all form 3×2 blocks. A model of this structure shows that, in addition to the main tunnels, smaller tunnels exist which are resolved in the 1 MeV image as white dots; these do not appear in the 100 kV image. The lower left-hand corner of each image is a computer-simulated image, produced by M. O'Keefe. Below the images are shown the instrumental transfer functions (see Section 4.2) which account for the improved resolution in (b), as discussed in Appendix 3.

The lattice imaging of small-unit-cell crystals using 100 kV machines presents special problems. In particular, for these specimens the image of highest contrast is generally not that which can be simply interpreted. This occurs from the following reasons. From Section 4.2 and Appendix 3 we see that the loss of contrast due to partial spatial coherence effects can be accounted for using a Gaussian damping envelope $\gamma(\nabla\chi/2\pi)$. Since $\gamma(u)$ is Gaussian, a high-contrast image is produced by adjusting the focus to make the slope of $\chi(u)$ (or zero) in the neighbourhood of important Bragg reflections. In small-unit-cell crystals such as gold or silicon, a single inner reflection (and those equivalent to it by symmetry) will dominate the form of the image seen through the binocular. There is thus a temptation to adjust the microscope focus control so the $\nabla\chi(u) = 0$ for $u = u_0$, where $u_0 = \theta/\lambda$ is the scattering vector for the dominating inner reflection. From eqn (3.24) this condition is

$$\Delta f = -C_s\lambda^2u_0^2 \qquad (5.76)$$

Since, for $C_s > 0.5$ mm at 100 kV and $u_0 > 3$ nm^{-1} this focus setting is greater than the Scherzer focus of eqn (6.16) (which allows a simple image interpretation), these images of high contrast in regions of perfect crystal will not correctly reveal the structure of any defects present. For example, in silicon, with $u_0 = 3.189$ nm^{-1} for the (111) reflections and $C_s = 2.2$ mm, eqn (5.76) gives $\Delta f = -306$ nm at 100 kV. This explains the common observation that sharp, high-contrast lattice fringes are found in metals and semiconductors at focus settings much larger than the Scherzer value. Indeed, the Scherzer image, which correctly reveals defect structure to the limited resolution available, may show a rather poor image in regions of perfect crystal, since one is attempting to image detail beyond the point-resolution of the instrument. The systematic use of 'passbands' or contrast transfer intervals derived from eqn (5.76) and (4.12) is exploited in image-processing studies, as further discussed in Appendix 3.

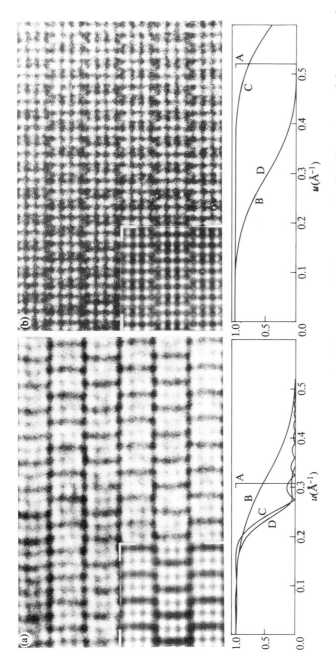

Fig. 5.30 (a) A structure image of $Nb_{12}O_{29}$ taken by S. Iijima on a 100 kV instrument. This is compared in (b) with an image of the same specimen recorded at 1 MeV, taken by S. Horiuchi. The improvement in resolution is clear. The insets show computed images for a thickness of 3.8 nm. Below the images are shown transfer functions for the conditions used to obtain the images. Curve A indicates the limit set by the physical objective aperture used, curve B includes only the damping effect of the energy spread from the electron gun (eqn (3.21b)), while curve C includes only the damping effect of using finite illuminating beam divergence (modulus of eqn (3.22a) with $\theta_c = 1.4$ mrad). The combined effect of all these factors (D) limits resolution to 0.38 nm at 100 kV and 0.25 nm at 1 MeV.

Table 5.2 Some early applications of high-resolution lattice imaging

Workers/Reference	Material and Spacings	Application and Comments
Komoda (1965)	copper (200) 0.181 nm palladium (200) 0.194 nm aluminium (111) 0.234 nm	instrumental resolution and stability test
Komoda (1966)	gold (200) + (020) + (020) 0.144 nm	three sets of crossed fringes from a simple metal
Yada and Hibi (1969)	palladium 0.097 nm nickel 0.088 nm	three-beam fringes resolution test
Sieber and Tonar (1975)	gold 0.072 nm	two beam with special aperture
Yada et al. (1971)	chrysotile asbestos 0.53 × 0.91 × 0.73 nm	structure determination of small, often faulted crystals
Heidenreich et al. (1968)	partially graphitized carbon black 0.3 nm	calibration and microscope testing
Krivanek et al. (1977)	germanium	grain boundary structure deter- mined at atomic resolution
Van Dyck et al. (1976)	YSeF	stacking sequence of polytypes
H.-R. Wenk (1976)	minerals	review of applications in mineralogy
Hutchinson et al. (1977)		review of applications in physical chemistry
Allpress and Sanders (1973)	niobium oxides	review of prospects for structure images
Iijima et al. (1974)	niobium oxides	observation of point defects in solids
Iijima (1975)	tungsten oxides	mechanism of crystallographic shear
Horiuchi and Matsui (1974)	niobium and vanadium oxides	HVEM many-beam high-resolution images
Bursill and Wilson (1977)	hollandite	image matching, and use of very low beam divergence to obtain 0.3 nm point resolution
O'Keefe and Sanders (1976)	Linde L	structure of zeolites, extraction of 'phase-contrast' image contribution
Bourret et al. (1979)	germanium	core structure of dislocations shown by lattice imaging
Cockayne et al. (1971)	germanium	two-beam dislocation lattice images
Uyeda et al. (1972)	copper hexadeca chlorophthalocyanine	organic crystal with low radiation sensitivity
Clarke (1978)	ZnO	use of lattice imaging to investigate varistor intergranular phases
Summerville et al. (1978)	$Pr_{24}O_{44}$	reaction mechanisms in fluorite related materials
Sinclair and Thomas (1975)	Cu_3Au	antiphase domains
Krivanek and Howie (1975)	Ge	structure of amorphous materials investigated by high-resolution CTEM
Spence and Kolar (1979)	Si	stacking fault energy from faulted dipole in silicon
Krivanek and Maher (1978)	Si	core structure of stacking faults in silicon
Matsuda et al. (1978)	Ni(220) 0.062 nm	current record for highest 'resolu- tion', (but see Section 8.10).

5.17 Early applications of lattice imaging

Table 5.2 lists some early publications in the field of high-resolution lattice imaging. Studies on biological crystals are not included and references to high-resolution work on these can be found in the section on minimum-exposure microscopy.

Specimen preparation details are given in many of the articles—more information can be found elsewhere (Hirsch *et al.* 1965; Goodhew 1972). More recent applications (since about 1978) are summarized in Chapter 12.

5.18 Lattice imaging in STEM

The purpose of this section is to summarize the important principles and features of high-resolution lattice imaging in scanning transmission electron microscopy, a topic which has received little attention in the research literature. STEM instruments were used to obtain the first images of isolated atoms and the interested reader can find an extensive review of this work in the article by Langmore (1978). The optimum choice of experimental conditions (detector and objective aperture sizes, choice of focus, etc.) for bright- and dark-field imaging of small molecules in STEM are discussed at length in the articles by Cowley (1976b) and Cowley and Au (1978).

The principles of STEM imaging are most easily understood through the reciprocity theorem. Figure 5.31(a) shows a simplified diagram of a single-lens conventional transmission electron microscope (CTEM) instrument in which S represents a single point source in the plane of the final condenser aperture, while P is an image point on the film. The reciprocity theorem of Helmholtz states that if a particular intensity† I_p is recorded at P due to a point source S at A, then this same intensity I_p would be recorded at A in Fig. 5.31(a) if the source S were transferred to the point P and the system were left otherwise unaltered. The instrument has been redrawn in Fig. 5.31(b) with the source and image points of Fig. 5.31(a) interchanged, and now corresponds to the ray diagram for a STEM instrument. The point source at A in Fig. 5.31(a) illuminates the CTEM specimen with a plane wave (A is a 'large' distance from the specimen) and lens L1 faithfully images all points near the optic axis simultaneously onto the film plane in the neighbourhood of P. This 'parallel processing' feature of CTEM imaging makes the equivalent STEM arrangement seem very inefficient, since, according to the reciprocity theorem, the intensity I_p of Fig. 5.31(b) gives the desired CTEM image intensity for a single point only of that image. In order to obtain a two-dimensional image from the STEM arrangement it is necessary to scan the focused probe at M across the specimen—the

---

† In fact the reciprocity theorem applies to complex amplitudes for elastic scattering, and may be extended to small inelastic losses if intensities are considered—see Pogany and Turner (1968).

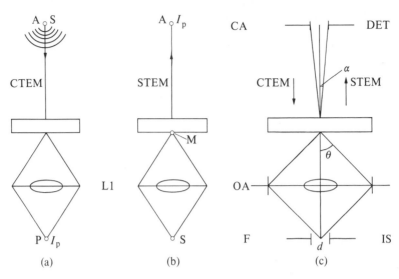

Fig. 5.31   (a) The ray diagram for a simplified CTEM instrument. Here S is a point within the final condenser aperture (strictly the 'effective source') which emits a spherical wave to illuminate the specimen. All points on the lower side of the specimen are imaged to conjugate points, such as P in the film plane, by the single lens L1. The intensity at P is I_p. (b) The source S and 'detector' (the image point P) have been interchanged, producing the ray diagram for STEM. (c) The condenser aperture (CA), objective aperture (OA) film plane (F) and film grain size (d) are defined for CTEM on the left, with the corresponding quantities for STEM shown on the right. Here DET is the STEM detector and IS refers to the incoherent STEM source plane, taken to be an electron emitter of size d.

reciprocity argument can then be applied to each scan point in turn. Conceptually, it may be simpler to imagine an equivalent arrangement in which the specimen is moved across over the probe in order to obtain the image signal in serial form (like a television image signal) from the STEM detector at A. The inefficiency in STEM arises from the fact that, in order to obtain an image equivalent to the bright-field CTEM image, most of the electrons scattered by the specimen in Fig. 5.31(b) must be rejected by the small detector at A. These scattered electrons carry information on the specimen and can be used by forming a dark-field image (see Cowley (1976b)). The serial-processing STEM arrangement has important advantages for analytical microscopy, and it should be noticed that, since there are no lenses 'down-stream' of the specimen in Fig. 5.31(b), the resulting STEM image does not suffer from chromatic aberration due to specimen-induced energy losses.

This reciprocity argument can be extended to cover the case of instruments which use extended sources and detectors. Thus, as shown in Fig. 5.31(c), consider the image formed by a CTEM instrument in which the illuminating aperture CA subtends a semi-angle α, the objective aperture subtends a semi-angle θ, and the image is recorded on film F with 'grain size' d. The illuminating aperture is taken to be perfectly incoherently filled.

The reciprocity theorem can be used to show that an identical STEM image can be obtained from a STEM instrument fitted with a finite incoherent source IS of size d in which the 'illuminating aperture' (called the STEM objective aperture OA) subtends a semi-angle θ, and in which the STEM detector DET subtends semi-angle α. The equivalence means that the CTEM-computed images described in Section 5.11 can be used to assist in the interpretation of STEM images. A computer program which calculates CTEM images will accurately reproduce STEM images of the same specimen so long as the incoherent sum over CTEM illumination angles (see Section 4.1) is re-interpreted as an incoherent sum over the STEM detector aperture.

These results can be used to find STEM analogues both for incoherent CTEM imaging and for the use of conical illumination in CTEM (see Section 4.4). We now consider the STEM analogue for many-beam lattice imaging in the symmetrical zone axis orientation using the three-beam case as a simple example. Figure 5.32, reading down the page, shows the ray-paths for CTEM three-beam lattice imaging. A spherical wave diverging from D illuminates the specimen, becoming a plane wave at O which is diffracted into three Bragg beams, passing through the objective aperture at A, B, and C to be synthesized by the lens into an image point at P. If we assume the use of a 'point' field-emission source in STEM (and no resolution limit due to the film used in CTEM) the reciprocity theorem (see

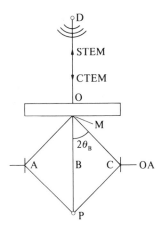

Fig. 5.32 Ray diagram for symmetrical three-beam lattice imaging in CTEM (reading down the page) and STEM (reading up the page). In CTEM, the optical paths MA, MB, and MC represents the directions in which three Bragg beams are scattered by the specimen which is illuminated by a plane wave in the direction DO. These are focused by the lens to interfere at the image point P. In STEM the specimen is illuminated from below simultaneously by three coherent plane waves in directions AM, BM, and CM. The Bragg condition permits scattering only through angles $2\theta_B$ (or zero), so that each of these incident plane waves can scatter in the direction OD of the detector, where they interfere to produce the STEM image signal. Note that for Bragg scattering, points A, B, and C are the only source points that can scatter into the detector D.

Fig. 5.31(c)) indicates that an identical STEM image can be formed (reading up the page) if a small STEM detector D is used together with a STEM objective aperture of semi-angle $2\theta_B$. This angle must be sufficiently large to accept the three Bragg beams of interest as diffracted from an imaginary point source placed at the STEM detector D. Thus, when setting up a STEM instrument in which the electron gun is below the specimen, the choice of objective aperture for lattice imaging with a small central detector can be made by imagining the specimen to be illuminated from above by an axial plane wave and using the same criteria which apply to the choice of objective aperture in CTEM (see Section 6.2). For a STEM instrument in which the electron gun is mounted above the specimen, the objective aperture is again selected by imagining the specimen to be illuminated from a point at the STEM detector (now below the specimen); however, the STEM image is now the same as that which would be obtained by placing the inverted specimen in a CTEM instrument. This inversion is rarely important.

The choice of STEM detector size for lattice imaging can be understood by drawing out Fig. 5.32 more fully in the STEM case. This is done in Fig. 5.33(a). The specimen is illuminated from below by a cone of radiation (semi-angle $2\theta_B$) which, for a point source at P, forms an aberrated spherical wave converging to the focused electron probe M on the specimen. Thus, around each scattered Bragg direction (for the illumination direction PM) we must draw a cone of semi-angle $2\theta_B$. The result is a set of overlapping 'convergent beam' diffraction discs as shown in Fig. 5.33(b). Three of these overlap at the detector D in the three-beam case shown. It is the interference between these three discs at D which produces the lattice image as the probe is scanned across the specimen. Figure 5.34 shows the arrangement which would be used to form the analogue of tilted-illumination, two-beam fringes in CTEM (see Fig. 5.1(b)). A detailed analysis of these arrangements (see Spence and Cowley (1978)) has established the following general results, which take account of all multiple scattering effects.

1. For crystalline specimens of any imperfections illuminated in a STEM instrument by an objective aperture of semiangle $\theta < \theta_B$ (θ_B is the Bragg angle), the set of convergent-beam diffraction discs observed in the detector plane shows an intensity distribution which is independent of the probe position, the aberrations of the probe-forming lens, and its focus setting. In this case the 'size' of the electron probe is much greater than that of the crystal unit cell.

2. The intensity within regions of disc overlap depends both on the probe position and the aberrations and focus setting of the probe-forming lens. The intensity at D of Fig. 5.34 varies sinusoidally as the probe moves across the specimen and is given by eqn (5.7) (for a small detector). The effect of enlarging the detector is exactly analogous to the effect of enlarging the final condenser aperture in CTEM. Outside the regions of overlap, the diffracted intensity does not depend on the probe position or lens aberrations.

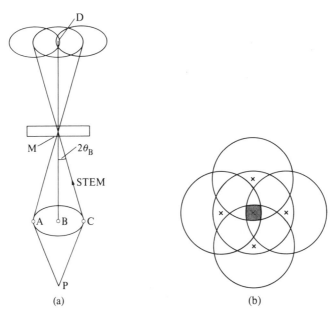

(a)   (b)

Fig. 5.33 The ray diagram of Fig. 5.32 drawn with a cone of semi-angle $2\theta_B$ around each Bragg direction. 'Bragg direction' means the scattered directions resulting from axial plane-wave illumination. These directions fall at the disc centres. In (b) the pattern is drawn out as it would appear in two dimensions for five-beam imaging; here the crosses denote Bragg directions while the hatched area indicates a possible STEM detector outline. Within this region, the intensity is sensitive to the probe position (scan coordinate). For thick crystals these discs will contain intensity variations characteristic of the crystal rocking curve.

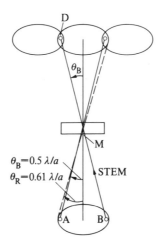

Fig. 5.34 The STEM analogue of tilted-illumination, two-beam CTEM lattice imaging described in Fig. 5.1(b). The CTEM conditions of Fig. 5.1(b) are obtained by placing the electron source at D, so that MA and MB become the two Bragg beams used for imaging. The figure also shows the illumination angle θ_R needed to make the electron probe 'width' approximately equal to the crystal unit-cell dimension a. As drawn, the diffraction discs subtend a semi-angle slightly larger than the Bragg angle and therefore overlap.

3. The intensity at the midpoint between overlapping discs is a special case. Here the intensity depends on the probe position, but not those lens aberrations which are an even function of angle, such as focus setting and spherical aberration.

Finally, a brief note concerning the frequently discussed question of probe size. In general the intensity distribution of the STEM probe incident on a specimen is a complicated and not particularly useful function. Near the Scherzer focus setting for the probe-forming lens it does, however, form a well defined peak with rather extensive 'tails'. The width of this peak (the F.W.H.M.) is given very approximately by $d = 0.61\lambda/\theta_R$ if we use the Rayleigh criterion and imagine the STEM lens to be imaging an ideal point field-emission source (see Fig. 5.34). If we set this probe size equal to the crystal unit-cell size a, then the illumination angle (STEM objective aperture semi-angle) needed to match the probe 'size' to the unit cell dimension is $\theta_R = 0.61\lambda/a$. But the Bragg angle is $\theta_B = 0.5\lambda/a$, and this is the condition that adjacent diffraction discs just begin to overlap. In this rather loose sense, lattice imaging becomes possible as the electron probe becomes 'smaller' than the crystal unit-cell dimension, and this is only possible if the diffraction discs are allowed to overlap. In fact, the important condition for lattice imaging in STEM is that the exit pupil of the probe-forming lens be coherent over an angular range equal to that covered by the Bragg beams one wishes to image. As the probe becomes very much smaller than the unit cell, information on the translational symmetry of the crystal is progressively lost (it becomes difficult to distinguish the reciprocal lattice using the overlapping convergent-beam pattern), leaving only information on the point-group symmetry of the unit cell contents and the probe position. The symmetry information contained in convergent-beam patterns is discussed in Section 11.4, which should be read in conjunction with this section.

5.19 Summary

This chapter provides the simplest theoretical background needed to understand the changes observed in the appearance of high-resolution lattice images with changes in experimental conditions. The important experimental conditions discussed include specimen thickness, defocus, objective aperture size, illuminating aperture size, and spherical aberration.

For two-beam lattice fringes formed using untilted illumination, we find that fluctuations in focus or electron wavelength cause an image shift normal to the legnth of the fringes. Using tilted illumination (see Fig. 5.1), the two-beam fringes become insensitive to changes in focus and electron wavelength and so to electronic instabilities. This method is, therefore, to be preferred when one is solely interested in obtaining the highest information-resolution limit possible. The method relies on accurately locating the objective lens optic axis, which is difficult in practice but should be done by the method of Section 10.2.

For three-beam fringes formed using axial illumination, the point is made that fringes are formed having either the lattice spacing or half this periodicity, depending on the focus setting. Frequently, the half-period fringes will not be resolved, owing to instrumental instabilities. The relationship between the size of illumination aperture used and depth of field is also investigated. This is the range of focus control settings over which lattice fringes can be seen, referred to the objective plane. We find that for two-beam fringes the smaller the illuminating aperture, the larger will be the range of focus values which will give fringes. That is, more accurate focusing is required when using a large illumination aperture. Depth of field is also proportional to the fringe spacing.

In Section 5.3, three mathematical approximations giving the strength of the Bragg reflections from a crystal of known structure and orientation are discussed. They confirm what is found in practice.

1. Two-beam fringes have a sinusoidal form whose maxima do not necessarily coincide with the atom positions in the specimen. The period of these fringes is equal to the atomic lattice spacing in the specimen (multiplied by the electron-optical magnification) if the Bragg condition is exactly satisfied.
2. Tilting the specimen will move a two-beam fringe image in a direction normal to the length of the fringes, as will a change in focus or a change in specimen thickness. Bent or wedge-shaped specimen produce two-beam images showing fringes of varying spacing.
3. The best thickness for obtaining two-beam fringes is approximately one-quarter of the extinction distance for the satisfied reflection. Here the contrast of two-beam lattice fringes is a maximum, since the two beams are of equal strength. Approximate values of extinction distances can be found in Hirsch *et al.* (1977).

In Section 5.4 the effects of variable lattice spacings are discussed. This is followed by sections on computing methods for both the Bloch wave and multislice techniques, together with a discussion of the effects of dynamically forbidden reflections, beam reduction, and sign conventions. The problem of errors in atomic positions as determined from experimental HREM images is also discussed.

The final sections of this chapter list some of the experimental parameters important for many-beam structure imaging. Examples of these images are given, several of which are readily interpretable in terms of specimen structure and one of which requires the use of computer-simulated images for its interpretation. Characteristic features of the images are described, and brief outline of the methods used for interpreting images of defects in crystals is also given in Section 5.15. The chapter ends with an historical bibliography of representative early lattice image applications and a review of the principles of lattice imaging in scanning transmission electron microscopy (STEM).

References

Allpress, J. G. and Sanders, J. V. (1973). The direct observation of the structure of real crystals by lattice imaging. *J. Appl. Crystallogr.* **6,** 165.

Anstis, G. R. (1977). The calculation of electron diffraction intensities by the multislice method. *Acta Crystallogr.* **A33,** 844.

Anstis, G. R., Lynch, D. F., Moodie, A. F., and O'Keefe, M. A. (1973). *n*-beam lattice images. II. Upper limits to ionicity in $W_4Nb_{26}O_{77}$. *Acta Crystallogr.* **A29,** 145.

Bethe, H. (1928). Theorie der Beugung von Elektronenan Kristallen. *Ann. Physik (Leipzig)* **87,** 55.

Blume, J. (1965). Die Kantenstreifung im elektronenmikroskopischen Bild Würfelförmiger MgO-Kristalle bei Durch strahlung in Richtung der Raumdiagonale. *Zeit. für Phys.* **191,** 248.

Booker, G. R. and Stickler, R. (1962). Method of preparing Si and Ge specimens for examination by transmission electron microscopy. *Br. J. Suppl. Phys.* **13,** 446.

Bourret, A. and Desseaux, J. (1979). The low angle [011] tilt boundary in germanium. *Phil. Mag.* **39,** 419.

Bracewell, R. (1968). *The Fourier transform and its applications.* McGraw-Hill, New York.

Burge, R. E. (1973). Mechanisms of contrast and image formation of biological specimens in the transmission electron microscope. *J. Microsc.* **98,** 251.

Bursill, L. A. and Wilson, A. R. (1977). Electron optical imaging of the Hollandite structure at 3Å resolution. *Acta Crystallogr.* **A33,** 672.

Buseck, P. and Iijima, S. (1975). High resolution electron microscopy of enstatite. II: Geological application. *Am. Mineral.* **60,** 771.

Buxton, B. F. (1976). Bloch waves and higher order Laue Zone effects in high energy electron diffraction. *Proc. R. Soc. Lond.* **A350,** 335.

Clarke, D. R. The microstructural location of the intergranular metal-oxide phase in a zinc oxide varistar. (1978). *J. Appl. Phys.,* **49,** 2407.

Cochran, W. (1952). The symmetry of real periodic two dimensional functions. *Acta Crystallogr.* **5,** 620.

Cockayne, D. J. H. and Gronsky, R. (1981). Lattice fringe imaging of modulated structures. *Phil. Mag.* **44,** 159.

Cockayne, D. J. H., Parsons, J. R., and Heolke, C. W. (1971). A study of the relationship between lattice fringes and lattice planes in electron microscope images of crystals containing defects. *Phil. Mag.* **24,** 139.

Cowley, J. M. (1959). The Electron Optical imaging of crystal lattices. *Acta Crystallogr.* **12,** 367.

Cowley, J. M. (1975). *Diffraction physics.* North-Holland/American Elsevier, Amsterdam.

Cowley, J. M. (1976a). The principles of high-resolution electron microscopy. In *Principles and techniques of electron microscopy; biological applications* Vol. 6. Van Nostrand Reinhold, New York.

Cowley, J. M. (1976b). Scanning transmission electron microscopy of thin specimens. *Ultramicroscopy* **2,** 3.

Cowley, J. M. (1978). The configuration of atomic defects from high-resolution transmission electron microscopy. *J. Nucl. Mat.* **69;** 70, 228.

Cowley, J. M. (1985). The future of high resolution electron microscopy. *Ultramicros.* **18,** 463.

Cowley, J. M. (1988). Electron microscopy of crystals with time-dependent perturbations. *Acta Cryst.* (1988) in press.

Cowley, J. M. and Au, A. Y. (1978). Image signals and detector configurations for STEM. In *Scanning electron microscopy/1978* (ed. Om Johari) Part 1, p. 53. IIT Research Institute, Chicago.

Cowley, J. M. and Moodie, A. F. (1959). The scattering of electrons by atoms and crystals, III. Single-crystal diffraction patterns. *Acta Crystallogr.* **12**, 360.

Cowley, J. M. and Moodie, A. F. (1960). Fourier images IV; the phase grating. *Proc. Phys. Soc.* **76**, 378.

Cowley, J. M. and Moodie, A. F. (1962). The scattering of electrons by thin crystals. *J. Phys. Soc. Japan* **17** (Suppl. B II), 86.

Cowley, J. M. and Iijima, S. (1972). Electron microcope image contrast for thin crystals. *Z. Naturforsch.* **27a**, 445.

Cracknell, A. P. (1968). *Applied group theory.* Pergamon, London.

Dederichs, P. H. (1972). Dynamical diffraction theory by optical potential methods. In *Solid state physics* Vol. 27, p. 135 (eds. H. Ehrenreich, F. Seitz, and D. Turnbull). Academic Press, New York.

Desseaux, J., Renault, A., Bourret, A. (1977). Multibeam lattice images from germanium in (011). *Phil. Mag.* **35**, 357.

Dowell, W. C. T. (1963). Das elektronenmikroskopische Bild von Hetzebenenscharen und sein Kontrast. *Optik* **20**, 535.

Doyle, P. A. and Turner, P. S. (1968). Relativistic Hartree-Fock X-ray and electron scattering factors. *Acta Crystallogr.* **A24**, 390.

Dryden, J. S. and Wadsley, A. D. (1958). The structure and dielectric properties of compounds with the formula $Ba_x (Ti_{8-x}Mg_x)O_{16}$. *Trans. Faraday Soc.* **54**, 1574.

Edington, J. W. (1976). *Practical electron microscopy in materials science.* Van Nostrand Reinhold, New York.

Fejes, P. L. (1977). Approximations for the calculation of high-resulution electron-microscope images of thin films. *Acta Crystallogr.* **A33**, 109.

Fejes, P. L., Iijima, S., and Cowley, J. M. (1973). Periodicity in thickness of electron microscope crystal lattice images. *Acta Crystallogr.* **A29**, 710.

Fields, P. and Cowley, J. M. (1978). Computer simulation of the imaging of atomic defects in F.C.C. metals. *Acta Crystallogr.* **A34**, 103.

Fujimoto, F. (1978). Periodicity of crystal structure images in electron microscopy with crystal thickness. *Phys. Stat. Sol.* **45**, 99.

Fujime, S., Watanabe, D., and Ogawa, S. (1964). On forbidden reflection spots and unexpected streaks appearing in electron diffraction patterns from hexagonal cobalt. *J. Phys. Soc. Japan* **19**, 711.

Fukahara, A. (1966). Many ray approximations in dynamical theory. *J. Phys. Soc. Japan* **21**, 2645.

Gevers, R. and David, M. (1962). Relativistic theory of electron and position diffraction at high and low energy. *Phys. Stat. Sol.* **113(b)**, 665.

Gibson, J. M. and Treacy, M. M. J. (1984). The effect of elastic relaxation on the local structure of lattice modulated thin films. *Ultramicroscopy* **14**, 345.

Gjønnes, J. and Gjønnes, K. (1985). Bloch wave symmetries and inclined surfaces. *Ultramicroscopy* **18**, 77.

Gjønnes, J. and Moodie, A. F. (1965). Extinction conditions in the dynamic theory of electron diffraction. *Acta Crystallogr.* **19**, 65.

Glaisher, R. W. and Spargo, A. E. C. (1985). Aspects of the HREM of tetrahedral semiconductors. *Ultramicroscopy* **18**, 323.

Goodhew, P. J. (1972). Specimen preparation in materials science. In *Practical methods in electron microscopy* (ed. A. M. Glauert) Vol. 1. North-Holland, Amsterdam.

Goodman, P. (1975). A practical method of three-dimensional space-group analysis using convergent-beam electron diffraction. *Acta Crystallogr.* **A31**, 804.

Goodman, P. and Lehmpfuhl, G. (1964). Verbotene Elektronenbeugungareflexe von CdS. *Z. Naturforsch.* **19a**, 818.

Goodman, P. and Moodie, A. F. (1974). Numerical evaluation of N-beam wave functions in electron scattering by the multislice method. *Acta Crystallogr.* **A30**, 280.

Hall, D. J., Self, P., Stobbs, W. M. (1983). The relative accuracy of axial and non-axial methods for the measurement of lattice spacings. *J. Micros.* **130**, 215.

Hashimoto, H. (1985). Achievement of ultra-high resolution by 400 kV analytical atom resolution electron microscopy *Ultramicroscopy* **18,** 19.

Hashimoto, H., Mannimi, M., and Naiki, T. (1961). Theory of lattice images. *Phil. Trans. R. Soc.* **253,** 459; 490.

Head, A. K., Humble, P., Clarebrough, L. M., Morton, A. J., and Forwood, C. T. (1973). *Computed electron micrographs and defect identification.* North-Holland, Amsterdam.

Heidenreich, R. D., Hess, W. M., and Ban, L. L. (1968). A test object and criteria for HREM. *J. Appl. Crystallogr.* **1,** 1.

Hirsch, P. B., Howie, A., Nicholson, R. B., Pashley, D. W., and Whelan, M. J. (1977). Electron microscopy of thin crystals. Krieger, New York.

Horiuchi, X. and Matsui, X. (1974). *Phil. Mag.* **30,** 777.

Hovmoller, S., Sjogren, A., Farrants, G., Sunberg, M., and Marinder, B.-O. (1984). Accurate atomic positions from electron microscopy. *Nature (Lond.)* **311,** 238.

Howie, A. (1971). The theory of electron diffraction image contrast. In *Electron microscopy in materials science* (eds. U. Valdre and A. Zichichi). Academic Press, New York.

Humphreys, C. J. (1979). The scattering of fast electrons by crystals. *Rep. Prog. Phys.* **42,** 1825.

Humphreys, C. J. and Spence, J. C. H. (1979). Wavons—A simple concept for high resolution microscopy. *Proc. EMSA, 1979* (ed. G. W. Bailey) pp. 554–555. Claitor's Publishing Division, Louisana.

Humphreys, C. J. and Spence, J. C. H. (1981). Resolution and illumination coherence in electron microscopy. *Optik* **58,** 125.

Humphreys, C. J., Drummond, R. A., Hart-Davis, A., and Butler, E. P. (1977). Additional peaks in the high-resolution imaging of dislocations. *Phil. Mag.* **35,** 1543.

Hutchinson, J. L., Jefferson, D. A. and Thomas, J. M. (1977). *Surface and defect properties of solids,* Vol. 6, Chap. 8. Chemical Society, London.

Iijima, S. (1975). High resolution electron microscopy of crystallographic shear structures in tungsten oxides. *J. Solid-State Chem.* **14,** 52.

Iijima, S., Kimura, S., and Goto, M. (1974). High-resolution microscopy of non-stoichiometric $Nb_{22}O_{54}$ crystals: point defects and structural defects. *Acta Crystallogr.* **A30,** 251.

Ishizuka, K. (1982a). Multislice formula for inclined illumination. *Acta Crystallogr.* **A38,** 773.

Ishizuka, K. (1982b). Translation symmetries in convergent-beam electron diffraction. *Ultramicroscopy* **9,** 255.

Ishizuka, K. (1985). Comments on the computation of electron wave propagation in the slice methods. *J. Microsc.* **137,** 233.

Jap, B. K. and Glaeser, R. M. (1978). The scattering of high-energy electrons. *Acta Crystallogr.* **A34,** 94.

Jauch, J. M. and Rohrlich, F. (1955). *The theory of photons and electrons.* Addison-Wesley, New York.

Kambe, K. (1982). Visualisation of Bloch waves of high energy electrons in HREM. *Ultramicroscopy* **10,** 223.

Kambe, K. and Moliere, K. (1970). Dynamical theory of electron diffraction. In *Advances in structure research by diffraction methods* Vol. 3, p. 83. Pergamon, London.

Kambe, K., Lehmpfuhl, G., and Fujimoto, F. (1974). Interpretation of electron channeling by the dynamical thoery of electron diffraction. *Z. Naturforsch.* **29a,** 1034.

Kogiso, M. and Takahashi, H. (1977). Group theoretical methods in the many-beam theory of electron diffraction. *J. Phys. Soc. Japan* **42,** 223.

Komoda, T. (1964). On the resolution of the lattice imaging in the electron microscope. *Optik* **21,** 94.

Komoda, T. (1965). Observation of lattice images of metal crystals. *Japan J. Appl. Phys.* **4,** 618.

Komoda, T. (1966). Observation of (220) lattice images of gold crystal. *Japan J. Appl. Phys.* **5,** 452.

Krivanek, O. and Howie, A. (1975). Kinematical theory of images from polycrystalline and random network structures. *J. Appl. Crystallogr.* **8**, 213.

Krivanek, O. and Maher, D. M. (1978). Krivanek, O. and Maher, D. M. (1978). The core structure of extrinsic stacking faults in silicon. *Appl. Phys. Lett.* **32**, 451.

Krivanek, O. L. and Rez, P. (1980). Imaging of atomic columns in [110] silicon. *Proc. 38th Ann. EMSA meeting* (ed. G. W. Bailey) Claitors, Baton Rouge.

Krivanek, O., Isoda, S., and Kobayashi, K. (1977). Lattice imaging of a grain boundary in crystalline Ge. *Phil. Mag.* **36**, 931.

Lang, A. R. (1983). The correct rules for determining the sign of fault vectors. *Phys. Stat. Sol.* **76(a)**, 595.

Langmore, J. (1978). Scanning transmission electron microscopy. In *Principles and techniques of scanning electron microscopy* Vol. 9. Van Nostrand Reinhold, New York.

Lewis, A. L., Villagrana, R. E. and Metherall, A. J. F. (1978). A description of electron diffraction from higher-order Laue zones. *Acta Cryst.* **A34**, 138.

Loretto, M. H. and Smallman, R. E. (1975). *Defect analysis in electron microscopy.* Chapman and Hall, London.

Lynch, D. F. (1971). Out of zone reflections in gold. *Acta Crystallogr.* **A27**, 399.

Lynch, D. F., Moodie, A. F., and O'Keefe, M. A. (1975). n-Beam lattice images V. The use of the charge density approximation in the interpretation of lattice images. *Acta Crystallogr.* **A31**, 300.

MacLagan, D. S., Bursill, L. A., and Spargo, A. E. C. (1977). Experimental and calculated images of planar defects at high resolution. *Phil. Mag.* **35**, 757.

Marks, L. D. (1984). Bloch wave HREM. *Ultramicroscopy* **14**, 351.

Marks, L. D. (1985). Direct observation of diffractive probe spreading. *Ultramicroscopy* **16**, 261.

Matusuda, T., Tonomura, A. and Komoda, T. (1978). Observation of lattice images with a field emission electron microscope. *Jap. J. Appl. Phys.* **17**, 2073.

Matsuhata, H., Van Dyck, D., Van Landuyt, J. and Amelinckx, S. (1984). A practical approach to the periodic continuation method. *Ultramicroscopy* **13**, 343.

Menter, J. W. (1958). The electron microscopy of crystal lattices. *Adv. Phys.* **7**, 299.

Metherall, A. J. E. (1975). Diffraction of electrons by perfect crystals. In *Electron microscopy in materials science,* Part II. Commission of the European Communities, Directorate General 'Scientific and Technical Information Management', Luxembourg.

Moodie, A. F. (1972). Reciprocity and shape functions in multiple scattering diagrams. *Z. Naturforsch.* **27a**, 437.

Nagakura, S. and Nakamura, Y. (1983). Forbidden reflection intensity in electron diffraction and structure image. *Trans. Japan Inst. Metals.* **24**, 329.

Nishida, T. (1980). Electron optical conditions for the formation of structure images of silicon oriented in (110). *Japan J. Appl. Phys.* **19**, 799.

Ohtsuki, Y. H. (1967). Normal and abnormal absorption coefficients in electron diffraction. *Phys. Lett.* **A24**, 691.

O'Keefe, M. A. and Sanders, J. V. (1975). n-Beam lattice images, V. Degradation of image resolution by a combination of incident-beam divergence and spherical aberration. *Acta Crystallogr.* **A31**, 307.

O'Keefe, M. A. and Sanders, J. V. (1976). The phase contrast component of lattice images of a zeolite crystal. *Optik* **46**, 421.

O'Keefe, M. A. and Saxton, W. O. (1983). The well known theory of electron image formation. *Proc. 41st Ann. EMSA Meeting,* p. 288.

O'Keefe, M., Spence, J. C. H., Hutchinson, J. L., and Waddington, W. G. (1985). *Proc. 43rd EMSA meeting* (ed. G. Bailey), San Francisco Press, San Francisco.

Olsen, A. and Spence, J. C. H. (1981). Distinguishing dissociated glide and shuffle set dislocations by high resolution electron microscopy. *Phil. Mag.* **43**, 945.

Pirouz, P. (1979). The effect of absorption on lattice images. *Optik* **54**, 69.

Priouz, P. (1981). Thin crystal approximations in structure imaging. *Acta Crystallogr.* **A37**, 465.

Pogany, A. P. and Turner, P. S. (1968). Reciprocity in electron diffraction and microscopy. *Acta Crystallogr.* **A24,** 103.

Portier, R. and Gratias, D. (1981). Diffraction symmetries for elastic scattering. In: *Inst. Phys. Conf. Ser.* No. 61 (I.O.P., Bristol) p. 275.

Radi, G. (1970). Complex lattice potentials in electron diffraction calculated for a number of crystals. *Acta Crystallogr.* **A26,** 41.

Reimer, L. (1984). *Transmission electron microscopy.* Springer-Verlag, Berlin.

Rez, P. (1977). PhD. Thesis. Oxford University, U.K.

Rez, P. (1985). The use of array processors attached to minicomputers for multislice image calculations. *Ultramicroscopy* **16,** 255.

Ritchie, R. and Howie, A. (1977). Electron excitation and the optical potential in electron microscopy. *Phil. Mag.* **36,** 463.

Saxton, W. O. and Smith, D. J. (1985). The determination of atomic positions from high-resolution electron micrographs. *Ultramicroscopy* **18,** 39.

Saxton, O., O'Keefe, M. A., Cockayne, D. J., and Wildens, M. (1983). Sign conventions in electron diffraction and imaging. *Ultramicroscopy* **12,** 75.

Self, P. G., O'Keefe, M. A., Buseck, P. R., and Spargo, A. E. C. (1983). Practical computation of amplitudes and phases in electron diffraction. *Ultramicroscopy* **11,** 35.

Serneels, R. and Gevers, R. (1969). Systematic reflection in transmision electron diffraction. *Phys. Stat. Sol.* **33,** 703.

Sieber, P. and Tonar, K. (1975). Test of electron microscopes by lattice imaging in the 0.1 nm domain. *Optik* **42,** 375.

Sinclair, R. and Thomas, G. (1975). Antiphase domains and superlattice spot splitting in Cu_3Au. *J. Appl. Crystallogr.* **8,** 206.

Skarnulis, A. J., Iijima, S., and Cowley, J. M. (1976). Refinement of the defect structure of 'GeNb$_9$O$_{25}$' by high-resolution microscopy. *Acta Crystallogr.* **A32,** 799.

Smith, D. J., Bursill, L. A., and Wood, G. J. (1985). Non-anomalous high-resolution imaging of crystalline materials. *Ultramicroscopy* **16,** 19.

Summerville, E., Tuenge, R. T. and Eyring, L. (1978). High resolution crystal structure images of Beta Phase $Pr_{24}O_{44}$. *J. Solid-State Chem.* **24,** 21.

Spence, J. C. M. (1978). Approximations for dynamical calculations of micro-diffraction patterns and images of defects. *Acta Crystallogr.* **A34,** 112.

Spence, J. C. H. and Cowley, J. M. (1978). Lattice imaging in STEM. *Optik* **50,** 129.

Spence, J. C. H. and Cowley, J. M. (1979). The effect of lens aberrations on lattice images of spinodally decomposed alloys. *Ultramicroscopy* **4,** 429.

Spence, J. C. H. and Kolar, H. (1979). *Phil. Mag.* **39,** 59.

Spence, J. C. H., O'Keefe, M., and Kolar, H. (1977). High-resolution image interpretation in crystalline germanium. *Optik* **49,** 307.

Spence, J. C. H., O'Keefe, M., and Iijima, S. (1978). On the thickness periodicity of atomic resolution dislocation core images. *Phil. Mag.* **38,** 463.

Stout, G. H. and Jensen, L. H. (1968). *X-ray structure determination,* p. 136. Macmillan, London.

Sturkey, L. (1950). Multiple diffraction of a scalar wave (electrons) in a periodic medium— Laue Case. *Proc. A.C.A. Summer Meeting,* New Hampton, New Hampshire.

Sturkey, L. (1962). The calculation of electron diffraction intensities. *Proc. Phys. Soc.* **80**(20), 321–354.

Taftø, J. (1979). Point symmetry reduction of the dispersion matrix. Internal report. University of Olso, Norway.

Tanaka, M. and Terauchi, M. (1985). *Convergent beam electron diffraction.* JEOL L.T.D., Tokyo.

Tinnappel, A. (1975) PhD. Thesis, Technische Universitat, Berlin.

Treacy, M. M. J., Gibson, J. M., and Howie, A. (1985). On elastic relaxation and long wavelength microstructures in spinodally decomposed $In_xGa_{1-x}As_yP_{1-y}$ expitaxial layers. *Phil. Mag.* **51,** 389.

Uyeda, N., Kobayashi, T., Saito, E., Harada, Y. and Watanabe, M. (1972). Molecular image resolution in electron microscopy. *J. Appl. Phys.* **43,** 5181.

Van Dyck, D. (1978). The path integral formalism as a new description for the diffraction of high-energy electrons in crystals. *Phys. Stat. Sol.* **72(b),** 321.

Van Dyck, D., Van Landuyt, J. and Amelincx, S. (1976). Direct imaging of polytypes in YSeF. *J. Solid-State Chem.* **19,** 179.

Van Dyck, D. and Coene, W. (1984). The real space method for dynamical electron diffraction calculations. *Ultramicroscpy* **15,** 29.

Vergasov, V. L. and Chuklovskii, F. N. (1985). Excitation of bound valence waves of energetic electrons and formation of crystal lattice images with atomic resolution. *Phys. Lett.* **110A,** 228.

Voss, R. Lempfuhl, G., and Smith, P. J. (1980). Influence of doping on the crystal potential in silicon. *Z. Naturforsch.* **35a,** 973.

von Laue, M. (1960). *Rontgenstrahlinterferenzen* (3rd edn.) Akademische Verlagsgesellschaft, Fankfurt-am-Main.

Yada, K. and Hibi, T. (1969). Fine lattice fringes around 1 Å resolved by the axial illumination. *J. Electron. Microsc.* **18,** 266.

Yada, K. (1971). Study of the microstructure of chrysotile asbestos by high resolution electron microscopy. *Acta Crystallogr.* **A27,** 659.

Wenk, R. (1976). *Electron microscopy in mineralogy.* Springer-Verlag, New York.

Yoshioka, H. (1957). Effect of inelastic waves on electron diffraction. *J. Phys. Soc. Japan* **12,** 618.

Zuo, J. M., Spence, J. C. H., and O'Keeffe, M. (1988). Bonding in GaAs. *Phys. Rev. Letts.* in press.

Note

The columns in eqn (5.54) represent different Bloch waves, with $i = 1$ to 7 from left to right. The rows number the beams, with $g = 1$ to 6 from top to bottom. The lowest row is the zero order beam.

6

NON-PERIODIC SPECIMENS

The unique advantages of high-resolution transmission electron microscopy are best utilized for the investigation of the atomic structure of non-periodic specimens and specimens composed of microcrystals. Many interesting and important structures do not form crystals large enough for analysis by X-ray diffraction methods; for these, electron images can provide a great deal of useful information. This chapter describes the simple phase-contrast theory, using which the images of non-periodic ultra-thin specimens, small molecules, and single atoms may be interpreted.

Many microscopists have published images which they claim show individual atoms. This is a promising step forward since it may make possible the structural analysis of small molecules, crystal defects, or suitably labelled biological molecules such as DNA. In this chapter the simplest theoretical expressions are given which allow the optimum experimental conditions to be determined for imaging these specimens. Particularly at high resolution, there are many instances in which a one-to-one correspondence does not exist between object and image, and it is important to understand these image artifacts and the experimental conditions which allow a simple intuitive image interpretation. The expressions given for the form of the image lead to a discussion of the methods used to compute electron microscope images of trial structures for comparison with experimental images. The inverse problem of deducing the specimen structure from an image is generally much more difficult and our interpretation must often be guided by experience gained in computing images of specimens whose structure is known.

This chapter also includes a discussion of the characteristics of images of single atoms and the dependence of single-atom image contrast on wavelength, atomic number, defocus, and objective aperture size. Most of the useful mathematical results for high-resolution single-atom imaging are included—the practical methods of single-atom imaging are discussed in Section 6.2 and Chapter 10. The techniques of minimum-exposure microscopy are outlined and some of the problems associated with reading electron micrographs into a computer are discussed, together with those of using a computer to simulate a micrograph.

6.1 Phase and amplitude contrast

At medium and low resolution the contrast in electron microscope images is understood to arise from the creation of an intensity deficit in regions of large scattering where the scattered rays are intercepted by the objective aperture. A clear account of this theory is given in Cosslett (1958). At high resolution where one seeks to examine distances comparable with the coherence width (Section 4.3) this incoherent theory is no longer adequate and phase contrast becomes the dominant contrast mechanism. The theory of phase contrast becomes important whenever detail in a micrograph is of interest which is smaller than the coherence width. Fresnel fringes, lattice fringes, and single-atom images are three examples of phase contrast. To discuss the transition from amplitude to phase contrast in a simple way it is necessary to use a highly simplified theory, which unfortunately is accurate only for the very thinnest specimens ($t < 5$ nm for light elements at 100 kV) or for rather thick specimens imaged at low resolution. There is no simple theory for the interpretation of image detail in the intermediate range—for example high-resolution detail (<1 nm) in a non-periodic specimen 20 nm thick imaged with coherent radiation ($X_c \gg 1$ nm). However, it is possible, though time-consuming, to compute such an image (see Section 6.7).

Aperture contrast (also known as amplitude or absorption contrast) can be incorporated into the transfer theory outlined in Chapter 3 in the following way. Since the amount of scattering excluded by a small aperture is approximately equal to the total atomic scattering, the transmitted intensity I for a specimen of thickness t and density ρ is

$$I = I_0 \exp(-\mu t)$$

with

$$\mu = \sigma_e \rho L / M \qquad (6.1)$$

where M is the local molecular weight, L is Avogadro's number, and σ_e is the local elastic scattering cross-section (see Fig. 6.6). Here the phase difference between waves scattered by successive atoms has been ignored, so the specimen exit-face wave amplitude becomes

$$\psi_e = \sqrt{I} = \psi_0 \exp(-\mu t / 2)$$

This expression can be used to describe the attentuation of the image wavefield in eqn (3.24) (Grinton and Cowley 1971; see also Appendix 2):

$$\psi_e = \exp(-i\sigma\phi_P(x, y)) = \exp(-i\sigma\phi_R(x, y))\exp(-\sigma\phi_i(x, y)) \quad (6.2)$$

where $\phi_P = \phi'_R - i\phi_i$ is the complex projected specimen potential with ϕ'_R now averaged over the resolution distance (Cowley and Pogany 1968). Here $\phi_i = \mu t/2\sigma$. This imaginary part of the potential may also be used to represent the depletion of the elastic wavefield by inelastic processes (Yoshioka 1957). Working through the transfer theory of Section 3.4 with this complex potential then gives for the bright-field image intensity

$$I = 1 + 2\sigma\phi_R(x, y) * \mathcal{F}\{A(u, v)\sin\chi(u, v)\}$$
$$- 2\sigma\phi_i(x, y) * \mathcal{F}\{A(u, v)\cos\chi(u, v)\} \quad (6.3)$$

where the $*$ again indicates the convolution or smearing of the specimen potential (the ideal image) with the point spread function of the instrument, shown in Fig. 3.6. This equation contains three terms. The first represents the uniform bright-field background, while the second represents the phase-contrast image. The third term represents the 'absorption' image. The relative importance of these last terms can be judged from Fig. 6.1, which shows the transfer functions at Scherzer focus (see Section 6.2). Most biological microscopy is limited by radiation damage and stain effects to a resolution far poorer than 1 nm. With an aperture cut-off at this point, Fig. 6.1 shows that we have, very approximately,

$$\cos\chi \approx 1 \qquad \text{and} \qquad \sin\chi \approx 0$$

so that the image intensity is, from eqn (6.3),

$$I \approx 1 - 2\sigma\phi_i(x, y) * \mathcal{F}\{A(u, v)\} \quad (6.4)$$

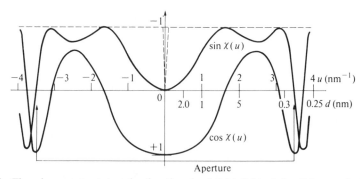

Fig. 6.1 The phase-contrast transfer function for bright field at the Scherzer focus $\Delta f = -1.2(C_s\lambda)^{1/2} = -61.0$ nm with $C_s = 0.7$ mm at 100 kV. This is about the lowest spherical aberration constant available on current instruments using a fixed high-resolution stage. The limits of a suitable bright-field aperture are indicated. The dotted line shows the transfer function required of an ideal phase-contrast microscope such as might obtain for a good quality optical phase-contrast microscope. In electron microscopy we approximate this ideal function by a suitable choice of focus and thereby 'convert' a phase object into an amplitude object. The abscissa is marked in units of $|K| = \theta/\lambda$ and resolution d. The effect of the damping envelope is shown in Fig. A3.2(a).

which predicts an image which is dark in regions of large 'mass-thickness' (ρt). This same image is obtained at Gaussian focus ($\Delta f = \chi = 0$) where there is no phase-contrast contribution, the effect of spherical aberration being negligible at this resolution. Thus ultra-thin specimens imaged with no objective aperture, so that $\phi_i = 0$, appear transparent at exact focus. This phenomenon is used to obtain a reference focus for high resolution, with some modification for the effects of the transfer function, as discussed in Chapter 9. Note that this analysis assumes perfectly coherent illumination. In practice with $\theta_{ap} = 1$ mrad, the 'incoherence width' is about 3.7 nm at 100 kV (eqn (4.17)) and so for resolution much poorer than 3.7 nm the imaging should be considered incoherent. In that case, as is the situation in most low-resolution biological microscopy, the expression corresponding to eqn (6.3) would be (Section 4.2),

$$I = 1 - 2\sigma\phi_i(x, y) * |\mathcal{F}\{A(u, v)\}|^2 * |\mathcal{F}\{\exp i\chi(u, v)\}|^2 \qquad (6.5)$$

where $A(u, v)$ describes the aperture used. For such an incoherent image, multiple-scattering effects will be small for thicknesses much less than the elastic scattering path length. For carbonaceous material at 100 kV this is about 130 nm. To assess the importance of multiple scattering for coherent imaging, see Section 6.3.

At high resolution, concentrating on spacings comparable with the coherence length between, say, 0.1 and 1 nm, we see from Fig. 6.1 that, very approximately,

$$\sin \chi \approx -1 \quad \text{and} \quad \cos \chi \approx 0$$

so the image is

$$I = 1 - 2\sigma\phi_R(x, y) * \mathcal{F}\{A(u, v)\} \qquad (6.6)$$

which is a pure phase-contrast image. This accurately reflects the specimen projected potential at this focus setting for sufficiently thin specimens. Like the absorption image, the image is dark in regions of high potential at this focus.

The above is a highly simplified theory and applies only to the very thinnest specimens for phase contrast ($t < 5$ nm for light elements). Nevertheless, eqn (6.3) provides the basis for much of the computer image processing presently undertaken (reviewed by Saxton (1978) and Frank (1973)). The relative simplicity of this equation has provided much of the stimulus for the preparation of ultra-thin specimens. Carbon support films can now be made with thicknesses down to about 2 nm. These are quite invisible in light (see Section 10.6).

To summarize, images of thick specimens formed with small apertures should be interpreted on the incoherent absorption-contrast model (eqn (6.5)), while high-resolution images are formed from phase-contrast effects and require larger (optimized) apertures and the thinnest possible specimens. Physically, absorption contrast arises from the exclusion of electrons from the image by the aperture. Phase contrast arises from the interference between waves within the aperture whose phase must be controlled (by accurate focusing) to produce contrast. Absorption contrast is increased at lower voltages and through the use of smaller apertures. The cut-off angle for absorption contrast can be reduced by increasing the lens focal length to provide more low-resolution contrast for stained biological specimens. On the other hand, phase contrast may increase with accelerating voltage (see Fig. 6.8). The maximum specimen thickness which allows a pure phase-contrast interpretation also increases with accelerating voltage. At 100 kV for stained biological specimens about 10 nm thick, both phase and amplitude effects are important if a resolution of 1 nm is expected. In practice the Fresnel fringe is also commonly used to enhance image contours. Since the width of these fringes is approximately equal to the transverse coherence width (Section 8.8) the approximate rule of thumb can be adopted that the more complicated phase-contrast theory is required to interpret detail smaller than the total width of any Fresnel fringes observed on a micrograph. The contrast for detail larger than this is proportional to the density of scattering matter in the object. The choice of focus under these conditions is discussed in Agar, Alderson, and Chescoe (1974).

6.2   Single atoms in bright field

In this section a brief review of the elementary theory of single-atom imaging is followed by a discussion of practical methods. In Chapter 3 an expression was given for the two-dimensional complex amplitude across the exit face of a phase object such as a single atom:

$$\psi_e(x, y) = \exp(-i\sigma\phi_p(x, y)) \tag{6.7}$$

where $\sigma = 2\pi m_0 e\lambda_r\gamma/h^2$ and $\gamma = m/m_0 = (1 - v^2/c^2)^{-1/2}$ incorporate the relativistic correction to the refractive index. A reasonable approximation for atoms of medium and low atomic number is given by the first-order expansion

$$\psi_e(x, y) = 1 - i\sigma\phi_p(x, y) \tag{6.8}$$

known as the weak-phase object approximation. An expression for the atomic potential can be obtained as follows. In the first Born approximation, the wave scattered by an atom is given by

$$\psi = \gamma(\exp(-i\mathbf{k}_i r)/r)f(\mathbf{K})$$

where

$$f(\mathbf{K}) = 2\pi m_0 e/h^2 \int_{-\infty}^{\infty} \phi(x, y, z) \exp(2\pi i\mathbf{K} \cdot \mathbf{r})\, dx\, dy\, dz \tag{6.9}$$

with $\mathbf{K}$ the scattering vector (see Fig. 6.2). Traditionally, the function $f(\mathbf{K})$ has been tabulated according to eqn (6.9) so that, regardless of the accuracy

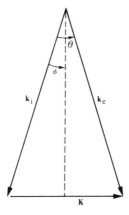

Fig. 6.2 Definition of the incident wavevector $\mathbf{k}_i$, the scattered wave vector $\mathbf{k}_e$, and the scattering vector $\mathbf{K}$ for elastic scattering. For elastic scattering $|\mathbf{k}_e| = |\mathbf{k}_i| = 1/\lambda_r$ and $\mathbf{k}_e - \mathbf{k}_i = \mathbf{K}$. The Bragg angle ϕ and the scattering angle θ are also shown.

of the first Born approximation, this publishing convention allows the atomic potential to be synthesized by inverse transform of this equation (Dawson, Goodman, Johnson, Lynch, and Moodie 1974). With the approximation of a flat Ewald sphere, equivalent in this case to the neglect of Fresnel diffraction or propagation within the atom, so that $K_z = 0$, we have, from eqns (6.8) and (6.9),

$$\psi_e(x, y) = 1 - i\lambda_r\gamma \int\int f(\mathbf{K}) \exp(-2\pi i \mathbf{K} \cdot \mathbf{r}) \, dK_x \, dK_y$$

The methods of Section 3.4 can be used to obtain the image complex amplitude as

$$\psi_i(x, y) = 1 - i\lambda_r\gamma \int\int f(\mathbf{K}) \exp(+i\chi(\mathbf{K})) \exp(-2\pi i \mathbf{K} \cdot \mathbf{r}) \, dK_x \, dK_y \quad (6.10)$$

where unimportant phase factors of unit modulus have been omitted for clarity. A magnificaton of unity is assumed and the function

$$\chi(\mathbf{K}) = \pi \Delta f \lambda_r \mathbf{K}^2 + \pi C_s \lambda^3 \mathbf{K}^4 / 2 = \pi [C_s \theta^4 / 2 + \Delta f \theta^2] / \lambda_r. \quad (6.11)$$

For elastic scattering (Fig. 6.2) we also have

$$|\mathbf{K}| = 2 \sin \phi / \lambda_r \approx \theta / \lambda = 1/d$$

with θ the angle between the incident beam and the scattered wave—this is twice the X-ray 'Bragg angle' ϕ. Scattering factors are usually tabulated as a function of ϕ/λ. For symmetric atoms,

$$f(\mathbf{K}) = f(|\mathbf{K}|)$$

and the image is radially symmetric. Equation (6.10) then becomes

$$\psi_i(r) = 1 - 2\pi i\gamma/\lambda_r \int_0^{\theta_{ap}} f(\theta)J_0(2\pi\theta r/\lambda_r) \exp(+i\chi(\theta))\theta \, d\theta \qquad (6.12)$$

which can be written

$$\psi_i(\mathbf{r}) = 1 - iz$$

where z is a complex function of $\mathbf{r}$.

The image intensity is then

$$I(r) = \psi_i\psi_i^* = 1 + 2\,\mathrm{Im}(z) + zz^* \approx 1 + 2\,\mathrm{Im}(z)$$

$$= 1 + 4\pi\gamma/\lambda_r \int_0^{\theta_{ap}} f(\theta)J_0(2\pi\theta r/\lambda_r)\,\theta \sin\chi(\theta)\,d\theta \qquad (6.13)$$

for a bright-field image of a single unsupported atom formed with a central objective aperture of semi-angle θ_{ap}. To a good approximation the integrated image intensity is proportional to the axial value

$$I(0) = 1 + 4\pi\gamma/\lambda_r \int_0^{\theta_{ap}} f(\theta) \sin\chi(\theta)\,\theta \, d\theta = 1 + A \qquad (6.14)$$

so that the image contrast becomes

$$C = |I(\infty) - I(0)|/I(\infty) = |A| \qquad (6.15)$$

The problem of maximizing single-atom contrast then becomes the problem of maximizing the integral in eqn (6.14), for which the experimental variables Δf, θ_{ap}, and $\chi(\theta)$ are available. The integral is dominated by the behaviour of the transfer function $\sin(\chi(\theta))$, shown in Fig. 6.1 with its characteristic rapid oscillation at high angles. Several workers have investigated the behaviour of eqn (6.14) and several criteria for choosing Δf and θ_{ap} have been proposed. All these lead to the approximate result that, for maximum contrast,

$$\Delta f = -1.2(C_s\lambda)^{1/2} \qquad (6.16a)$$

and

$$\theta_{ap} = 1.5(\lambda/C_s)^{1/4} \qquad (6.16b)$$

a result first obtained by Scherzer (1949). This is the $n = 0$ case in eqn (4.12). The diffraction limit set by this aperture size corresponds to a point resolution of

$$d = 0.66C_s^{1/4}\lambda^{3/4} \qquad (6.17)$$

in an ideal phase-contrast microscope fitted with the objective aperture of eqn (6.16b). Note, however, that since the imaging is coherent, the resolution cannot strictly be defined in this simple way, since the ability to distinguish neighbouring point objects depends on their scattering properties under coherent illumination (see Chapter 3). The sign of Δf indicates that a slightly under-focused lens is required for optimum contrast (Scherzer focus). Here the lens is slightly weakened from the Gaussian focus condition and is focused above the specimen, while the first Fresnel fringe would appear bright. This value of Δf gives the best approximation to an ideal lens for phase contrast, shown dotted in Fig. 6.1 and giving $\chi(\theta) = -\pi/2$ for all θ except at the origin where $\chi = 0$. This would lead to a darkened image on an otherwise bright background. Alternatively, $\chi(\theta) = +\pi/2$ would lead to a bright image. Notice that the low spatial frequencies are severely attenuated so that the pure phase-contrast image of, say, a labelled DNA molecule would be synthesized from the middle-range spatial frequencies only. In practice there is a maximum contribution from absorption contrast at low spatial frequencies which may complement this lack of information from the real potential (Section 6.1). All spatial frequencies beyond $1/d = \theta_{ap}/\lambda$ are excluded by the aperture, corresponding to detail smaller than $d_c = \lambda/\theta_{ap}$. Resolution for coherent images must be understood in this sense of an allowed band of spatial frequencies with differing weights given by the instrumental transfer function. Focusing the microscope can be thought of as 'tuning' the instrument to the spatial frequencies of interest.

Figure 6.3 shows this idealized bright-field image for a gold atom using the optimum values of Δf and θ_{ap}. The image was computed from eqn (6.13) using the relativistic Hartree–Fock scattering factors of Doyle and Turner (1968). The use of a complex scattering factor has been investigated by Hall (1971) and found to have little effect on contrast. A value of $C_s = 0.7$ mm has been used corresponding to a commercial CTEM instrument (JEM-100C) with fixed high-resolution stage. An image of lower contrast can be obtained with a small positive value of Δf which produces a bright image more intense than the uniform background. This transition from bright to dark image can be observed experimentally with small clusters of evaporated gold atoms on thin amorphous carbon films as described in Section 1.3, or using the resolution standards referred to in Section 8.10.

While the ability to image single isolated atoms is an impressive instrumental achievement, the usefulness of these experiments has yet to be established. Generally it is the arrangement of atoms on an atomic scale which is of interest. Nevertheless, isolated single atoms have been imaged in bright-field transmission microscopy by several workers (Formanek, Muller, Hahn, and Koller 1971; Iijima 1977a) and it is important to establish that these are indeed individual atoms. Perhaps the most convincing evidence that one is seeing individual atoms comes from experiments in which images of the same substrate area are recorded before and after the addition of a heavy atom. A method for adding atoms inside the microscope ('*in situ*') has been developed by Iijima (1977a), who installed a small evaporation furnace in the microscope. The most important practical problem in all single-atom studies so far has been the preparation of an 'invisible' substrate to support the atoms of interest. Iijima has used thin graphite flakes which are etched in the microscope to reduce their thickness still further by the

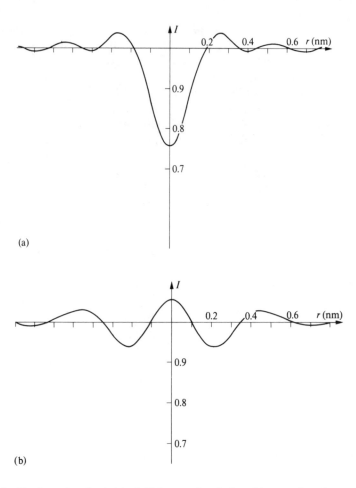

Fig. 6.3 The intensity of a bright-field image of a single gold atom plotted as a function of radial distance from the centre of the atom. The contrast is about 25 per cent, which well exceeds the threshold of detectability (5 per cent). The experimental conditions simulated are $C_s = 0.7$ mm, $\Delta f = -61$ nm, accelerating voltage 100 kV. This theoretical calculation based on eqn (6.13) does not take account of the image background due to a substrate supporting film. In (b) the same image is plotted for $\Delta f = -100$ nm to show the effect of an error in estimating the optimum focus used in (a). The two-dimensional intensity distribution is obtained by rotating the figure about the vertical axis and the intensity distribution for two neighbouring atoms consists of the sum of two displaced functions such as these. For such a pair of atoms the resulting image depends both on the focus condition and the atomic separation—it is possible to obtain a single deep intensity minimum at the point where subsidiary minima from adjacent atoms overlap. This could lead to the false identification of a single atom where two actually exist. The complexities of image interpretation are compounded in dark field where complex amplitudes must be added.

action of the electron beam on tungsten trioxide particles. The thickness of these ultrathin specimen supports has been estimated from the chance observation of graphite crystals whose edges curl up, giving a view down between the graphite crystal layers. Since individual layers are resolved, the thickness of the substrate can be found by counting the atomic layers. Crystals of about 2 nm thickness were frequently used. The question of substrate films is further discussed in Section 10.6, and full details of Iijima's method can be found in the papers by Iijima (1977a,b). Examples of Iijima's work are shown in Figs. 6.4–6.7. In these figures, taken at the optimum (Scherzer) focus given by eqn (6.16), we note the following. Owing to the use of an extremely thin graphite substrate there is almost no discernible contrast change between the region outside the graphite support film and the region inside the graphite flake. In addition, image noise due to thickness variations of the support film is not seen. (2) The first Fresnel fringe at the edge of the graphite flake appears bright, indicating that the image was recorded slightly under-focus (weakened lens). (3) The atoms and clusters of atoms appear dark, as expected from Fig. 6.3 at the focus setting used.

Three tests can be applied to confirm the interpretation of these images as showing single atoms. (1) The *size* of an isolated atom image should agree with the width of the function shown in Fig. 6.3(a) computed from eqn (6.13). This size depends mainly on the objective aperture used, the value of C_s and the focus setting. If no objective aperture is used, an effective

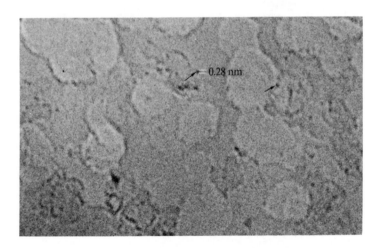

Fig. 6.4 Single atoms and clusters of atoms of tungsten on the surface of a few atomic layers of graphite. The black areas are larger tungsten oxide crystals. Fresnel edge fringes can be seen surrounding the region where a single atomic layer of graphite has been removed. Individual atoms are marked by arrows. The experimental conditions are $\Delta f = -61$ nm, $C_s = 0.7$ mm, $\lambda = 0.0037$ nm, illumination semi-angle 1.4 mrad, objective aperture semi-angle 16 mrad, electron optical magnification 500 000. Th objective aperture used excludes the Bragg beams diffracted by the graphite substrate from the image. No image processing, background subtraction, or contrast enhancement techniques have been used in printing these micrographs.

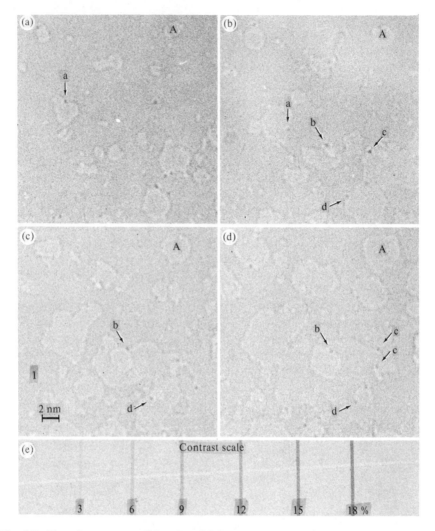

Fig. 6.5 These images, recorded under slightly different conditions from those of Fig. 6.4 ($C_s = 1.8$ mm, Scherzer focus) show individual atoms (a, b, d), one of which (a) can be seen to move between exposures. All the atoms cling to rough spots on the surface where the electrostatic potential 'well' is deepest. Atom c is believed to be a pair of atoms which separate in the last image. A contrast scale is given below to allow comparison with Fig. 6.3.

aperture must be assumed whose size is determined by the magnitude of the incoherent instabilities (see Sections 3.3 and 4.2 and Appendix 3). The size of a single-atom image does not depend on the type of atom to any appreciable extent, rather, it depends on the impulse response of the objective lens (see Section 3.4). (2) The *contrast* of a single atom image should also agree with calculations. Figure 6.5 shows three optimum focus images of tungsten atoms on a graphite substrate recorded at two-minute

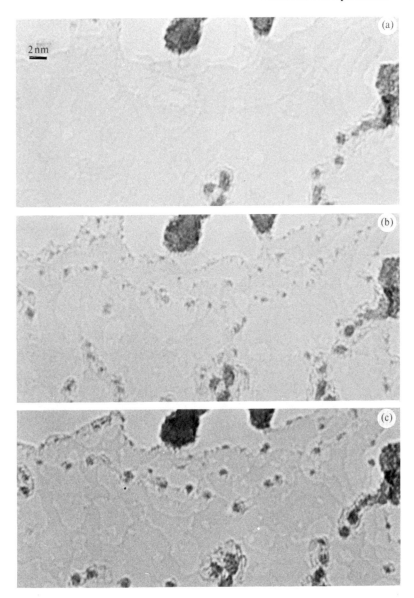

Fig. 6.6   Graphite flake shown (a), before and (b), after *in situ* evaporation of tungsten atoms. Again the atoms stick to surface steps. (c) The same region after five minutes' beam irradiation when the heavy atoms have migrated together to form metallic tungsten.

intervals. Atoms can be seen to have moved about between exposures, while the contrast of single atom blobs is between 6 and 9 per cent, in agreement with calculations. (3) Images showing the same region before and after the addition of a known number of atoms provide the best confirmation of image interpretation.

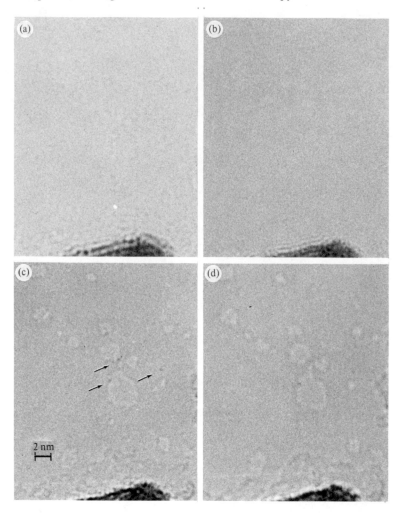

Fig. 6.7 Through-focus series showing the phase-object nature of single-atom images. Focus settings are (a) $\Delta f = +42.5$ nm, (b) $\Delta f = -25.0$ nm, (c) $\Delta f = -92.5$ nm, (d) $\Delta f = -160.0$ nm. Atoms are arrowed. Same experimental conditions as Fig. 6.4. Note the changing appearance of the Fresnel fringes around the clump of tungsten on the lower edge of the images and the absence of contrast in (b) taken under conditions similar to those of Fig. 1.2(b). Owing to the presence of spherical aberration, the focus condition for minimum contrast occurs at a small under-focus value rather than for $\Delta f = 0$ (see Section 10.4). (c) The optimum (Scherzer) focus image.

Figure 6.6(a) and (b) shows such a pair of images, while Fig. 6.6(c) shows the effect of beam-induced atomic mobility. The atoms use some of the energy available from the electron beam to hop about on the surface, finally collecting in the small clusters shown.

It should be emphasized that an ultra-low-spherical-aberration objective lens is not needed to image well-separated atoms. A low spherical aberration constant (resulting in reduced width of the atom image) only becomes important when one wishes to distinguish closely separated atoms, such as are found in the neighbourhood of defects in solids.

Figure 6.7 beautifully illustrates the 'phase object' nature of single-atom images (see Sections 1.1 and 3.4). In accordance with theoretical predictions, the image recorded at 'exact' focus shown in Fig. 6.7(b) shows little or no contrast. As further predicted by eqn (6.13), the over-focus image (Fig. 6.7(a)) shows bright fluctuations against the bright-field background, changing to dark atom images of high contrast at optimum focus (Fig. 6.7(c)). The reason why the image of minimum contrast (Fig. 6.7(b)) occurs at a slight under-focus value rather than for $\Delta f = 0$ is discussed in Section 10.5. It can be seen from a comparison of Fig. 6.7(c) and 6.7(d) that for single atom work a focusing accuracy of a few tens of nanometres is needed. Even greater accuracy (perhaps ± 4 nm) is needed if the atoms are closely spaced so that their images overlap (see Fig. 6.3(b)).

The usefulness of single-atom imaging remains to be determined. The likely applications of these methods arise in the following fields. (1) *Surface physics.* By observing the movement of single adatoms on surfaces the mechanisms of chemisorption and physisorption by which atoms stick to surfaces can be investigated. These mechanisms are fundamental to our understanding of solid-state chemistry, corrosion, and catalysis. Figure 6.6 makes it clear that newly arriving atoms are more likely to stick to exposed corners and edges of a substrate where a larger number of bonds can be formed than on the atomically smooth regions of the substrate. The edges of additional layers of graphite are outlined by the fine Fresnel fringes seen in this figure. (2) *Molecular biology.* The hope is that specific heavy atoms (or groups of heavy atoms of known structure) may be used to label nucleotides, thus allowing direct sequencing from a high-resolution image. Success in this field depends on the development of techniques for the minimization of radiation damage (see Section 6.9) however; some limited success has been achieved in this difficult area (Whiting and Ottensmeyer 1972; Beer 1978). (3) Many problems in materials science could be resolved by the development of reliable methods for structure determination at the atomic level. The distribution of segregant at grain boundaries and the detailed atomic structure of dislocation cores are two such outstanding problems of current research.

6.3   The use of higher accelerating voltage

In this section the advantages of higher accelerating voltage for high-resolution work are briefly reviewed. Several review articles covering the use of high-voltage machines for biological applications have appeared (Glauert 1974; Humphreys 1976). In what follows I have tried to emphasize aspects of high-voltage microscopy which are important for high-resolution phase-contrast microscopy, that is, from the point of view of coherent imaging theory (see Section 3.4).

Leaving aside the formidable engineering problems associated with the construction of high-energy electron microscopes, the wavelength reduction resulting from an increase in accelerating voltage has several generally beneficial effects on high-resolution phase-contrast images. On the experimental side, evidence is also slowly accumulating in favour of higher voltage. The advantages are considered in turn.

The dependence of single-atom contrast in bright-field on wavelength can be obtained from eqns (6.15), (6.14), and (6.16) using the relativistic wavelength and mass correction factor

$$\gamma = m/m_0 = (1 - v^2/c^2)^{1/2} = (1 + eV_0/m_0c^2) \tag{6.18}$$

The small additional effects of electron spin are discussed in Fujiwara (1962). The result of such a calculation is shown in Fig. 6.8 for a gold atom using Turner–Doyle scattering factors. The results of electron-optical calculations and experiment suggest that the product of C_s and λ is approximately constant. The smooth curve in Fig. 6.8 is drawn for

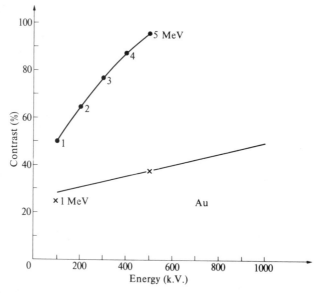

Fig. 6.8 Contrast of the bright-field image of a single gold atom as a function of accelerating voltage. Unlike the normal aperture 'absorption' contrast, this phase contrast increases with voltage. The Scherzer focus has been assumed and the spherical aberration constant used is an extrapolation from the value applying to Dr Kobayashi's instrument ($C_s = 1.2$ nm at 500 kV) assuming the product of C_s and λ to be independent of wavelength. Through the reciprocity theorem, this plot can be taken to apply to scanning transmission microscope images taken under reciprocal experimental conditions with a small, central detector. Note that the assumption that the product of C_s and λ is constant also suggests that the Scherzer 'optimum focus' condition is also independent of accelerating voltage (eqn (6.16)). The crosses on this diagram correspond to C_s values taken from commercial instruments now available.

$C_s\lambda = 11.3 \times 10^3$ nm^2, while the crosses represent the performance of commercial instruments now available. With increasing voltage the atomic scattering becomes more narrowly forward-peaked; however, the dependence of the scattering factor on θ differs from that of the transfer function and both effects must be considered. The main cause of the large increase in contrast with accelerating potential is the increase in relativistic electron mass. This calculation assumes that a phase contrast image *can* be formed at very high voltages, that is, that the resolution is not limited by the incoherent image smearing due to chromatic aberration (high-voltage instability) and mechanical vibration. High-quality many-beam phase-contrast images have indeed been obtained on 1 MeV and 3 MeV instruments by both French and Japanese groups (Jouffrey, personal communication; Horiuchi, Matsui, and Bando 1976). This increase in phase contrast with increasing accelerating voltage should be distinguished from the conventional (incoherent) aperture absorption contrast of interest at low resolution. In fact, for a fixed objective aperture size, the absorption contrast decreases with accelerating voltage as the electron scattering becomes more strongly forward-peaked.

The resolution of electron images also improves at higher voltage, both in theory and in practice, as shown in Fig. 5.30. Combining the empirical wavelength dependence of spherical aberration with eqn (6.17) suggests that the resolution should improve as the square-root of the wavelength. Some experimental work in support of this has recently appeared (Horiuchi and Matsui 1974) in which an improvement in resolution from 0.35 to 0.2 nm is reported in going from 100 kV to 1 MeV. This would make, very approximately, the resolution

$$d = 6.8\lambda^{1/2} \qquad (6.19)$$

with λ in nanometres. This would give a resolution of 0.1 nm at 4 MeV and 0.07 nm at 10 MeV. At present it appears unlikely that machines which operate at voltages above 3 MeV will be built, in view of their very high cost. In addition, problems of electronic and mechanical stability have yet to be resolved for instruments operating much above 1 MeV. A more detailed analysis of the dependence of resolution on electron wavelength can be found in Appendix 3.

The effects of chromatic aberration become less severe at high voltage. The incoherent image blurring is (eqn (2.38))

$$\Delta r = C_c \theta_{\mathrm{ap}} \Delta V_0 / V_0$$

which decreases inversely as the accelerating voltage for a fixed energy loss ΔV_0. The focal plane separation is also correspondingly reduced.

We now consider the importance of multiple scattering as a function of wavelength. In the incoherent imaging theory used at low resolution, multiple-scattering effects can be neglected for specimens much thinner than the path length for elastic scattering (see Section 6.1). However, when we consider high-resolution coherent imaging there are two quite distinct physical mechanisms which must be considered: the strength of the electron scattering interaction and the effects of Fresnel propagation through the specimen. The strength of the electron scattering interaction is measured by σv_g, where v_g are the crystal potential Fourier coefficients (see Section 5.3.2). Relativistic effects are correctly incorporated if the relativistic mass m, accelerating voltage, and wavelength λ_r (see eqn (2.5)) are used in the definition of σ (see Section 2.1). The relativistic exctinction distance is then given by equation (5.17). A rule of thumb which expresses the spreading of a wave from a point source due to Fresnel propagation can be obtained by setting the maximum phase shift allowed by Fresnel propagation equal to π (see Section 3.1), or, equivalently, by restricting the resolution to those reflections for which the Ewald sphere passes through the central maximum of the kinetmatic shape transform shown in Fig. 5.2. Thus the curvature of the Ewald sphere expresses physically the process of the propagation of the Huygen's wavelets in the crystal. Setting the phase of eqn (3.16) equal to π gives

$$u^{-1} = d = \sqrt{\lambda t} \qquad (6.20)$$

That is

$$t < d^2 \lambda$$

gives the approximate conditions that Fresnel broadening of the diffracted waves may be neglected in a specimen of thickness t if one is interested in detail of size d or larger. Note that eqn (6.20) also gives the approximate width of the first Fresnel edge fringe (see Section 8.8). One approximation which neglects the effect of Fresnel diffraction (or Ewald sphere curvature) is the phase-grating approximation (eqn (6.7)) and this belongs to a class of approximations known generally as 'projection approximations'. Using them, it is not possible to distinguish differences in the height of atoms within the specimen. Resolution in the direction of the electron beam is only possible using a theory which takes full account of Fresnel diffraction and the Ewald sphere curvature.

Thus eqn (6.20) may be used to estimate the maximum thickness for which eqn (6.7) can be used for non-periodic specimens. From eqn (6.20) we see that while this expression accurately describes images of 0.35 nm resolution in specimens 30 nm thick at 100 kV, for the same resolution this thickness is increased to 400 nm at 4 MeV. For crystalline specimens the orientation is important, and detailed calculations must be used to assess the accuracy of the phase-grating approximation (see Section 5.5).

While the radiation sensitivity of electron microscope specimens varies widely, there is evidence of a reduction in certain types of damage at higher accelerating voltage. The three important types of radiation damage at high voltage are displacement damage (in which an atom is removed from its site by the incident fast electron), ionization damage, and beam heating. Displacement of atoms is not possible below a certain high-voltage threshold, which depends on the atomic number of the displaced atom. This threshold is slightly greater than 100 kV for carbon atoms. Except for metals imaged at high voltages, displacement damage is normally a small effect in HREM work, compared with ionization damage. Ionization damage, which in covalently bonded structures can lead to structural rearrangement of the atoms, is expected to decrease inversely as the square of the fast electron velocity according to the Bethe law. This result has been confirmed experimentally, and the evidence for this and other types of damage has been reviewed extensively by Glaeser (1975). The effects of beam heating have not been extensively studied; however, a reduction is expected with increasing voltage. A more detailed discussion of radiation damage effects can be found in the article by Isaacson (1976), and in Section 6.9.

An important problem in high-voltage microscopy concerns the reduced sensitivity of photographic emulsions and viewing phosphors at high accelerating voltages. Methods of minimizing these effects, using thicker viewing phosphors and fast medical X-ray film have now been developed (Iwanaga, Ueyanagi, Hosoi, Iwasa, Oba, and Shiratsuchi 1968; Cosslett 1974). One final consequence of using higher accelerating voltage should also be mentioned: the increase in electron source brightness (see eqn (7.2)).

To summarize, there seems little doubt that the trend in high-resolution electron microscopy will be toward machines operating at several hundred kilovolts (Kobayashi, Suito, Uyeda, Watanabe, Yanaka, Etoh, Watanabe, and Moriguchi 1974; Cosslett 1974). For reasons of cost and performance these 'medium energy' machines have now become the most popular instruments for HREM work, with a point-resolution of about 0.17 nm. Appendix 3 and Fig. 5.30 provide instructive comparisons in more detail on the effects of accelerating voltage on HREM performance.

6.4 Contrast and atomic number

Experimental images of heavy atoms show more contrast than those of light atoms (Crewe, Langmore, Isaacson, and Retsky 1974; Whiting and Ottensmeyer 1972). This is expected theoretically as shown in Fig. 6.9, where the bright-field, single-atom contrast is plotted as a function of atomic number. Equation (6.15) has been used with the optimum values of Δf and θ_{ap} at 100 kV and 3 MeV. The steeper slope of the graph at higher voltage suggests that the discrimination of atomic species may be easier at higher voltage. In practice the effect of the substrate must be allowed for; however, experimental single-atom images have been obtained in bright field (Formanek *et al.* 1971; Iijima 1977a) and work continues in an effort to find a substrate that produces lower noise than amorphous carbon (see Section 10.6).

It is of interest to compare the expression for bright-field single-atom contrast with the total elastic scattering cross-section to clarify the distinc-

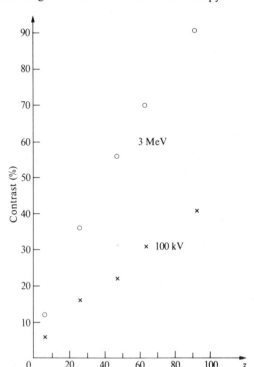

Fig. 6.9 The contrast of bright-field images of single atoms as a function of atomic number at two different accelerating voltages. The contrast increases more rapidly with atomic number at higher accelerating voltage. Here the Scherzer conditions are assumed with $C_s = 0.5$ mm at 100 kV and $C_s = 4.7$ mm at 3 MeV, again an extrapolation from Dr Kobayashi's instrument.

tion between a scattering experiment and a phase-contrast image. For an ideal phase-contrast image in bright field ($\chi(\theta) = \pi/2$ in eqn (6.10)), with no objective aperture, the contrast would be

$$C_{BF} = \lambda\gamma \int f(\mathbf{K}) \, d\mathbf{K} \tag{6.21}$$

For comparison the elastic scattering cross-section is defined as

$$\sigma_e = \int |f(\mathbf{K})|^2 \, d\Omega \tag{6.22}$$

which can be expressed as an integration over $\mathbf{K}$.

Values of σ_e are indicated in Fig. 6.10, obtained from Turner–Doyle scattering factors (Humphries, Hart-Davis, and Spencer 1974). On the same plot is shown the expression due to Lenz (1954),

$$\sigma_e = \frac{\lambda^2 Z^{4/3}}{\pi(1 - v^2/c^2)} \quad \text{Å}^2 \tag{6.23}$$

which provides a very rough approximation.

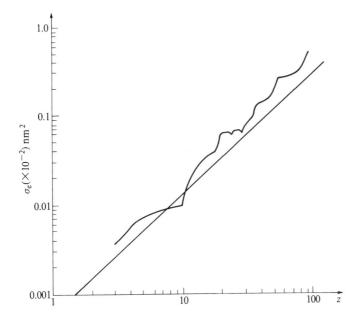

Fig. 6.10 The elastic scattering cross-section σ_e plotted as a function of atomic number. This shows how strongly 100 kV electrons are scattered by atoms, and how the strength of the scattering increases with atomic number. In transmission electron microscopy only a small part of the total scattering shown here is used to form the image. The straight line is a simple expression due to Lenz, the other curve is the result of detailed calculations based on Doyle–Turner scattering factors. Both scales are logarithmic. (Courtesy of C. J. Humphreys.)

The contrast in a dark-field atom image depends critically on the substrate used; however, the intensity relative to unit incident intensity at the centre of a dark-field image of a single atom formed in an ideal phase-contrast microscope ($\chi(\theta) = 0$ in eqn (6.25)) would be

$$I_{DF} = \lambda^2 \gamma^2 \sigma_e \qquad (6.24)$$

if an unsupported central beam stop were used within a large objective aperture.

Accurate calculations must take account of the substrate, the geometry of the beam stop, and the effects of defocus and spherical aberration as in eqn (6.25) (see also Section 6.5).

Through the reciprocity theorem (Cowley 1969), the results given in Figs. 6.9 and 6.10 also apply to the case of single atoms imaged in STEM under reciprocal aperture conditions—that is, a small central detector and optimum illuminating aperture and focus condition. The dark-field mode has many advantages in STEM (Langmore and Wall 1973); however, there are complications in the interpretation of these images (Cowley 1976b; see also Section 5.10 on STEM imaging).

6.5 Dark-field methods

A large increase in contrast is obtained by eliminating the bright-field background represented by the first term in eqn (6.13), the eye being sensitive to ratios of intensity rather than their absolute magnitude. The unscattered wave is focused to a small spot in the objective lens back-focal plane and may be eliminated using a suitably shaped aperture or tilted illumination. Against this important advantage must be set the disadvantage that dark-field images are less easily interpreted than bright-field images, since the effect of the aperture geometry on the image Fourier synthesis must be considered.

First consider a single atom imaged using an unsupported central beam stop to exclude the central diffraction maximum. Then eqn (6.8) allows the image to be calculated as

$$I(x, y) = \left| \lambda \gamma \int_{-\infty}^{\infty} f'(\mathbf{K}) e^{i\chi(\mathbf{K})} e^{2\pi i \mathbf{K} \cdot \mathbf{r}} \, d\mathbf{K} \right|^2 \tag{6.25}$$

where $f'(\mathbf{K})$ is the atomic scattering amplitude modified by the aperture. For an ideal central beam stop, $I(x, y)$ is a smooth peak, so that a bright spot will be seen as the dark-field image of an isolated single atom. For more complicated apertures, the function $f'(\mathbf{K})$ must be set to zero in the regions excluded by the beam stop. For groups of atoms and more realistic apertures, the situation is complicated and false image detail is possible in dark field. In practice the diffraction conditions may not be symmetric, so that a two-dimensional image simulation is required to predict the image of a known structure. Computed images for a variety of dark-field modes (wire beam stop, displaced aperture, tilted illumination) have been published (Krakow 1976; Chiu and Glaeser 1974). Beam stops have been made using fine tungsten wire across the diameter of the objective aperture or a gold-plated spider's web in the same position. Fine quartz fibres have also been used. The effect of electrostatically charged contaminant forming on these wires is an important experimental problem. The effect of the wire thickness on the appearance of the image of an atom is discussed in Chiu and Glaeser's paper. Figure 6.11 shows the image distortion possible under various dark-field conditions for a small molecule.

The optimum defocus and aperture size which must be used in dark field differ from those used in bright field. Unlike the bright-field case, both the real and imaginary parts of eqn (3.23) are equally influential for dark field, as can be seen by comparing eqns (6.25) and (6.13). The optimum focus for dark field is one which makes $\chi(\theta) \approx 0$ to the highest resolution possible. The condition $\sin(\chi(\theta_{ap})) = 0.3$ has been suggested (Cowley 1976a), which gives an optimum focus for dark field of

$$\Delta f = -0.44\sqrt{C_s \lambda} \tag{6.26}$$

With $C_s = 0.7$ mm, this gives $\Delta f = -22$ nm (under-focus). $\sin(\chi(u))$ and

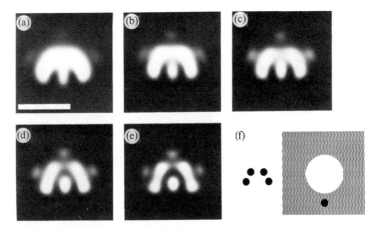

Fig. 6.11 Five computed dark-field electron micrographs of the mercury-thiophene molecule whose atoms are arranged in the pattern shown in (f). The simulated conditions are $\Delta f = -75$ nm (under-focus), $C_s = 1.35$ mm, accelerating voltage 100 kV, objective aperture semi-angles: (a) 6.5 mrad, (b) 7 mrad, (c) 7.5 mrad, (d) 8.0 mrad. The optic axis passes through the centre of the objective aperture shown in (f), while the central 'unscattered' beam falls at the position of the black dot. In practice any fine structure seen in these images would be lost if a large illuminating aperture were used and the contrast would be further degraded by scattering from the necessary supporting film. Nevertheless, these images would clearly lead to the identification of five atoms (see (e)) where only four actually exist. A further problem arises because one is not certain that the molecule will lie 'flat' on the substrate, in order to give a favourable projection. (Images courtesy of Dr W. Krakow.)

$\cos(\chi(u))$ are shown under these conditions at 100 kV in Fig. 6.12. Since the form of the image is mainly determined by the $\cos(\chi(\theta))$ function at small Δf, the image change with defocus is less rapid than for bright field, as is readily seen experimentally. We emphasize that since the image in dark field is no longer linear in specimen potential, the convenient product representation and envelope functions of the bright-field theory cannot be used (Hanssen and Ade 1978). Thus Fig. 6.12 should not be thought of as representing a 'transfer function'. Equation (6.26) also gives the minimum-contrast bright-field reference focus (see Section 10.5).

In practice the best dark-field high-resolution images have been obtained using tilted illumination with the unscattered beam intercepted on the rim of an aperture placed centrally about the optic axis, or using annular illumination (see Section 4.4). The optimum defocus for this mode is the same as that required for the ideal beam stop and the best aperture size to use can be found by drawing out the transfer function once the electron wavelength and spherical aberration constant have been found by the methods of Chapter 8. Notice that it is possible to include almost twice the range of spatial frequencies that could be included using axial illumination and a translated aperture (Fig. 6.12). The semi-angle subtended by an aperture is easily measured by taking a double exposure of the objective aperture and the diffraction pattern of a crystal of known structure.

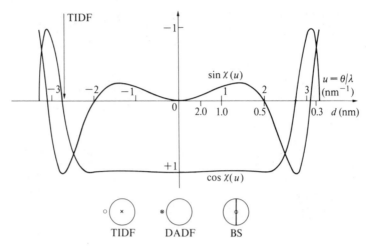

Fig. 6.12 The sin and cosine of $\chi(u)$ at the optimum conditions for high-resolution dark-field imaging. The phase shift due to aberrations has been kept close to zero to the highest resolution possible. Here $C_s = 0.7$ mm, and $\Delta f = -22$ nm (under-focus). The position of the optic axis ($\times$) and the 'unscattered' beam ($\bigcirc$) are shown in relation to the objective aperture for three dark-field imaging modes below. These are TIDF (tilted-illumination dark-field), DADF (displaced-aperture dark-field) and BS (beam stop). The vertical arrow on the left suggests the position of the unscattered beam in the tilted-illumination mode most commonly used for high-resolution work. This is not a 'transfer functon' (see text).

A clue to the kind of false contrast which may appear in dark-field images of non-periodic specimens may be obtained from eqn (6.8) (Cowley 1973). The image intensity formed with an ideal central beam stop is

$$I(x, y) = \sigma^2 \, |\phi_p(x, y) - \bar{\phi}|^2 \tag{6.27}$$

where $\bar{\phi}$ is the average value of the projected potential (equal to the excluded zero-order Fourier component) with the average taken over a coherently illuminated region. This equation suggests that both positive and negative deviations from the mean potential could produce the same contrast in dark field. That is, both a void and a heavy-atom inclusion would appear bright. It is also possible to show that a sinusoidal object produces an image whose periodicity is half that of the object, if the dark-field image is formed with a central beam stop. This is similar to the formation of 'half-spacing' fringes when three beams are used for lattice imaging (Section 5.1). Physically, dark-field images are formed by interference between Fourier components across the full width of the aperture. In bright field the important contribution is from interference between the central beam and a particular Fourier component or spatial frequency. Thus, image detail may appear in dark-field images on a scale which is finer than that seen in a bright-field image taken with the same aperture.

More realistic aperture arrangements, such as tilted illumination dark field, introduce an additional distortion in the form of an elongation in the image in the direction of a line between the unscattered beam and the optic axis. This 'Schlieren' distortion is similar to the effect of astigmatism and can be corrected accordingly. Figure 6.13 shows the individual atoms of the thorium oxide lattice imaged using this method (Hashimoto, Kumao, Hino, Endoh, Yotsumoto, and Ono 1973). Thus, in practice it does appear to be possible to make small molecules and atoms appear 'round' by careful adjustment of the stigmator, and a great deal of trial and error. Leaving the distortion aside, we may say that *the intensity variation in a high-resolution*

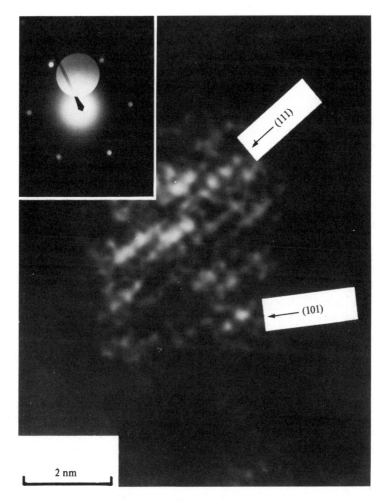

Fig. 6.13 Experimental image of the atoms in the thorium oxide lattice obtained by Dr Hashimoto using the tilted illumination dark-field method. The Miller indices of two identified planes of atoms are shown. Also shown is the diffraction pattern of the graphite substrate used to support the atoms, indicating the position of the objective aperture between the substrate reflections which is therefore excluded from the image. Inelastic scattering around the central beam is also excluded by this method. Note that the objective aperture size must be matched to the substrate used and the objective lens focal length.

dark-field image is proportional to the square of the deviation of the specimen projected potential from the mean projected potential.

At the magnification used for single-atom images (200 000–500 000) there is insufficient contrast to see these atoms through the viewing binocular and one is focusing on blobs about the size of a small molecule. It remains to be seen whether it is practical to include molecules of known structure with the unknown specimen as a focusing and stigmating guide, thereby allowing the microscopist to compensate for the image distortion which otherwise appears in high-resolution dark-field images formed with tilted illumination.

6.6 Inelastic scattering

Electrons which lose energy within the specimen suffer a wavelength change and therefore contribute an out-of-focus background to molecular images, generally incoherent with the elastic wavefield. These image intensities can be summed in an image calculation once the energy loss distribution for the specimen is known (Misell 1973). If the microscope is focused for electrons of energy V_0, the focus defect for other energies can be described by an additional term in the transfer function, as mentioned in Section 2.8.2. The overall effect of this blurred, incoherent background is then to reduce the elastic image contrast. Note that the expression for image broadening given in Section 2.8.2 deals with the incoherent images of neighbouring point sources. For most thin biological specimens the only way the microscopist can control chromatic aberration is by choosing a specimen position which minimizes C_c (see Fig. 2.17); however, this can only be done by increasing C_s with a consequent degradation in resolution.

Inelastic scattering makes the dominant contribution to the image background when amorphous carbon or graphite are used to support small molecules. The experimental work which has been done using energy-selecting microscopes to measure the ratio of inelastic to elastic scattering for amorphous carbon suggests that the inelastic scattering dominates by a factor of about 3 if all angles and energies are included (see Fig. 6.14). The objective aperture will increase this ratio to perhaps 10, since the important inelastic processes for amorphous carbon are confined to small angles around the central beam. Practically all the inelastic scattering occurs within a milliradian of the central beam under single-scattering conditions. The mean free path for inelastic scattering in amorphous carbon, which gives an indication of the average distance between scattering events, is about 70 nm at 100 kV. With a substrate much thinner than this, most of the inelastic scattering is therefore excluded in the tilted-illumination dark-field technique. A calculation of the inelastic and elastic scattering from amorphous materials, including coherence effects, is difficult, since the atomic coordinates must be known (Howie, Krivanek, and Rudee 1973). Incoherent calculations, accurate for large thickness, have appeared (Misell 1973); however, it is the coherent Fresnel diffraction and possibly any micro-

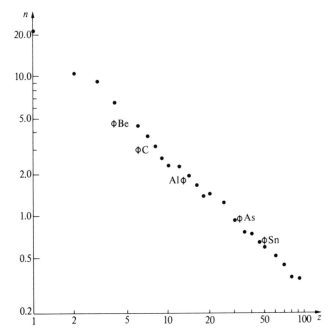

Fig. 6.14   The ratio n of total inelastic to total elastic scattering as a function of atomic number for 80 kV electrons. The open circles (with error bars) are experimental points and include scattering angles up to 0.13 rad. Less than 3 per cent scattering occurs outside this range. The experimental points were obtained from amorphous materials similar to biological specimens. The points give the predictions of the theory due to Lenz (1954), which neglects the small plasmon contribution. The figure is intended as an approximate guide, based on the work of Egerton (1975).

crystalline structure in 'amorphous' carbon when used as a substrate which gives rise to the unwanted background 'noise'. Experimentally it is a simple matter to measure the image background intensity from a bare substrate under the proposed diffraction conditions.

The partial ordering of atoms in amorphous carbon gives rise to broad ring maxima in the diffraction pattern. When using this substrate to support small molecules or strained biological specimens, some advantage can be gained by tilting the illumination so that the central aperture falls between these rings, thereby reducing the image background and enhancing the image contrast. Some experimental measurements on inelastic scattering in carbon can be found in Egerton (1975).

For crystalline substrates where the objective aperture is placed between the sharp Bragg reflections of the substrate, scattering will occur into the aperture owing to the inelastic processes of phonon excitation, plasmon excitation, and single-electron excitations. The intensity of this scattering can be computed if the substrate thickness is known but again an experimental measurement is simpler. An example of an attempt to fit theoretical estimates of diffuse inelastic scattering in a crystal to experimental measurements can be found in Howie and Gai (1975). There will also be an elastic contribution to the image background in these experiments from any non-periodic crystal structure such as surface contamination or crystal defects.

For isolated atoms, the proportion of elastic and inelastic scattering depends on atomic number. At low atomic number inelastic scattering dominates, while at high atomic number the scattering is predominantly elastic. The important early work of Lenz (1954) gives the ratio of inelastic to elastic scattering as inversely proportional to atomic number with a value of unity for copper. Figure 6.14 shows this result compared with experimental measurements. More accurate calculations are now possible and these calculation methods are reviewed by Burge (1973); however, Fig. 6.14 gives a sufficiently accurate indication for most biological work. Again the inelastic scattering is narrowly forward-peaked with a half-width of much less than a milliradian. Thus, for a single atom imaged in bright field all the inelastic scattering is included within the aperture, while only a portion of the elastic scattering is included. As an example, the bright-field image of a single aluminium atom is found to include only 3 per cent of the elastic scattering but about 80 per cent of the inelastic scattering (optimum aperture) at 100 kV (Whelan 1972).

The precise form of an image formed only from inelastically scattered electrons depends on the particular inelastic process. Images formed from plasmon-scattered electrons are not expected to contain high-resolution detail, owing to the non-localized plasmon interaction. However, images formed from electrons which have lost energy owing to K-shell or core electron excitations should show high-resolution detail, thereby allowing the possibility of image formation from a particular atomic species using a selective energy filter 'tuned' to this loss peak (Harauz and Ottensmeyer (1984), see also Fig. 6.15). This prospect of 'coloured' electron images (one 'colour' or fast electron wavelength for each atomic species in the specimen) is a subject of active research in high-voltage microscopy where the unwanted loss peak background is reduced. There are, however, two fundamental difficulties which must be considered in any proposal to develop this deceptively simple technique. Firstly, as mentioned above, only large-energy-loss processes are expected to provide high spatial resolution, for reasons based essentially on the uncertainty principle (see Craven, Gibson, Howie, and Spalding (1978) and Spence (1988) for a fuller discussion). Using a STEM instrument and energy filter it has proved possible, however, to detect a variation in the barium M edge intensity as a sub-nanometer probe is moved within a single unit cell of a barium aluminate crystal (Spence and Lynch 1982). On the other hand, smaller-energy-loss processes (such as plasmon excitation) preserve diffraction contrast effects (such as stacking fault fringes) but not high-resolution detail. The loss of diffraction contrast as one forms images at higher energy loss can be understood by considering the progressive overlap of the diffraction

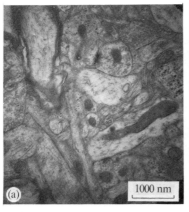 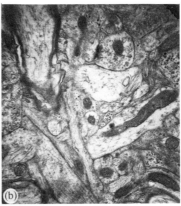

Fig. 6.15   Energy-filtered electron imges. The image on the left (a) has been formed in the normal way using all electrons. That on the right (b) was formed using only those electrons which are elastically scattered in the specimen and lose no energy. Chromatic aberration due to specimen energy losses is thereby eliminated. The specimen is visual cortex tissue (stained). The importance of these inelastic electrons depends strongly on the specimen thickness. In this case appreciable inelastic scattering has been removed leading to an improvement in image contrast. The inelastic image contribution produces a low-resolution background which reduces the image contrast in unfiltered images. (Image courtesy of R. Egerton.)

discs laid down at each of the satellite reflections responsible for the fault fringes, whose angular width is proportional to the energy loss. As with lattice imaging, one does not expect to see fringes in TEM from overlapping diffraction orders. A second important problem results from the fact that one normally has a strong mixture of multiple inelastic and elastic scattering present in the specimen, and it is rarely possible to devise experimental conditions which truly separate the two. Thus, for example, if one were to form an image in the characteristic core loss channel of a single uranium atom lying on an amorphous carbon substrate, it would be difficult to exclude from this image the contribution of electrons elastically scattered in the substrate in addition to their inelastic scattering at the uranium atom. (This elastic substrate image contribution can, however, be greatly reduced in STEM by suitable choice of detector geometry.) Thus, the attempts to form 'inelastics only' electron images should be clearly distinguished from lattice images formed from inelastically scattered electrons which are also Bragg scattered. Dynamical inelastic calculations (Spence and Lynch 1982) show, for example, that in images filtered for a particular species, other species may also appear as a result of coupled elastic–inelastic scattering. Using this filtering technique in scanning transmission electron microscopy it is possible to form a lattice image using electrons which are Bragg-scattered 'following' an inelastic event (Craven and Colliex 1977).

   The contrast improvement which results when the inelastic image contribution is removed from the image by an energy-selecting electron microscope is shown in Fig. 6.15. The image formed by elastically scattered electrons is compared with that formed using all electrons scattered by the specimen. We can conclude that for high-resolution work the inelastic background provides a slowly varying low-resolution image which has the effect of reducing the high-resolution elastic image contrast.

It is this improvement in the quality of energy-filtered images, particularly at high resolution, which has motivated the construction of imaging energy-selecting microscopes for transmission microscopy (Ottensmeyer and Henkelman 1974) based on the work of Castaing and Henry (1962). A much simpler device can be used for scanning transmission microscopes, where information is processed serially and the filter need not act as a lens. An important recent development has been the commercial availability of efficient magnetic-sector energy-loss analysers with parallel detection (Krivanek, Ahn and Kenney (1988)).

6.7   Molecular image simulation, processing and hardcopy devices

The purpose of this brief section is to put new workers in touch with the important sources of software and hardware needed to create computer-simulated electron micrographs, and to review the recent literature on image processing. Section 5.11, which emphasizes dynamical effects, should be read in conjunction with this section.

The image of a non-periodic specimen can be computed by the method of periodic continuation. A new potential function is constructed, equal to the specimen potential in some large cell and infinitely periodic outside it. This new potential can then be represented by a Fourier series and manipulated by a computer. For example the evaluation of eqn (6.10) by computer using integration steps $k_x = h/a$, $k_y = l/a$, (h, l, integers) implies a two-dimensional periodic extension with one atom in each unit cell of a plane rectangular lattice of side a. Thus the methods of Chapter 5 can be used to calculate the image of a small organometallic molecule or positively-stained biological specimens if the dimensions of the unit cell are made large enough. To avoid artifacts in the diffraction pattern (Lanczos 1966), the potential must join smoothly to its continuation in adjacent cells. Abrupt boundary discontinuities need not introduce artifacts into the interior of the image cell, however, since the forward-scattering nature of electron diffraction allows the method of image patching to be used (see Section 5.11). Even for single atoms, these two-dimensional calculations require considerable computing facilities. They allow the details of the objective aperture shape and position to be included. For specimens sufficiently thin that their images are described by eqn (6.7) the procedure is broadly as follows. Using either tabulated atomic potentials or relation (5.18), the cell potential is synthesized and the complex exit-face wave amplitude is evaluated over a grid of points in the unit cell using eqn (6.7). The Fourier coefficients of this must then be obtained and these diffracted amplitudes, which sample a plane in reciprocal space, must be multiplied by the complex transfer function (eqn (3.23)) and aperture function. Finally the image is synthesized by Fourier summation, usually using the fast Fourier transform library routines. These may require the number of points to be equal to 2^n (n integer). Thicker specimens can be handled by the dynamical calculations described in Chapter 5. Such a calculation for DNA is reported by Lynch and Moodie (1974) and shown in Fig. 6.16. The important effects of partial coherence are included by an integration over incident beam directions

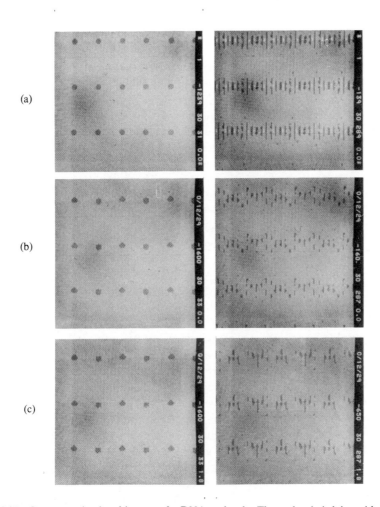

(a)

(b)

(c)

Fig. 6.16 Computer-simulated images of a DNA molecule. The molecule is lying with its long axis parallel to the support film and is not stained. Images are shown on the left as calculated using an objective aperture allowing 1 nm resolution and on the right for 0.33 nm resolution. In (a) the image predicted by the PCD approximation is shown; in (b) the full effects of dynamical scattering have been included with no spherical aberration; (c) shows the effect of introducing an aberration constant $C_s = 1.8$ mm. The atomic coordinates for DNA are given in Langridge, Wilson, Hooper, and Wilkins (1960). (Images courtesy of D. Lynch and A. F. Moodie.)

(Chapter 4) if the illuminating aperture is incoherently filled. In these dynamical calculations the potential within each slice varies with depth in the specimen, allowing the scattering from, say, a crystal substrate to be included in the calculation (Spence 1975; see also Section 5.11).

A variety of methods have been devised for the presentation of these images. The cheapest method and the most useful for a quick impression of an image needed, for example, while debugging a program, is that based on line-printer overprinting. A grey scale is obtained from the normal alphanumeric character set available on the computer's line printer using, where necessary, two superimposed characters. A photoreduction of the resulting two dimensional half-tone image gives a surprisingly good image representation. For example, the inset images in Fig. 5.30 were obtained by this method. Best results are obtained by printing the image on a large scale, then defocussing the enlarger slightly during photoreduction. Full details of the method, including Fortran program listings and the experimental reflectance measurements used to obtain the grey scale for popular character sets are contained in Head, Humble, Clareborough, Morton, and Forwood (1973). Since only 'ASCII' characters are used, this method has the important advantage of output device independence—the software will work on most computers. (The 'ASCII' code is the only successful international standard in computing). By avoiding the cost and delay of photographic processing, and by providing a good writing surface for annotation, the line-printer method is most useful for a quick search through a large number of images while experimental parameters are varied.

Higher quality images can also be produced by one of the many video hardcopy units now available (e.g. Sony UP-811, Mitsubishi P-60U). These are ideal for final publication-quality images, however specialised 'device-driver' software will be needed to interface a particular computer to the hardcopy unit. By printing on special paper, these units also avoid the need for photographic processing. They can be strongly recommended.

Laser printers may also be used to produce half-tone images. However the transfer of a bit-mapped image from the computer is likely to be prohibitively time consuming. It is therefore important to ensure that the grey-scale elements can be generated by the hardware of the laser-printer. The laser printer must therefore have enough memory to store an entire page of (at least) 300 dot-per-inch graphics.

More sophisticated image presentation methods are based on optical film writing equipment. The Optronics Photomation P-1700, for example, uses a small light spot (12.5–200 μm diameter) to scan unexposed film mounted on a rotating drum. The intensity of the light beam is modulated by the calculated image intensities read from magnetic tape. In the read mode the instrument can also be used as a two-dimensional scanning microdensitometer to 'read' electron micrographs onto magnetic tape suitable for a computer. The machine uses panchromatic film, so that all processing must be done in complete darkness. This machine, and others like it such as the Perkin–Elmer PDS Scanner (Model 1010A) and the Syntex AD-1 densitometer are ideally suited both to digital *a posteriori* image improvement projects (Frank 1973; Andrews 1970; Saxton 1978) and for the simulation of images by computer. The video hard-copy method provides a useful complement to this method because of its speed.

Oscilloscope displays have also been used to present two-dimensional computed images, although the number of grey-scale levels is reduced with this method. A specially prepared phosphor may provide up to 32 levels at 1024 by 1024 points. Up to 64 grey levels are commonly obtained with the film-writing systems. An example of the oscilloscope display system for computed images is described in Billington and Kay (1974), and images produced by this system are shown in Fig. 6.16. The most important practical problem which arises when using computed images for a comparison with experimental micrographs concerns the choice of contrast and mean background for the display of the computed images. To some extent these parameters can be used to compensate for an incorrect choice of other image-matching parameters $(C_s, \theta_{ap}, \Delta f, \text{etc.})$. Their choice must be based on a comparison with images of a known structure recorded under standardized experimental conditions (including all photographic conditions) and these standard conditions must be used when recording experimental images of unknown structures.

A vast literature exists on the subject of image processing, and there seems little point in reiterating here the excellent review of Hawkes (1982), in which over six hundred papers are indexed by topic. These cover just two years! Three popular software packages (amongst many) for HREM image processing are the SEMPER system (Saxton 1978), the SPIDER system (Frank, Shimkin, and Dowse 1981), and the INTELLECT system (Skarnulis 1982). Multiple-unit-cell images of copper phthalocyanine are analysed, including multiple scattering, partial coherence and noise effects, in Kirkland, Siegel, Uyeda, and Fujiyoshi (1985). Here a 30–50% increase in resolution is claimed as a result of image processing. The three-dimensional reconstruction of non-periodic objects from projections is exhaustively reviewed in Frank and Radermacher (1986). The use of digital Fourier 'transforms' of small regions of HREM images for the purposes of microanalysis is described in Tomita, Hashimoto, Ikuta, Endoh, and Yokota (1985) (see also Section 3.5 for background). Image processing of HREM images of $K_7NB_{15}W_{13}O_{80}$ is described in Kihlborg, Sundberg, and Sarborg (1985), who claim an accuracy in atomic coordinate determination of ±0.015 nm (see Section 8.10 for a discussion of the principles and limitations of this method). Noise filtering for periodic images is discussed in Tanji, Hashimoto, Endoh, and Tomioka (1982), and the image filtering of single-atom images is analysed in Dorignac and Jouffrey (1980). Volume 127, Part 1 of the *Journal of Microscopy* (1982) is devoted entirely to computer techniques in electron microscopy.

6.8   Noise and information

The intelligent use of an optical microdensitometer requires some familarity with the concepts of information theory. This field is also important when one wishes to obtain the maximum amount of structural information from a radiation-sensitive specimen while administering the smallest possible radiation dose. The following is intended to give a brief introduction to some useful ideas in this well-established field.

The grainy appearance of electron micrographs is due not to the film grain but to statistical fluctuations in the number of electrons contributing to each picture element. Experiments with photomultipliers show that the statistical distribution of electron arrivals at a small image area is Poisson. Figure 6.17 shows an electron emulsion exposed to a uniform intensity of electrons and this is compared with the same emulsion exposed to approximately the same optical density under light. Whereas several grains of emulsion are rendered developable by a single electron, the cooperative action of many photons is required for each grain with correspondingly improved statistics producing a smoother optical image. A measure of the noise in an electron image is given by the standard deviation of the distribution of electrons contributing to each image element. For Poisson statistics this is $\sqrt{N}$ where N is the mean number of electrons per picture element, so that the signal-to-noise ratio S is $N/\sqrt{N} = \sqrt{N}$ in the rather

Fig. 6.17 The response of photographic emulsion to light (a) and electrons (b). The grainy appearance in (b) is due to statistical fluctuations in the number of electrons contributing to each image element. This graininess represents an irreducible minimum of background for high-resolution images.

unrealistic case of a dark-field image recorded with no background noise contribution.

While electron noise is the dominant source of noise in an electron image, there is a further contribution to image noise from the photographic emulsion itself. Detective quantum efficiency (DQE) is defined as the ratio of the square of the signal-to-noise ratio measured from the emulsion to the square of this ratio for the electron noise in the electron beam (Hamilton and Marchant 1967). It thus provides a measure of the noise introduced by the recording medium. A DQE of 0.75 indicates that the signal-to-noise ratio in the recorded image is 87 per cent of that in the electron beam. Most modern emulsions have DQEs in the range 0.5–1.

An impression of the severity of electron noise can be obtained by analysing a typical bright-field phase-contrast image. Taking the brightness of a hair-pin filament to be about $3 \times 10^5 \, A \, cm^{-2} \, s^{-1}$, the object current density becomes about $1 \, A \, cm^{-2}$ for an illuminating aperture semi-angle of 1 mrad (eqn (7.1)). The total charge passing a $0.3 \, nm^2$ image element (referred to object space) after a two-second exposure is then 1.8×10^{-15} coulombs, corresponding to about 11 250 electrons. With 10 per cent contrast we would expect a reduction of $\Delta N = 1125$ electrons in the vicinity of a heavy-atom image (see Fig. 6.18). If a densitometer trace were taken of this image, the error (standard deviation) in an estimate of the peak area would be given by the square root of the variance (eqn (2.40))

$$\sigma^2(\Delta N) = \sigma^2(N_b - N_a) = \sigma^2(N_b) + \sigma^2(N_a) \approx 2N_b \qquad (6.28)$$

since $\sigma^2(N) = N$ and $N_a \approx N_b$. Thus $\sigma(\Delta N) = \sqrt{2N_b}$ and the electron signal-to-noise ratio for a bright-field single-atom image is

$$S = \Delta N / \sqrt{2N_b}. \qquad (6.29)$$

Applying this result to the given example gives $S = 7.5$; $S = 5$ is

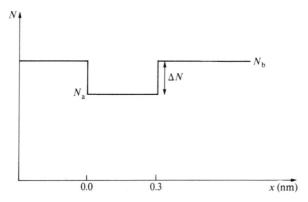

Fig. 6.18 Simplified representation of an atom image. A total number N of electrons contribute to the bright-field image within the image area. The background, for the same area, is made up from N_b electrons.

commonly accepted minimum value. Notice that the signal-to-noise ratio is improved by increasing the image element size, so that a trade-off between resolution and noise is expected.

The question of electron noise becomes important for radiation-sensitive materials for which the object current density must be limited. The relationship between contrast, resolution, and noise for biological specimens is discussed in more detail in Glaeser (1971).

A good deal of attention has been given to the problem of *a posteriori* correction of micrographs for image defects (see Frank 1973; Saxton 1978). In this field the approximation of Gaussian, additive noise is frequently made to simplify the analysis of filtering operations. This approximation is justified for bright-field images where N_b is large so that the variation of $\sqrt{N_b}$ with N_b is slow. The Poisson distribution becomes Gaussian when the mean is large. Electron noise is also stationary in the sense that its average value is independent of the choice of origin, so that many of the techniques of stationary time series analysis are applicable to electron images (Jenkins and Watts 1968).

Statistical fluctuations are responsible for the noise 'contrast' apparent in Fig. 6.16, and this has an important consequence for instruments used to read electron micrographs onto magnetic tape for storage and analysis by computer. As with all analogue-to-digital conversion, the number of discernible discrete levels (in this case the grey levels on the micrograph) are limited by noise superimposed on the analogue signal. Thus there is no point in attempting to distinguish grey levels within the noise 'contrast'. A relationship can be obtained for the average number of electrons per image element required for p grey levels at a signal to noise ratio S (Van Dorsten 1971):

$$N = p(p + 1)S^2/2 \qquad (6.30)$$

An acceptable image has $p = 5$, $S = 6$. The assumption has been made that there is no correlation between different image elements. The effect of correlations, which always exist between different parts of the image in a partially coherent aberrated optical system, on the information content of an image is discussed in Fellget and Linfoot (1955).

To obtain the maximum spatial resolution from a micrograph as it is read into a computer, the scanning microdensitometer light spot must be matched in size to the microscope point spread function. The electron-optical magnification must be chosen accordingly and the limit of resolution set by the film grain size must also considered (Section 10.8). Computer storage limitations usually mean that only a small portion of a micrograph can be read into a computer. Scanning spot sizes down to one micrometre are available on some two-dimensional scanning microdensitometers, and the practical considerations involved in digitizing electron micrographs are discussed in more detail in the text by Misell (1979). The direct electronic recording of images is discussed in Section 11.3.

The information contained in a microscope image is measured in bits and is given by the logarithm to the base 2 of the total number of possible pictures. If the electron noise limits the number of grey levels to p per image element with n image elements, the maximum information content is

$$I = \ln_2(p^n) = n \ln p \qquad (6.31)$$

For a micrograph of area A recorded by an instrument with a point spread function of size a in the image space, $n \approx A/a$. The value of p is seen from eqn (6.30) to depend only on the electron exposure. At 0.35 nm resolution an 11×5 cm micrograph recorded at 200 000 magnification is thus seen to contain about 1.8×10^{12} bits of information.

Equation (6.31) is obtained under the assumption that all images are equally probable, which must be made in the absence of *a priori* information. The probability that a particular image is recorded is then p^{-n}. The information content of a message or picture is an 'inverse' measure of its probability of occurrence—unlikely messages are supposed to contain a lot of information, while those which are certain to occur (probability of occurrence unity) contain no information or 'surprise'. By taking the negative logarithm of the probability of an event occurring, a suitable additive measure of information is obtained. The quantity so defined depends only on the likelihood of the event (or image) occurring, obtained from past observations. It tells us nothing about the image's usefulness to the observer. The definition of information can, however, be extended to the case where *a priori* probabilities can be assigned to individual image elements (Brillouin 1962).

A recent emphasis has been on techniques which use all the information which can be collected in an electron imaging experiment. This is particularly important for biological specimens and other radiation-sensitive materials where the best use must be made of every electron passing through the specimen. The concepts of information theory are thus likely to become increasingly important in high-resolution electron microscopy of non-periodic biological specimens, in which one wishes to extract the maximum amount of structural information from the specimen while delivering the smallest possible radiation dose (Cowley and Jap 1976). The use of entropy and information concepts for the analysis of HREM images of amorphous carbon is described in Fan and Cowley (1988).

6.9 Radiation damage and minimum-exposure microscopy

With the growing popularity of 'medium-energy' machines operating at 300–400 kV, the problem of radiation damage has recently taken on a renewed significance for HREM work. With rare exceptions (notably in some minerals and heavy metals), radiation damage effects cannot be neglected with these machines, and methods for minimizing damage are therefore outlined below. The study of radiation damage mechanisms (briefly reviewed at the end of this section) is a large subject, and the interested reader can find more information in the articles by Glaeser (1975), Isaacson (1976), Zietler (1982), and Cosslett (1978). We commence with a brief review of work on biological materials at lower voltages, before discussing high- and medium-voltage work in materials science. Many important but easily damaged biological specimens form thin crystals. The formation of an acceptably intense image at high magnification would destroy these crystals, but they can be imaged at low magnification. If the many unit cells in such an image are added together in a computer, an average unit-cell structure is obtained with greatly improved signal-to-noise ratio. A composite image can thus be formed with an acceptable signal-to-noise ratio from a very low-dose recording, showing the undamaged average projected cell structure. In addition, using a modified electron-optical system, it is made possible for the microscopist to focus on a region of specimen other than that which will be recorded for the final image. This is the principle of minimum-exposure microscopy (Williams and Fisher 1970; McLachlan 1958). A detailed description of this technique lies outside the scope of this book, and has been given in two recent texts where the problems of extracting high-resolution detail from micrographs of radiation-sensitive specimens are discussed at greater length (Saxton 1978; Misell 1979). In addition, the review article by Glaeser (1985), which contains an extensive bibliography, can be highly recommended as in introduction to the field of the electron crystallography of biological macromolecules. Rather than aligning and adding together the individual unit-cell images photographically, a common technique is to work with the digital Fourier 'transform' of the low-magnification image, and it is instructive to consider the way in which these two techniques are related.

As a result of radiation damage, the individual unit-cell images will differ slightly, with the result that the transform $F_0(u)$ of the true image $I(x)$ will consist of a set of slightly broadened Bragg peaks, as shown in Fig. 6.19(a). We wish to investigate the form of a new image $I'(x)$ formed using only those transform values within a narrow angular range around each Bragg peak. To do this we modify $F_0(u)$, multiplying it (in the computer) by the mask $M(u)$ shown in Fig. 6.19(b). The result is the function $G(u)$ of Fig. 6.19(c), in which a 'window' has been placed around each Bragg peak. From the convolution theorem of Fourier analysis, the transform $I'(x)$ of $G(u) = F_0(u)M(u)$ is then given by the true image $I(x)$ convoluted with the transform of $M(u)$. This transform is shown in Fig. 6.19(d) for a window width $1/b = 1/2a$ (a is the period of the one-dimensional 'crystal'). The effect of convoluting the true image $I(x)$ with the function shown in Fig. 6.19(d) is thus to add the image to $I(x)$ to itself after translation by a. This process, called 'periodic averaging', is repeated many times, with each new image added in with a weighting given by the Fourier transform of the window function. As the window 'top-hat' function becomes narrower, so the envelope in Fig. 6.19(d) becomes broader, resulting eventually in the unweighted addition of many translated and superimposed copies of the

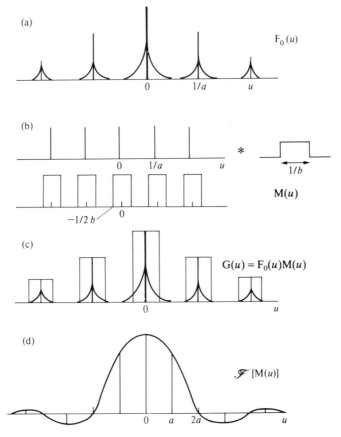

Fig. 6.19 The steps used to synthesize a single composite 'average' unit-cell image from the images of many damaged cells. In (a) is shown the transform $F_0(u)$ of the damaged crystal image $I(x)$. The mask function $M(u)$ used to select only values of $F_0(u)$ near Bragg peaks is shown in (b), represented as the convolution of a comb function with a 'top-hat' function. The effect of multiplying $M(u)$ by $F_0(u)$ is shown in (c), while (d) shows the transform of $M(u)$ for $b = 2a$, with a the lattice period and $1/b$ the window width. A similar analysis can also be applied to optically processed electron micrographs of crystals in which a mask is used in the back-focal plane which transmits only the Bragg beams.

original image $I(x)$. Thus the method of adding together photographically all the individual, aligned unit-cell images in a crystal is in principle exactly equivalent to the Fourier synthesis of an image from the original in which only the Bragg intensities are used in the Fourier series. In practice the major advantage of this technique is the improvement in signal-to-noise ratio seen in the periodically averaged image. The many subtleties involved in the digital processing of electron micrographs are discussed in greater detail in the book by Misell (1979).

In an application of this method (Unwin and Henderson 1975) the structure of purple membrane and that of catalase have been determined to resolutions of 0.7 and 0.9 nm respectively, using the weak-phase object theory of Section 3.4. Their method is a good example of image reconstruction methods which are becoming increasingly popular (see, for example, Erickson and Klug (1971), Saxton 1978; Frank 1973). The principle is as follows. Specimens are prepared using non-volatile glucose to replace the water normally present. We imagine an experiment in which low-magnification ($M \simeq 40\,000$), low-dose recordings have been made of a periodic low-atomic-number-specimen a few nanometres thick in a zone-axis orientation. A set of Fourier coefficients is computed from the image intensity. From eqn (3.29) we see that these are proportional to

$$A_g = 2\sigma v_{h,k,0} \sin \chi(u)$$

where $v_{h,k,0}$ is in general complex. If the defocus is known (for example from the optical diffractogram of a higher-dose image recording taken subsequently) together with the spherical aberration constant (Section 8.2), the function $\sin \chi(u)$ is known (eqn (3.24)). This function affects the amplitude of A_g and the sign of its phase. Knowing this function enables the complex $v_{h,k,0}$ to be found, from which only the phase is retained. The amplitudes of the reflections are obtained from a recording of the electron diffraction pattern. From eqn (3.27) we see that electron diffraction pattern intensities are proportional to $|v_{h,k,0}|^2$ from which the amplitude of $v_{h,k,0}$ is easily obtained. A Fourier summation of the complex $v_{h,k,0}$ finally produces the unaberrated projected specimen potential. The neglect of Fresnel diffraction within the specimen (curvature of the Ewald sphere—see Section 5.3.3) is frequently justified for these extremely thin specimens, but the detailed conditions under which the projection approximation fails for biological specimens has been analysed in detail by Cohen, Schmid, and Chiu (1984b). In practice the image dose used with this method may be so low that no observable image contrast can be seen, yet the computed transform will show well-defined Bragg peaks. The symmetry properties of a specimen may provide a check on calculated phases—for example, for a centrosymmetric crystal the $v_{h,k,0}$ are real with a phase ambiguity of sign only.

Generous computing facilities are required for this and similar methods in view of the enormous amount of information contained in an electron micrograph. Digital fast Fourier transforms are discussed in the article by Bantz and Zwick (1974). An even greater amount of storage is required for three-dimensional structure analysis. This is possible using images taken over a range of orientations (DeRosier and Moore 1970), since each projected image corresponds to a plane in reciprocal space.

This and similar methods based on the collection of information from both the image and diffraction pattern together with subsequent computer analysis, appears to be the most promising direction for the development of non-destructive high-resolution structure analysis, both for transmission and scanning transmission electron microscopy of radiation-sensitive materials.

A method for identifying radiation-sensitive biological molecules in STEM using the Paterson function and the techniques of pattern recognition has been proposed by Cowley and Jap (1976), and these functions have been used to assess damage in 5-iodouracil by Misra and Egerton (1985). Electron energy loss spectroscopy also forms a powerful tool for the study of damage (Egerton 1980).

The choice of emulsions for high-resolution work with radiation-sensitive specimens is discussed in Section 10.8, that the use of nuclear-track emulsions may have advantages for this work. These workers have also stressed the importance of minimizing stray magnetic fields in the vicinity of the viewing window when recording high-resolution detail at moderate magnification ($M \simeq 60\,000$).

We now turn to a review of some more recent applications of low-dose and image-processing techniques applied to the high-resolution electron imaging of biological and organic crystals and molecules. Section 12.5 contains additional material on this topic. A study of negatively-stained DNA packing in the T4 capsid can by found in Fujiyoshi, Yamagishi, Kunisada, Sugisaki, Kobayashi, and Uyeda (1982), and of monoclinic yeast tRNA embedded in glucose at 0.4 nm resolution in Fujiyoshi, Uyeda, Morikawa, and Yamagishi (1984). Dark-field imaging and image reconstruction from projections of the poly-L-lysine α-helix at 0.37 nm resolution are described in Harauz, Andrews, and Ottensmeyer (1983). The use of dark-field STEM techniques for the quantitative spatial mapping of molecular mass has been reviewed by Wall and Hainfeld (1986). Work on sensitive crystalline materials included an extensive study of the crotoxin complex by Jeng, Chiu, Zemlin, and Zeitler (1984). Here crystals embedded in glucose have been imaged at 0.35 nm resolution using image-processing and low-dose methods. The use of ice-embedding for high-resolution work is also described in Cohen, Jeng, Grant, and Chin (1984a). Practical design details of a minimum-dose control device are given in Fujiyoshi, Kobayashi, Ishizuka, Uyeda, Ishida, and Harada (1980), where it is applied to the imaging of zinc–phthalocyanine at doses of 0.3 and 1 coulomb cm^{-2}, and to Ag–TCNQ. Beam-induced specimen movement has been identified as a major resolution-limiting factor in biological work by Henderson and Glaeser (1985). A summary of work on purple membrane, and a study of lipid-depleted material can be found in Glaeser, Jubb and Henderson (1985). The use of correlation functions to align and superimpose similar images of a prosthecate halophilic microorganism is described by Kessel, Radermacher, and Frank (1985), where overlapping lattices have been separated by this technique. The three-dimensional reconstruction of a subunit of *Escherichia coli* ribosomes from randomly oriented particles is demonstrated in Verschoor, Frank, Radermacher, Wagenknecht, and Boublik (1984). Here non-crystalline particles lying in random orientations are used to obtain the necessary number of projections needed for three-dimensional image reconstruction. This powerful technique would appear to have application to the problem of catalyst particle morphology (see Section 12.11). The effects of a reduction in temperature on damage mechanisms are summarized in Glaeser and Taylor (1978).

It would appear that research in materials science by HREM would greatly benefit from the application of these minimum exposure techniques, which are now routinely used in much biological work (Glaeser 1985). The most popular method of minimizing dosage in materials science at present appears to be the use of rapid video recording. We now briefly review the radiation damage mechanisms which are important for HREM work on inorganic crystals. It is convenient to distinguish between the effects of ionization, which predominate at accelerating voltages below the atomic displacement threshold, and those above it. Above the threshold, direct collisions between the beam electron and an atom may displace it. A further useful classification is according to type of bonding present in the material, such as metallic, covalent, ionic, or Van der Waals. A sizeable literature exists on each of these topics. Two excellent reviews (Urban 1980; Hobbs 1979) contain extensive references to particular materials and processes—we restrict the following to a qualitative review of principles, and to recent work which has a particular bearing on high-resolution imaging. A third distinct area of study is that of radiation-induced defect reactions, based on the chemical reaction rate theory.

The displacement threshold for metals varies from about 150 kV for aluminium (at 300 K) to about 1 MeV for gold, and has a weak temperature dependence. Below these accelerating voltages, direct atomic displacement due to collisions between the fast electron and an atomic nucleus ('knock-on') is unlikely; however, ionization of an atom may lead, through secondary processes, to the ultimate rearrangement of the atoms in the crystal. This process is known as radiolysis. The Bethe theory for single-electron excitation, described in Section 6.6, gives an ionization cross-section inversely proportional to the square of the fast electron velocity, indicating that ionization damage should decrease at higher voltages. This cross-section (for carbon) is of the order of 10^6 barns at 100 keV (1 barn = 10^{-24} cm^2). Unfortunately the elastic cross-section falls off in a similar way, resulting in a reduction in image contrast at higher voltages which largely offsets the reduction of radiation damage for a specimen of given thickness. The essential requirements for the energy of excitation to appear as an atomic displacement are (1) that the excitation is localized (this excludes plasmon excitation); (2) that there is sufficient energy in the excitation to move a nucleus through the crystal potential saddle-point; (3) that the lifetime of the excitation exceeds typical atomic vibration times (about a picosecond). Radiolysis is an unimportant process in metals because of screening and the delocalized nature of excitations; however, it may occur in semiconductors (particularly at point defects), and is very important in organic materials, in the alkali halides and in many oxides, minerals, and other insulating crystals of ionic or covalent character. The energies needed to break bonds in these materials varies from about a tenth of an electron volt for Van der Waals bonding to tens of electron volts for ionic crystals.

For the simplest ionic materials the criterion for displacement that the electron–hole recombination energy should exceed the lattice binding energy is a useful guide, but this simple rule fails for many covalent solids such as silicates. The most extensive studies of radiation damage by this mechanism have been for the alkali halides. Note that although the cross-section for inner-shell ionization is small, the corresponding energy loss is large, so that the rate of energy transfer to the specimen may be comparable to that from valence-band excitations.

The rate at which interstitials and vacancies are created owing to the 'knock on' process at accelerating voltages above the displacement threshold is given by the product of the displacement cross-section and the local electron-current density. (The total dose may be more important for ionization damage.) Since this last factor is influenced by electron channelling effects and diffraction conditions, increased damage has been seen inside bend contours (Fujimoto and Fujita 1972). For most materials, these cross-sections are in the range of 1–100 barn at 100 kV. The displaced atom and remaining vacancy are known as a Frenkel pair. Displacement energies for many crystals are spread throughout the literature—typical values are 20–40 eV for metals, 80 eV for diamond, about 60 eV for magnesium and oxygen in MgO, and 11–22 eV for silicon. Thus, whereas in earlier HREM instruments operating below 120 kV only first- and second-row elements were subject to the knock-on mechanism, a much larger range of materials is now vulnerable with modern HVEM instruments operating between 200 kV and 1 MeV. A wide variety of reaction products and mechanisms have been observed and studied, including the recombination of interstitials and vacancies, their interactions with surfaces, dislocations and grain boundaries, and the formation of vacancy clusters and dislocation loops. A great deal of work has been devoted to problems associated with 'swelling' in nuclear reactor materials. It must be emphasized, however, that the displacement threshold energy quoted for a perfect crystal is rarely relevant, since damage occurs initially at the weakly bonded atoms on surfaces or in line or point defects. Thus, for example, it is found that whereas 'float-zone' silicon remains 'undamaged' (as evidenced by the HREM image) for extended periods in 400 kV microscopes, CZ silicon at the same voltage is damaged heavily within a few minutes, presumably owing to the presence of oxygen-related defects. Similarly, deformed crystals, containing a high concentration of point defects, are found to damage more readily than perfect crystals as one approaches the displacement threshold. In some high-voltage machines, the complication also arises of distinguishing electron-beam damage from that due to ions generated in the gun (Werner and Pasemann 1982). Some machines are fitted with a suitable ion trap.

We now consider specifically the direct evidence for radiation damage seen in HREM structure images of inorganic materials. Figure 6.20 shows a structure image of $4Nb_2O_5.9WO_3$ both before and after about 20 minutes of irradiation by 1 MeV electrons at typical microscope beam currents (perhaps 1 cm^{-2}) (Horiuchi 1982). In Fig. 6.20(a), the dark spots are the

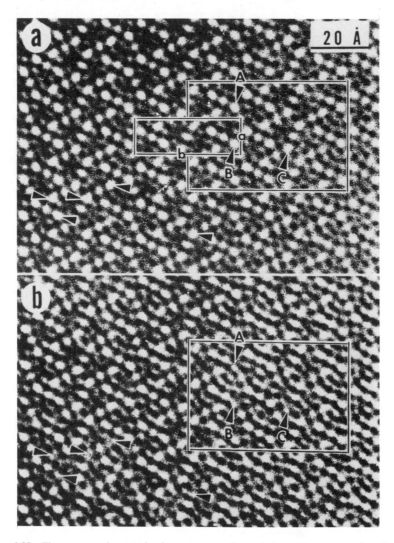

Fig. 6.20 The upper micrograph shows an atomic-resolution micrograph of undamaged $4Nb_2O_5.9WO_5$ with the unit cell indicated. Below is seen the altered image after 20 minutes' irradiation of 1 MeV electrons. Arrows indicate pentagonal tunnels filled with metal atoms which reverse contrast as the atom is 'knocked on' to a neighbouring pentagonal tunnel. (From Horiuchi (1982).)

cations, bright areas are pentagonal or square tunnels through the structure. The structure consists of corner-sharing octahedra, and the beam direction is parallel to the c-axis. The contrast changes at the arrowed tunnel positions are interpreted as knock-on events at the pentagonal tunnel sites, and these are not seen at the square tunnel sites. Some pentagonal tunnels contain a string of alternating metal–oxygen atoms, and it is assumed that oxygen is initially released, leaving a weakened metal–metal bond. Subsequent knock-on transfers one of these metal atoms into a neighbouring empty pentagonal tunnel.

This study, and the many others like it which have appeared since, reveals the power of high-resolution imaging for the determination of the atomic mechanisms involved in radiation damage in favourable cases. In a similar study of Zr_3Al, Au_3Cd, and Au_3Mn alloys, Shindo, Hiraja, Hirabayashi, and Aoyagi (1984) have used direct structure imaging to show that replacement disordering in these alloys due to 1 MeV irradiation occurs in the atomic rows parallel to the beam. Dislocation loops are also observed with a diameter of less than 10 Å at higher intensities, and lighter atoms are observed to be displaced preferentially. Damage in silicon at 400 kV is analysed under HREM conditions by Kuwabara, Endoh, Tsubokawa, Hashimoto, Yokota, and Shimizu (1985).

The study of radiation damage in quartz is important in both the geological and nuclear waste storage fields. Cherns, Hutchison, Jenkins, Hirsch, and White (1980) have found that dissociated dislocations in quartz become vitrified after irradiation, while Pascucii, Hobbs, and Hutchison (1981) have observed a two-stage damage process. At low doses, ionization damage produces a heterogeneous nucleation of amorphous inclusions. Higher doses produce an additional homogeneous loss of correlation in the surrounding matrix. The role of water or OH radicals (which causes a marked softening of quartz) has yet to be fully elucidated. Other studies of radiation damage in inorganic materials by HREM include those of sodium β'' alumina by Hull, Cherns, Humphreys, and Hutchison (1981), and the spectacular sub-nanometer 'hole drilling' of Mochel, Humphreys, Mochel, and Eades (1983). There seems little doubt that the now well-established method of video recording HREM images (see Section 11.3) will greatly assist in this work, since replay and freeze-frame reproduction of the first images of a fresh area of crystal frequently reveal details of undamaged material which cannot be seen in micrographs exposed for a longer period.

Much current high-resolution work is concerned with on the effects of radiation damage at surfaces. Thus Sinclair, Smith, Erasmus, and Ponce (1982) were able to observe radiation-induced partial dislocation motion in CdTe using structure images and other dynamic effects of atoms in surface positions. More recently, Bovin and co-workers (Bovin, Wallenberg, and Smith 1985; Iijima and Ichihashi 1985) have obtained some remarkable atomic-resolution video recordings of the radiation-induced motion of small metal particles (less than 100 Å in diameter) on silicon. Individual atoms are seen performing a 'random walk' on the crystal surface. Radiation-induced effects are clearly seen in most of the recent HREM profile images (see Section 12.8), and the usefulness of this effect for generating sub-micrometre metallic layers on dielectric metal oxide crystals has been investigated (Smith and Bursill 1985). The limitations which radiation damage imposes on the microanalysis of sub-nanometer regions by energy-loss spectroscopy are beautifully illustrated in the work of Bourret, Colliex, and Trebbia (1983) on oxygen segregation to dislocation cores in germanium. The considerable amount of work on damage in the rare-earth compounds can be traced through the recent study of vacancy ordering in scandium sulphide by Franzen, Tuenge, and Eyring (1983).

6.10   Summary

The non-periodic and nearly-periodic specimens described in this chapter provide some of the most useful applications of high-resolution microscopy. In the first section the distinction between phase and amplitude contrast is developed and we see that, for the very thinnest specimens, a phase-contrast image intensity can be obtained which is proportional to the specimen's atomic potential projected in the direction of the incident beam. This phase contrast, unlike 'absorption' or aperture contrast, increases with accelerating voltage and is obtained by using coherent illumination, a large objective aperture, and the correct choice of focus. The choice of aperture size and focus defect are given in eqns (6.16) together with a simple formula for the approximate microscope resolution under these conditions (eqn (6.17)). Microscopists interested in small-molecule imaging can gain useful experience by examining clusters of gold atoms evaporated onto thin carbon films. Groups of several atoms will be clearly seen through the binocular at an electron-optical magnification of 400 000 if a sufficiently thin substrate has been prepared. This can be done by using the indirect evaporation method described in Section 10.6. The through-focus behaviour of these atom clusters should be studied carefully and practice should be obtained in astigmatism correction and the identification of drift in images (see Sections 1.3 and 8.7). A bright annular fringe will be seen around these clusters on the under-focus side (weakened lens) changing to a dark fringe or the over-focus side. The minimum-contrast condition can be found on a neighbouring region of carbon film (see Section 10.5). Optical diffractograms obtained from this will reveal the presence of drift, the accuracy of the astigmatism correction, and the focus setting (see Section 8.8). By comparing the focus setting judged at the microscope to give images of highest contrast with that measured from the optical diffractogram of the best micrograph, the microscopist will soon learn the art of focusing non-periodic detail at high resolution.

In Section 6.3 the advantages of higher accelerating voltage are reviewed. These include increased contrast and resolution together with reduced chromatic aberration and specimen ionization damage. An impression of the increase in contrast with atomic number expected and found experimentally for single atoms is given in Section 6.4, while Section 6.5 outlines the principles of high-resolution dark-field microscopy. The image distortion which is commonly seen in dark-field images is discussed and the important point is made that both positive and negative deviations from the specimen's average potential will show up as bright areas in dark-field images. It is not possible therefore to distinguish areas of very high specimen potential from those of low potential from a dark-field image—both appear bright.

The image contribution of electrons which lose energy in their passage through the specimen is discussed in Section 6.6. Any loss of energy results in a small change in wavelength. The imaging then becomes analogous to image formation with sunlight in optics, where a range of wavelengths are also present. Since the focal length of the objective lens depends on the electron wavelength, these energy-loss images will be out of focus, typically by several hundred nanometres (see Section 2.8.2). They therefore contribute only a blurred background to high-resolution images resulting in a slight loss of contrast. The ratio of inelastic scattering to elastic scattering is approximately inversely proportional to atomic number for amorphous materials and the scattering is concentrated within a narrow cone (semi-angle less than 1 mrad) around the central beam and any Bragg reflections. It is therefore possible to exclude these energy-loss electrons from dark-field images by a judicious choice of objective aperture position. An image from which inelastically scattered electrons have been excluded using a special energy filter is shown in Fig. 6.15.

A great deal has been learnt about the interpretation of high-resolution images from the study of computer-simulated images of specimens of known structure. Simulation methods are discussed generally in Section 6.7 and the interested reader will find references there to more detailed information. The next section makes the important point that the grainy appearance of an electron micrograph is not due to the film grain, but rather to electron noise (see Fig. 6.17). This noise can only be reduced by increasing the number of electrons contributing to the image, so that high-resolution exposures should be as long as possible subject to the limitations of radiation damage and specimen drift. A trade-off between resolution and noise is seen to exist and some simple expressions are used to find the maximum number of discernible grey-levels in a micrograph formed with a given number of electrons. The final section in this chapter describes the way in which image processing methods may be used to reduce the amount of radiation damage caused while recording high-resolution detail from biological and other crystalline specimens.

References

Agar, A. W., Alderson, R. H., and Chescoe, D. (1974). Principles and practice of electron microscope operation In *Practical methods in electron microscopy* (ed. A. M. Glauert). North-Holland, Amsterdam.

Andrews, H. C. (1970). *Computer techniques in image processing*. Academic Press, New York.

Bantz, D. and Zwick, M. (1974). The use of symmetry with the fast Fourier transform. *Acta Crystallogr.* **A30**, 257.

Beer, M. (1978). Structural analysis of macromolecular assemblies with STEM. In *Scanning electron microscopy/1978* (ed. O. Johari), p. 69. AMF, Illinois.

Billington, C. and Kay, N. R. (1974) Pictorial presentation of two-dimensional calculations. *Aust. J. Phys.* **27**, 73.

Bovin, J. O., Wallenberg, R., and Smith, D. (1985). Imaging of atomic clouds. *Nature (Lond.)* **317**, 47.

Bourret, A., Colliex, C., and Trebbia, D., (1983). Oxygen segregation on a dislocation core. *J. de Physique Lettres* **44**, p. L33.

Brillouin, L. (1962). *Science and information theory.* Academic Press, New York.

Burge, R. E. (1973). Mechanisms of contrast and image formation of biological specimens in the transmission electron microscope. *J. Microsc.* **98**, 251.

Castaing, R. and Henry, L. (1962). Filtrage magnetique des vitesses en microscopie electronique. *C. R. hebd. Seanc. Acad. Sci., Paris* **255**, 76.

Cherns, D., Hutchison, J. L., Kenkins, M. L., Hirsch, P. B., and White, S. (1980). Electron irradiation induced vitrification at dislocations in quartz, *Nature* (Lond.) **287**, 314.

Chiu, W. and Glaeser, R. M. (1974). Single atom contrast: Conventional dark field and bright field electron microscopy. *J. Microsc.* **103**, 33.

Cohen, H. A., Jeng, T.-W., Grant, R. A., and Chiu, W. (1984a). Specimen preparative methods for electron crystallography of soluble proteins. *Ultramicroscopy* **13**, 19.

Cohen, H. A., Schmid, M. F., and Chiu, W. (1984b) Estimates of the validity of the projection approximation for three-dimensional reconstructions at high resolution. *Ultramicroscopy* **14**, 219.

Cosslett, V. E. (1958). Quantitative aspects of electron staining. *J. R. Microsc. Soc.* **78**, 18.

Cosslett, V. E. (1974). Perspectives in high voltage electron microscopy. *Proc. R. Soc. Lond.* **A338**, 1.

Cosslett, V. E. (1978). Radiation damage in the high resolution electron microscopy of biological materials: a review. *J. Microsc.* **113**, 113.

Cowley, J. M. (1969). Image contrast in a transmission scanning electron microscope. *Appl. Phys. Lett.* **15**, 58.

Cowley, J. M. (1973). High resolution dark field electron microscopy I. Useful approximations. *Acta Crystallogr.* **A29**, 529.

Cowley, J. M. (1976a). The principles of high resolution electron microscopy. In *Principles and techniques of electron microscopy, biological applications* (ed. M. P. Hayat), Vol. 6. Van Nostrand Reinhold, New York.

Cowley, J. M. (1976b). Scanning transmission electron microscopy of thin specimens. *Ultramicroscopy* **2**, 3.

Cowley, J. M. and Jap, B. K. (1976). The use of diffraction information to augment STEM imaging. In *Scanning electron microscopy/1976* (ed. O. Johari) Part 1, p. 377. IIT Research Institute, Chicago.

Cowley, J. M. and Pogany, A. O. (1968). Diffuse scattering in electron diffraction patterns. I. General theory and computational methods. *Acta Crystallogr.* **A24**, 109.

Graven, A. and Colliex, C. (1977). The effect of energy loss on lattice fringe images of dysprosium oxide. In *Proc. 35th Ann. Meet. Electron Microsc. Soc. Am.* (ed. G. W. Bailey), p. 242. Claitor's, Baton Rouge.

Craven, A., Gibson, J. M., Howie, A., and Spalding, D. R. (1978). Study of single-electron excitations by electron microscopy I. Image contrast from delocalized excitations. *Phil. Mag.* **38**, 519.

Crewe, A. V., Langmore, J., Isaacson, M., and Retsky, M. (1974). Understanding single atoms in the STEM. In *Proc. 8th Int. Cong. Electron Microsc.* Canberra, Vol. 1, p. 30.

Dawson, B., Goodman, P., Johnson, A. W. S., Lynch, D. F., and Moodie, A. F. (1974). Some definitions and units in electron diffraction. *Acta Crystallogr.* **A30**, 297.

DeRosier, D. J. and Moore, P. (1970). Reconstruction of three dimensional images from electron micrographs of structures with helical symmetry. *J. Mol. Biol.* **52**, 355.

Dorignac, D. and Jouffrey, B. (1980). Digital processing of atomic images. *J. Microsc. Spectrosc. Electron.* **5**, 671.

Doyle, P. A. and Turner, P. S. (1968). Relativistic Hartree-Fock X-ray and electron scattering factors. *Acta Crystallogr.* **A24**, 390.

Egerton, R. F. (1975). Inelastic scattering of 80 kV electrons in amorphous carbon. *Phil. Mag.* **31**, 199.

Egerton, R. F. (1980). Measurement of radiation damage by electron energy loss spectroscopy. *J. Microsc.* **118**, 389.

Erickson, H. P. and Klug, A. (1971). Measurement and compensation of defocusing and aberrations by Fourier processing of electron micrographs. *Phil. Trans. R. Soc.* **B261**, 105.

Fan, G.-Y. and Cowley, J. M. (1988). Assessing the information content of HREM images. *Ultramicros.* **24**, 49.

Fellget, P. B. and Linfoot, E. H. (1955). On the assessment of optical images. *Phil. Trans. R. Soc.* **247**, 369.

Formanek, H., Muller, M., Hahn, M. H., and Koller, T. (1971). Visualization of single heavy atoms with the electron microscope. *Naturwissenschaften* **58**, 339.

Frank, J. (1973). Computer processing of electron micrographs. In *Advanced techniques in biological electron microscopy* (ed. J. K. Koehler). Springer-Verlag, Berlin.

Frank, J. and Radermacher, M. (1986). Three-dimensional reconstruction of nonperiodic macromolecular assemblies from electron micrographs. *Advanced Techniques in biological electron Microscopy* (ed. J. Koehler). Springer-Verlag, Berlin pp. 1–72.

Frank, J., Shimkin, B., and Dowse, H. (1981). SPIDER-a modular software system for electron image processing. *Ultramicroscopy* **6**, 343.

Franzen, H., Tuenge, R. T., and Eyring, L. (1983). Vacancy ordering in scandium sulphide. *J. Solid-State Chem.* **49**, 206.

Fujimoto, F. and Fujita, H. (1972). Radiation damage induced by channeling. *Phys. Stat. Sol.* **11(a)**, K103.

Fujiwara, K. (1962). Relativistic dynamical theory of electron diffraction. *J. Phys. Soc. Japan* **17**, BII, 118.

Fujiyoshi, Y., Kobayashi, T., Ishizuka, K., Uyeda, N., Ishida, Y., and Harada, Y. (1980). A new method for optimal-resolution electron microscopy of radiation-sensitive specimens. *Ultramicroscopy* **5**, 459.

Fujiyoshi, Y., Yamagishi, H., Kunisada, T., Sugisaki, H., Kobayashi, T., and Uyeda, N. (1982). Visualization of the DNA thread packing within bacteriophage T4 heads. *J. Ultrastruct. Res.* **79**, 235.

Fujiyoshi, Y., Uyeda, N., Morikawa, K., and Yamagishi, H. (1984). Electron microscopy of tRNA crystals II. 4 Å Resolution diffraction pattern and substantial stability to radiation damage. *J. Mol. Biol.* **172**, 347.

Glaeser, R. M. (1971). Limitations to significant information in biological electron microscopy as a result of radiation damage. *J. Ultrastruct. Res.* **36**, 466.

Glaeser, R. M. (1975). Radiation damage and biological electron microscopy. In Physical aspects of electron microscopy and microbeam analysis (ed. B. M. Siegel and D. R. Beaman). Wiley, New York.

Glaeser, R. M. (1985). Electron crystallography of biological macromolecules. *Ann. Rev. Phys. Chem.* **36**, 243.

Glaeser, R. M. and Taylor, K. A. (1978). Radiation damage at low temperature. A review. *J. Microsc.* **112**, 127.

Glaeser, R. M., Jubb, J. S., and Henderson, R. (1985). Structural Comparison of Native and Deoxycholate-treated purple membrane. *Biophys. J.* **48**, 775.

Glauert, A. M. (1974). The high voltage electron microscope in geology. *J. Cell. Biol.* **63**, 717.

Grinton, G. R. and Cowley, J. M. (1971). Phase and amplitude contrast in electron micrographs of biological materials. *Optik* **34**, 221.

Hall, C. R. (1971). Contrast calculations for small clusters of atoms. In Electron microscopy in materials science (ed. U. Valdre and A. Zichichi). Academic Press, New york.

Hamilton, J. F. and Marchant, J. C. (1967). Image recording in electron microscopy. *J. Opt. Soc. Am.* **57**, 232.

Hanssen, K. J. and Ade, G. (1978). Phase contrast transfer with different imaging modes in electron microscopy. *Optik* **51**, 119 [and references therein].

Harauz, G. and Ottensmeyer, F. P. (1984). Nucleosome reconstruction via phosphorus mapping. *Science* **226**, 936.

Harauz, G., Andrews, D. W. and Ottensmeyer, F. P. (1983). Electron microscopic visualization of the sidechains of the poly-L-lysine alpha helix. *Ultramicroscopy* **12**, 59.

Hashimoto, H., Kumao, A., Hino, K., Endoh, H., Yotsumoto, H., and Ono, A. (1973). Visualization of single atoms in molecules and crystals by dark field electron microscopy. *J. Electron Microsc.* **22**, 123.

Hawkes, P. W. (1982). Electron image processing, 1978–80. A survey. *Computer graphics and image processing,* **18**, 58.

Head, A. K., Humble, P., Clareborough, L. M., Morton, A. J., and Forwood, C. T. (1973). *Computed electron micrographs and defect identification.* North-Holland, Amsterdam.

Henderson, R. and Glaeser, R. M. (1985). Quantitative analysis of image contrast in electron micrographs of beam-sensitive crystals. *Ultramicroscopy* **16**, 139.

Hobbs, L. W. (1979). In *Introduction to Analytical Electron Microscopy* (ed. J. Hren, J. Goldstein, and D. C. Joy). Plenum, New York.

Horiuchi, S. (1982). Detection of point defects accommodating nonstochiometry in inorganic compounds. *Ultramicroscopy* **8**, 27. [See also *Acta Crystallogr.* **15**. 323 (1982).]

Horiuchi, S. and Matsui, Y. (1974). Lattice images of $Nb_{22}O_{54}$ and V_6O_{13} in the 100 kV electron microscope. *Phil. Mag.* **30**, 777.

Horiuchi, S., Matsui, Y., and Bando, Y. (1976). A high resolution lattice image of $Nb_{12}O_{29}$ by means of a high voltage electron microscope newly constructed. *Japan J. Appl. Phys.* **15**, 2483.

Howie, A. and Gai, P. (1975). Diffuse scattering in weak beam images. *Phil. Mag.* **31**, 519.

Howie, A., Krivanek, O. L., and Rudee, M. L. (1973). Interpretation of electron micrographs and diffraction patterns of amorphous materials. *Phil. Mag.* **27**, 235.

Hull, R., Cherns, D., Humphreys, C. J., and Hutchison, J. L. (1981). Electron irradiation damage mechanisms in sodium β'' alumina. *Inst. Phys. Conf. Ser. No. 61,* p. 23. Institute of Physics, Bristol.

Humphreys, C. J. (1976). High voltage electron microscopy. In *Principles and techniques of electron microscopy: biological applications* (ed. M. A. Hayat) Vol. 6, p. 1. Van Nostrand Reinhold, New York.

Humphreys, C. J., Hart-Davis, A., and Spencer, J. P. (1974). Optimizing the signal/noise in the dark field imaging of single atoms. *Proc. 8th Int. Congr. Electron. Microsc.,* Canberra, p. 248.

Iijima, S. (1977a). Observation of single and clusters of atoms. *Optik* **48**, 193.

Iijima, S. (1977b). Thin graphite support films for high resolution electron microscopy. *Micron* **8**, 41.

Iijima, S. and Ichihashi, T. (1985). Motion of surface atoms on small-gold particles. *Japan J. Appl. Phys.* **24** L125.

Isaacson, M. (1976). Radiation damage to biological specimens. In *Principles and techniques of electron microscopy: biological applications* (ed. M. A. Hayat) Vol. 7. Van Nostrand Reinhold, New York.

Iwanaga, M., Ueyanagi, H., Hosoi, K., Iwasa, N., Oba, K., and Shiratsuchi, K. (1968). Energy dependence of photographic emulsion sensitivity and fluorescent screen brightness for 100 kV through 600 kV electrons. *J. Electron Microsc.* **17**, 203.

Jeng, T.-W., Chiu, W., Zemlin, F., and Zeitler, E. (1984). Electron imaging of crotoxin complex thin crystal at 3.5 Å. *J. Mol. Biol.* **175**, 93.

Jenkins, G. M. and Watts, D. G. (1968). *Spectral analysis and its applications.* Holden Day, New York.

Kessel, M., Radermacher, M., and Frank, J. (1985). The structure of the stalk surface layer of a brine pond microorganism: correlation averaging applied to a double layered lattice structure. *J. Microsc.* **139**, 63.

Kihlborg, L., Sundberg, M., and Sarborg, O. (1985). Structure analysis by HREM with online and offline image processing. *Ultramicroscopy* **18**, 191.

Kirkland, E. J., Siegel, B. M., Uyeda, N., and Fujiyoshi, Y. (1985) Improved high resolution image processing of bright field electron micrographs II. Experiment. *Ultramicroscopy* **17**, 87.

Kobayashi, K., Suito, E., Uyeda, N. Watanabe, M., Yanaka, T., Etoh, T., Watanabe, H., and Moriguchi, M. (1974). A new high resolution electron microscope for molecular structure observation. *Proc. 8th Int. Congr. Electron Microsc.*, Canberra, p. 30.

Krakow, W. (1976). Computer experiments for tilted beam dark-field imaging. *Ultramicroscopy* **1**, 203.

Krivanek, O. L., Ahn, C. C., and Keeney, R. B. (1987). A parallel detection electron spectrometer using quadropole lenses. *Ultramicroscopy* **22**, 103.

Kuwabara, M., Endoh, H., Tsubokawa, Y., Hashimoto, H., Yokota, Y., and Shimizu, R. (1985) In situ observation of the formation process of radiation damages in atomic level. *International Symposium on In Situ Experiments with HVEM* Osaka University, 1985 p. 341.

Lanczos, C. (1966). *A treatise on Fourier series*. Oliver and Boyd, Edinburgh.

Langmore, J. P. and Wall, J. (1973). The collection of scattered electrons in dark field electron microscopy. *Optik* **38**, 335.

Langridge, R., Wilson, H. R., Hopper, C. W., and Wilkins, M. H. F. (1960). The molecular configuration of deoxyribonucleic acid. *J. Mol. Biol.* **2**, 19.

Lenz, F. (1954). Zur Streung mittleschneller elektronen in klienste Winkel. *Z. Naturforsch.* **9A**, 185.

Lynch, D. F. and Moodie, A. F. (1974). Image contrast of DNA calculated for a real electron microscope. *Proc. 8th Int. Congr. Electron Microsc.*, Canberra, p. 224.

McLachlan, D. (1958). Crystal structure and information theory. *Proc. Nat. Acad. Sci. USA* **44**, 948.

Misell, D. L. (1973). Image formation in the electron microscope with particular reference to the defects in electron optical images. *Adv. Electron. Phys.* **32**, 63.

Misell, D. L. (1979). Image analysis, enhancement and interpretation. In *Practical methods in electron microscopy* (ed. A. M. Glauert). North-Holland, Amsterdam.

Misra, M. and Egerton, R. F. (1985). Assessment of electron irradiation damage to biomolecules using Patterson functions. *J. Microsc.* **139**, 197.

Mochel, M. E., Humphreys, C. J., Mochel, J. M., and Eades, J. A. (1983). Cutting of 20 Å holes and lines in β-Aluminas. *Appl. Phys. Lett.* **42**, 392.

Moodie, A. F. (1972). Reciprocity and shape functions in multiple scattering diagrams. *Z. Naturforsch.* **27a**, 437.

Ottensmeyer, F. P. and Henkelman, R. M. (1974). An electrostatic mirror. *J. Phys. E* **7**, 176.

Pascucci, M. R., Hobbs, L. W., and Hutchison, J. L. (1981). Lattice image of the metamict transformation in synthetic quartz. *Proc. 39th EMSA*, p. 110. San Fransicso Press, San Francisco.

Saxton, W. O. (1978). Computer techniques for imaging processing in electron microscopy. *Adv. Electron. Electron Phys.*, Suppl. 10.

Scherzer, O. (1949). The theoretical resolution limit of the electron microscope. *J. Appl. Phys.* **20**, 20.

Shindo, D., Hiraja, K., Hirabayashi, M., and Aoyagi, E. (1984). HREM of radiation defects in ordered alloys. *Science Reports of the Research Institutes, Tohoku University*, A-Vol. 32, No. 1. (Sendai, Japan).

Sinclair, R., Smith, D. J., Erasmus, S. T., and Ponce, F. A. (1982). Lattice resolution movie of defect modification in cadmium telluride. *Nature*, **298**, 127.

Skarnulis, A. J. (1982). A computer system for on-line image capture and analysis. *J. Microsc.* **127**, 39.

Smith, D. J. and Bursill, L. A. (1985). Metallisation of Oxide surfaces in HREM. *Ultramicroscopy* **17**, 387.

Spence, J. C. H. (1975). Single atom contrast. In *Electron microscopy and analysis* (ed. J. A. Venables), p. 257. Academic Press, New York.

Spence, J. C. H. (1988). Localisation in inelastic scattering. In *High resolution electron microscopy* (ed. P. Buseck, J. M. Cowley and L. Eyring). Oxford University Press, Oxford.

Spence, J. C. H. and Lynch, J. (1982). STEM microanalysis by ELS in crystals. *Ultramicroscopy* **9**, 267.

Tanji, T., Hashimoto, H., Endoh, H., and Tomioka, H. (1982). Theory and application of noise filter for periodic objects. *J. Electron Miscrosc.* **31**, 1.

Tomita, M., Hashimoto, H., Ikuta, T., Endoh, H., and Yokota, Y. (1985). Improvement and application of the Fourier-transformed pattern from a small area of high resolution electron microscope images. *Ultramicroscopy* **16**, 9.

Unwin, P. N. T. and Henderson, R. (1975). Molecular structure determination by electron microscopy of unstained crystalline specimens. *J. Mol. Biol.* **94**, 425.

Urban, A. (1980). Radiation damage in inorganic materials in the electron microscope. *Electron microscopy 1980* Vol. 4, p. 188. E. M. Congress Foundation, Leiden.

Van Dorsten, A. C. (1971). Contrast phenomena in electron images of amorphous and macromolecular objects. In *Electron microscopy in materials science* (ed. U. Valdre and Z. Zichichi). Academic Press, New York.

Verschoor, A., Frank, J., Radermacher, M., Wagenknecht, T., and Boublik, M. (1984). Three-dimensional reconstruction of the 30S Ribosomal subunit from randomly oriented particles. *J. Mol. Biol.* **178**, 677.

Wall, J. S. and Hainfeld, J. F. (1986). Mass mapping with the scanning transmission electron microscope. *Ann. Rev. Biophys. Biophys. Chem.* **15**, 355.

Werner, P. and Paseman, M. (1982). Generation of radiation induced defects in silicon. *Ultramicroscopy* **7**, 267.

Whelan, M. J. (1972). Elastic and inelastic scattering. *Proc. 5th Eur. Congr. Electron Microsc.*, p. 430.

Whiting, R. F. and Ottensmeyer, F. P. (1972). Heavy atoms in model compounds and nucleic acids imaged by dark field transmission electron microscopy. *J. Mol. Biol.* **67**, 173.

Williams, R. C. and Fisher, H. W. (1970). Disintegration of biological molecules under the electron microscope. *Biophys. J.* **10**, 53a.

Yoshioka, H. (1957). Effect of inelastic waves on electron diffraction. *J. Phys. Soc. Jap.* **12**, 618.

Zeitler, E. (1982). Cryomicroscopy and radiation damage. *Ultramicroscopy* **10**, 1.

7

ELECTRON SOURCES AND THE
ILLUMINATION SYSTEM

The four common electron sources are tungsten hair-pin filaments, tungsten pointed filaments, lanthanum hexaboride sources, and field-emission sources. This chapter gives brief notes on the suitability of each for high-resolution transmission electron microscopy and summarizes their characteristics. For this application the main requirements are as follows.

1. High brightness (current density per unit solid angle). This is important in high-resolution phase-contrast experiments where a small illumination aperture is required together with sufficient image current density to allow accurate focusing at high magnification.
2. High current efficiency (ratio of brightness to total beam current). This is achieved through a small source size. Reducing the area of specimen illuminated reduces unnecessary specimen heating and thermal movement during the exposure.
3. Long life under the available vacuum conditions.
4. Stable emission. Exposures of up to a minute are not uncommon for high-resolution dark-field work.

An ideal illumination system for high-resolution CTEM would give the operator independent control of the area of specimen illuminated, the intensity of the illumination, and the coherence conditions. Only the field-emission source approaches this ideal; however, for most laboratories the pointed tungsten filament or the lanthanum hexaboride sources offer the best compromise between cost and performance for high-resolution CTEM. A promising development is the laser-heated filament, where an increase in brightness of 3000 has been reported over that of the hair-pin filament with an effective source diameter of about 10 nm. These filaments operate at moderate vacuum (10^{-4} Torr) (see van der Mast, Barth, and LePoole 1974).

Before discussing particular electron sources, the properties of condenser lenses and triode electron guns are briefly reviewed.

7.1 The illumination system

The two condenser lenses of an electron microscope illumination system are shown in Fig. 7.1. The properties of simple lenses are discussed in Section 2.2, which provides some background for this section. The microscopist usually has independent control of the focal lengths of both these lenses (C1 and C2). The first condenser lens excitation is sometimes called 'spot size'. The instrument is normally used with these excitations arranged so that

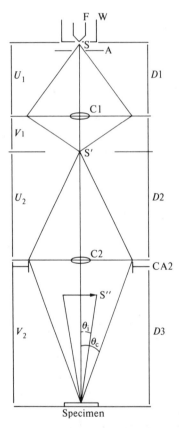

Fig. 7.1 The illumination system of a modern electron microscope. There are two condenser lenses C1 (a strong lens) and C2 (a weak lens). F-filament electron source; W = Wehnelt cylinder; S = virtual electron source, with images S' and S''; CA2 = second condenser aperture. The distances U_1, U_2, V_1, V_2 are electron optical parameters, while distances $D1$, $D2$, $D3$ are easily measured from the microscope column. Values of these distances are given for a particular microscope in Table 2.1.

planes S, S' and the specimen plane are conjugate, that is, with a focused image of the source formed at the specimen (focused illumination).

For a hair-pin filament the source size is about 30 μm. To prevent unnecessary specimen heating and radiation damage, a demagnified image of this is required at the specimen. The working distance $D3$ must also be large enough to permit specimen-change manoeuvers. These conflicting requirements (magnification M_1 small, $D3$ large) would necessitate an impractically large distance $D1$ if a single condenser lens were used. The usual solution is to use a strong first condenser lens C1 to demagnify the source by between 5 and 100, followed by a weak second lens C2 providing magnification of about 3 and a long working distance. Demagnification values for the two lenses and the corresponding focal lengths are given in Table 2.1 for a modern instrument. Since $M_1 = V_1/U_1$ is small the image of S formed by C1 is situated close to the focal plane of C1.

As discussed in Section 2.9, that portion of the objective lens field on the illuminating side of the specimen (the pre-field) also acts as a weak condenser lens. The total demagnification of the source at the specimen is

$$M = M_1 M_2 M_3$$
$$= \left(\frac{V_1}{U_1}\right)\left(\frac{V_2}{U_2}\right)M_3$$
$$\approx M_3\left(\frac{f(C1)D3}{D1D2}\right)$$

where M_1, M_2, M_3 are the magnifications of C1, C2, and the objective pre-field respectively. A typical value of M_3 is $\frac{1}{3}$ (JEM-100C). The minimum value of $f(C1)$ is the minimum projector focal length discussed in Chapter 2, and may be about 2 mm. If the condition of focused illumination is maintained, the size of the illumination spot on the specimen can be reduced from a large value by moving the image S' towards S, making V_1 smaller and U_2 larger.

A simple method of measuring the source size is to image S using lens C2 alone with lens C1 switched off. The magnification $M_2 = D3/(D1 + D2)$ is then about unity and can be obtained from measurements taken from the microscope column. If the pre-field demagnification is not known, a shortened specimen holder should be used on top-entry stages to bring the specimen out of the lens field.

If the aperture CA2 is incoherently filled (as it is when using a hair-pin filament), the field of view or illuminated specimen region is limited by the size of source S and the lens demagnification, while the coherence at the specimen is solely determined by the semi-angle θ_c. This method of focused illumination is known in light optics as critical illumination. Another method, where the image S' is placed in the focal plane of C2 ($U_2 = f(C2)$) is also possible and known as Köhler illumination in light optics. Here the coherence is limited by the source size, while the field of view depends on the size of CA1. On many electron microscopes the pole-piece of C1 forms the aperture CA1.

Finally, the use of unfocused illumination must be considered. This is treated in most of the electron optics texts mentioned in Chapter 2 (e.g., Grivet (1965), p. 453), where it is shown that the beam divergence θ_i (see Fig. 7.1) for unfocused illumination is given by the semi-angle subtended by the source image at the specimen if this angle is less than the angle subtended by the illumination aperture. Otherwise, the beam divergence θ_c is given by the angle subtended by the illumination aperture (see Fig. 7.1).

7.2 Brightness and its measurement

The electron current I passing through area A is proportional both to A and to the solid angle subtended by any filled illuminating aperture at the source. The constant of proportionality β is known as the beam brightness and is measured in amps per square centimetre per steradian. The brightness is strictly defined for vanishingly small areas and angles as

$$\beta = I/\pi A \theta^2 \qquad (7.1)$$

for an aperture of semi-angle θ. The effect of a lens is to reduce the current density (proportional to M^2) and increase the angular aperture (proportional to $1/M^2$) for a magnification M, leaving the brightness constant at conjugate planes in the microscope if aberrations are neglected. A full discussion of the signficance of brightness is beyond the scope of this text and can be found in the electron-optics texts mentioned in Chapter 2. Values of β for four modern sources are given in Table 7.1. For a hair-pin filament with $\beta = 5 \times 10^5$ A cm^{-2} sr^{-1}, a 400 μm final condenser aperture in the plane of CA2 (Fig. 7.1) gives a current density in the illumination spot

Table 7.1 The properties of four modern electron sources. These figures are approximate and intended as a rough guide only. The effect of source size on coherence at the specimen is discussed in Section 4.5

Source	Virtual Source Diameter	Measured Brightness at 100 kV (A cm^{-2} sr^{-1})	Current Density at Specimen with 0.8 mrad Illuminating Aperture (A cm^{-2})	Energy FWHM (eV)	Melting point (°C)	Vacuum Required (T)	Emission Current (μA)
Heated field-emission	5–10 nm	10^7–10^8	20	0.3	3370	10^{-8}–10^{-9}	50–100
Pointed filament	1–5 μm	2×10^6	4	2	3370	10^{-5}	10
Hair-pin filament	30 μm	5×10^5	1	See Fig. 7.5	3370	10^{-5}	100
LaB$_6$	5–10 μm	7×10^6†	14	1.0	2200	10^{-6}	50

† Measured at 75 kV (Broers 1976).

focused on the specimen of

$$j_0 = \pi\beta(r/D3)^2 = 1 \text{ A cm}^{-2}$$

where r is the aperture radius. This is a typical value for high-resolution electron microscopy of radiation-insensitive materials. The corresponding image current density on the viewing phosphor at a magnification of 500 K is

$$j_i = j_0/M^2 = 5 \times 10^{-12} \text{ A cm}^{-2}$$

which is sufficient to expose the Ilford EM4 emulsion to an optical density of unity after 6 seconds, assuming no specimen or pre-field effect. Corresponding figures for a pointed filament are $j_0 = 4$ A cm^{-2} and $j_i = 1.6 \times 10^{-11}$ A cm^{-2} with an exposure time of 1.7 seconds at the same magnification.

The theoretical upper limit to brightness is given in many texts as

$$\beta_m = \rho e V_r / \pi k T$$
$$= 3694 \rho e V_r / T \quad \text{A cm}^{-2} \text{sr}^{-1} \tag{7.2}$$

where ρ is the emission current density at the cathode (filament) in A cm^{-2}, $T(\text{K})$ is the filament temperature, and k is Boltzmann's constant. For thermionic sources the brightness increases with temperature, since ρ increases exponentially with temperature. Note that the brightness is proportional to accelerating voltage. Brightness measurements at high voltage have been given by Shimoyama, Oshita, Maruse, and Minamikawa (1972a), who found $\beta = 1.6 \times 10^6$ A cm^{-2} sr^{-1} at 500 kV. Note also that V_r is the relativistic high voltage (Shimoyama, Ohshita, and Maruse 1972b).

Since the early work of Haine and Einstein (1952), many workers have shown that the maximum theoretical brightness could be achieved in practice for suitable combinations of filament height (h), bias voltage (V_b), and filament temperature (T). For $T = 2800$ K, a temperature giving reasonable filament life, eqn (7.2) gives $\beta = 4.4 \times 10^5$ A cm^{-2} sr^{-1} at 100 kV, in rough agreement with measured values ($\rho = 3$ A cm^{-2} at this temperature).

Higher values of brightness can be obtained using pointed filaments, discussed in a later section. In designing a gun one aims for the highest current efficiency, that is the highest ratio of brightness to total beam current, as well as for high absolute brightness. Current efficiency is increased by making the Wehnelt hole small and h small. A typical beam current for a tungsten hair-pin filament is 150 μA, while many excellent high-resolution lattice images have been taken using pointed filaments with a beam current as low as 1 μA.

In order to compare electron sources or determine the optimum conditions for operating a source, it is necessary to be able to measure the source brightness. This may be done by the following method, which is useful for determining the best height for a pointed filament. A direct-reading image current electrometer is needed, connected to a small electron collector in the plane of the viewing phosphor. Some instruments are fitted with a differential image current screen which is suitable. The solid angle $\Omega \approx \pi\theta^2$ subtended by the second condenser aperture at the specimen must be measured from the microscope. The aperture size should be carefully measured using either a calibrated optical microscope or the optical diffraction pattern of the aperture, since these aperture sizes vary considerably within each nominal 'size'. The electron-optical magnification M must be known and the current I passing through a small circular area of radius r at the centre of the focused illumination spot should be noted. The brightness is then

$$\beta = M^2 I^2 / \pi^2 \theta^2 r^2$$

The following points are important.

1. The procedure should be repeated for successively smaller condenser apertures until the value of β obtained does not change with aperture size. This ensures that the aperture is filled with radiation for the measurement.

2. The current measuring area must be small compared with the focused spot size. If the full spot size is used the result will be a small fraction of the true brightness.

3. A correction for electron backscattering from the collection screen must be made, using published coefficients (Badde, Drescher, Krefting, Reimer, Seidel, and Buhring 1971). Alternatively, a specially shaped collector with low backscattering coefficient can be used.

4. A specimen such as a holey carbon film must be used to define the object plane. Bear in mind that the screen intensity (not β) will depend on the strength of the objective pre-field, that is, on the specimen height and objective focus (Section 2.9). To ensure that the condenser aperture actually limits the incident beam divergence, it will be necessary to work with the specimen outside the objective lens field and a correspondingly weakly excited objective. Alternatively, with the object plane immersed within the lens field, the angle θ_c can be found from the size of diffraction spots as described in Section 8.8.

Brightness measurement is of great assistance when setting up a pointed filament and it is worth spending some time on this since the accurate focusing of radiation-insensitive materials at high magnification is greatly facilitated by using the brightest possible source. Plots of filament height and gun-bias against brightness should both show maxima. It may also be worth experimenting with the size of the Wehnelt hole and its effect on current efficiency. The hole size can be made variable using interchangeable recessed apertures.

An example of a modern theoretical calculation of the brightness of a triode electron gun can be found in Kamminga (1972).

7.3 Biasing and high-voltage stability

A common problem in recording images at high resolution is high-voltage fluctuation during the exposure. Tests for high-voltage stability are de-

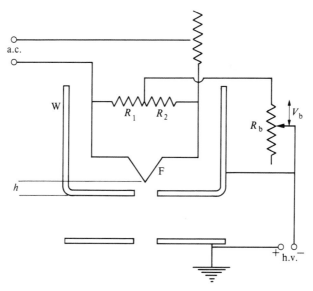

Fig. 7.2 The self-biasing gun. The alternating current filament heating supply is connected to a.c. (a d.c. supply is sometimes used). Balancing resistors R_1 and R_2, which keep the filament tip at a constant potential, are also shown. A bias resistor R_b is shown with bias voltage V_b. The anode is labelled A, the Wehnelt W, and the filament F. The feedback circuit shown provides improved current and temperature stability over a fixed-bias arrangement.

scribed in Section 9.2. On modern instruments the high voltage is stabilized against fluctuations by the negative feedback circuit shown in Fig. 7.2. The Wehnelt is biased a few hundred volts negative with respect to the filament and to tends to reduce the electron beam current. The circuit is most easily understood by analogy with the thermionic triode, and can be redrawn as shown in Fig. 7.3. While the anode of an electron microscope is kept at

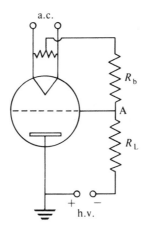

Fig. 7.3 Equivalent triode circuit for an electron gun. In the conventional cathode biased triode amplifier point A, rather than the anode, would be approximately at earth potential. R_L is an equivalent load resistor. The similarity with Fig. 7.2 should be noted.

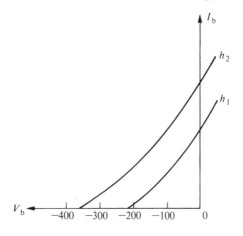

Fig. 7.4 Approximate total beam current I_b as a function of gun bias for two different filament heights h_1 and h_2. Here h_2 is greater than h_1. The bias required for optimum conditions (high brightness) is less for the h_1 case than the h_2 case.

ground potential for convenience, point A of the triode circuit is usually grounded. Figure 7.4 shows the relationship between grid (Wehnelt) voltage and plate (beam) current obtained by experiment. The presence of condensed oil vapour of evaporated tungsten on the gun chamber walls will encourage micro-discharge in the gun. If the discharge is small, the resulting increase in beam current will increase the bias voltage developed across R_b which in turn, from Fig. 7.4 will tend to oppose the initial increase in beam current, the result being a stabilization in high voltage and electron wavelength.

The three triode constants (Brophy 1972) can be defined and measured for an electron gun (Munakata and Watanabe 1962), and the stabilization analysed by applying Maxwell's loop equation to the small signal triode equivalent circuit for Fig. 7.3. This treatment shows that the stabilization depends both on the amplification factor of the equivalent triode and on the value of the bias resistor R_b.

In summary, the effects of an increase in R_b are as follows.

1. Reduced total beam current as shown in Fig. 7.4. The bias required for cut-off depends on the filament height h, becoming larger as h is reduced.
2. Improved high voltage stability.
3. Reduced chromatic aberration. Figure 7.5 shows the measured energy distributions for a tungsten filament operated at high and low bias. The energy width decreases with increasing bias. For a field-emission gun the measured half-width is about 0.3 eV.
4. By treating the gun as an electrostatic lens the size and position of the virtual electron source can be found (Lauer 1968). This is shown in

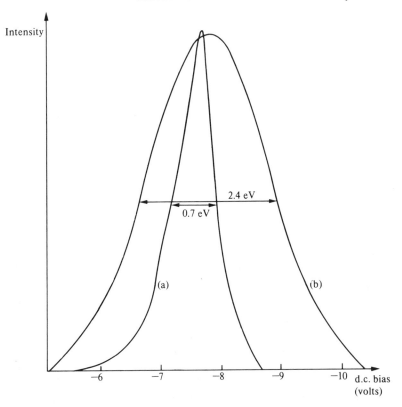

Fig. 7.5 An experimental plot of the energy spread of electrons leaving the filament of the gun of a JEM-6A electron microscope for maximum gun bias (a) and for minimum gun bias (b). The intensity scale is arbitrary and the absolute values of the abscissa have been subject to an origin shift, arising from the design of the energy-selecting microscope used. The experimental data have been smoothed. We see that the range of energies, and hence the range of wavelengths present in the electron beam is reduced when using maximum gun bias. In practice this results in a reduction in chromatic aberration when using very thin specimens (see Sections 4.2 and 2.8.2). A discussion of the expected form of this data can be found in Andersen and Mol (1970).

Fig. 7.6—its size varies only slightly with gun bias, while its position moves toward the anode with decreasing bias.

5. For a particular filament height and temperature there is a weakly defined maximum in brightness as a function of bias for self-biased guns.

The bias resistor R_b (several tens of megohms) can be adjusted by the operator. For pointed-filament operation a continuously variable servo-potentiometer is a great advantage since the operating conditions are sensitively dependent on gun bias. On many microscopes a clockwise rotation of the bias control (turning it 'up') actually decreases the bias, resulting in increased beam current.

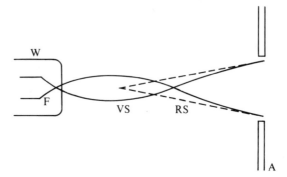

Fig. 7.6 Formation of a virtual source (VS) in the region between the Wehnelt (W) and the anode (A). A real crossover is formed at RS.

An important feature of the self-biasing system is that the bias is controlled by the filament temperature, since this limits the beam current. Electrons are emitted only within the circular zero equipotential on the hair-pin cap, the size of this area grows smaller with increasing bias, vanishing to a point at the filament tip at cut-off bias. This explains the changing appearance of the focused illumination spot as the filament current is increased, increasing both the filament temperature and gun-bias. The spot contracts as electrons are drawn from a progressively smaller region of the filament tip. A full treatment of the operating conditions for self-biased guns is contained in the article by Haine, Einstein, and Borchards (1958).

The bias setting may have an important bearing on the quality of high-resolution images of the thinnest specimens, where energy losses in the specimen are less important than the energy spread of the beam leaving the filament. A fuller discussion of this question is given in Section 5.8, where techniques for minimizing the thermal energy spread of electrons leaving the filament are outlined. In Section 4.2 and Appendix 3 the effect of gun-bias on image quality is also assessed (see, in particular, Fig. 4.3).

7.4 Hair-pin filaments

Most manufacturers supply pre-centred, annealed filaments at a pre-set height with replacement instructions. The tungsten wire thickness and length have been chosen with care since these determine the load applied to the filament supply. The filament life is limited by the effects of gas attack (mainly water vapour) and thermal evaporation. With a vacuum poorer than 10^{-5} Torr the first failure mechanism is more likely. On many modern instruments fitted with two diffusion pumps, filament lives well in excess of 50 hours are common. An hour meter actuated by the high-voltage relay can be used to measure the filament life and provides an inexpensive monitor of the microscope vacuum. As small leaks develope in the microscope column, the filament life rapidly deteriorates. The two failure mechanisms can be distinguished as follows. Most filaments fail when a break develops near the top of either arm where the temperature is greatest. If the wire has retained its original thickness with a blunt rounded appearance in the region of the break, the likely cause of failure is evaporation. Gas attack is characterized by a narrower tapering toward the break.

The brightness and coherence properties of hair-pin filaments are indicated in Table 7.1. Bear in mind that the intensity of the final image depends on a large number of factors apart from the source brightness. Three of these, often overlooked, are the age of the viewing phosphor (which should be replaced at least once a year for high-resolution work), the correction of condenser lens astigmatism, and the extent of objective lens pre-field used. A far more intense final image is obtained as the specimen is immersed deeper into the objective-lens field (Section 2.9).

While these filaments are adequate for routine high-resolution lattice imaging of radiation-insensitive materials, their chief disadvantage is the large specimen area which must be illuminated leading to unnecessary heating and specimen movement.

7.5 Pointed filaments

The pointed tungsten filament offers a significant increase in brightness and current efficiency over conventional filaments at a fraction of the cost of a field-emission source (see Table 7.1). In the hands of skilled users, pointed filaments have been used to produce many fine high-resolution CTEM images. The small specimen area illuminated reduces specimen heating and drift while the increased brightness enables more accurate focusing at high magnification for the same illumination aperture size. Alternatively, the image current density obtainable with a conventional source can be achieved using a pointed filament with a smaller illumination aperture (eqn (7.1)). The resulting improvement in the quality of lattice images with reduced beam divergence is discussed in Chapter 5. An improvement in the quality of biological images when using pointed filaments has been reported by Hibi and Takahashi (1971). These filaments also convey an important advantage for the difficult task of focusing accurately on point detail (e.g., a small molecule) in non-periodic specimens. Finally, astigmatism correction is a great deal easier using pointed filaments (see Section 10.1).

Pointed filaments are available commercially, and the suppliers will fit these filaments to used conventional filament bases, supplied by the customer. Techniques for manufacturing pointed filaments have been given by several workers, the simplest method being to grind a point on the apex of a hair-pin filament (Bradley 1961). A much finer point can be obtained by the method used for making field-emission tips. Tungsten wire of 0.1 mm diameter is electrolytically etched to a fine point in a solution of sodium hydroxide (about 1M strength). A copper cathode can be used and a polishing voltage between 5 and 40 V is required. Both the strength of the solution and the voltage must be varied to obtain the best conditions. One should aim for a tip diameter of a few micrometres, though larger tips will give some improvement over hair-pin performance. A narrowing occurs in the wire at the liquid surface, leading to a break. The sudden drop in current when this occurs can be used as an indication to terminate the supply current. Figure 7.7 shows the way in which this wire stub is spot-welded before etching to the top of a hair-pin filament, which is used to heat the tip. The tip temperature depends on the length of wire used—good results have been obtained with the dimensions shown. Etching techniques are discussed in the article by Dyke, Trolan, Dolan, and Barnes (1953).

Setting up a pointed filament takes considerably more time and care than does a conventional filament. Once operating correctly, a pointed filament should give a life of between 30 and 50 hours. Too high an operating temperature is a common cause of shorter life. These filaments frequently fail when being turned off—the filament heating current should be turned down quickly to extend filament life (Iijima, personal communication). Fernandez-Moran (1966) gives details of a modified Wehnelt in which interchangeable apertures allow the size of the hole to be varied. The hole size shown in Fig. 7.7 has been found to give high brightness and current efficiency. A Wehnelt suitable for use with pointed filaments must allow the filament height and the centring of the filament tip within the hole to be adjusted. If the Wehnelt is fitted to the gun with a fine thread bearing against a light spring, the filament height can be varied by rotating the Wehnelt. The change in height with rotation can be calibrated using a dial micrometer. For a small hole, the centring accuracy required places serious demands on the machining accuracy of this thread, which must not allow

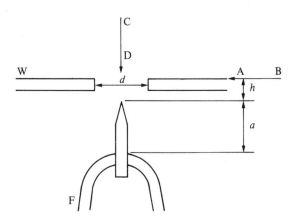

Fig. 7.7 Construction and dimensions of a pointed filament. Here d should be between 0.5 and 1 mm, and a should be about 1 mm or less. A trial value of h is 0.7 mm. The pointed stub has been spot-welded to the top of a conventional hair-pin filament. The filament is aligned by sighting along AB and CD.

any sideways movement of the Wehnelt. Using an optical microscope to sight along AB (Fig. 7.7), the Wehnelt can be rotated until the filament tip just coincides with the top of the Wehnelt. Using this as a reference level, the Wehnelt should be screwed out until the correct filament height (h) is obtained. A starting value for h is 0.7 mm, but the optimum value must be determined by experiment and may be as low as 0.3 mm.

Having set the filament height, the tip must be centred within the hole using, say, three set-screws let into the side of the Wehnelt base. This should be done under an optical microscope at a magnification of about 10, viewing down CD (Fig. 7.7), with a light source to one side of and above the Wehnelt. Adjust the lamp until a bright point of light is seen reflected from the filament tip. It is difficult to judge the centring sufficiently accurately by eye; however, time spent on this job will reduce the number of microscope trials required. It will usually be necessary to re-centre the filament after a few minutes of operation have distorted the filament shank, but this should only be necessary once. On examining the unsaturated filament image, a symmetrical image should be observed if the tip has been correctly centred (Fig. 10.2). A bright central spot with a comet-like tail of intensity indicates an off-centre filament. A continuously variable bias supply greatly facilitates the operation of pointed filaments. The filament temperature can be monitored by metering the filament heating current or a voltage proportional to it, which will ensure reproducibility of operating conditions. These three measured variables—gun-bias, filament tempera- ture, and filament height then determine the brightness and life of the filament. A systematic investigation is generally not necessary to determine the best operating conditions. Set the filament temperature to the value used for conventional filaments and vary the filament height until a saturation beam current of between 1 and 10 microamperes is obtained for a reasonably large bias setting. Then adjust the filament excitation and bias setting until the filament image coalesces into a small intense spot at saturation (magnification about 20 000), using the least possible filament excitation to prolong life. The conditions can be further optimized using image current-density measurements (Section 10.9). Depending on the top orientation, characteristic patterns are seen within the illumination spot. The resulting intensity variation within the spot will cause unevenly exposed plates to be recorded if a focused spot is used at magnifications below about 100 000.

High-voltage instability is sometimes experienced when using pointed filaments. The likely causes are poor gun vacuum (see Section 9.2) or dirt or evaporated tungsten in the gun chamber or around the Wehnelt hole. When changing pointed filaments the Wehnelt must be cleaned with metal-polish, alcohol, and finally ultrasonic agitation. Acetone is a health hazard and should be used sparingly. The Wehnelt hole must be examined optically for traces of dust or dirt before fitting the filament. These filaments seem to be particularly sensitive to the cleanliness of this hole.

Pointed filaments are also more sensitive to stray a.c. magnetic fields than conventional filaments (see Fig. 10.2). Tests for magnetic fields are discussed in Section 9.1. These act to enlarge the focused spot size and so reduce the image current density by periodic deflection of the electron beam. The effect is most serious where the electron velocity is lowest. A common cause is a mains earth-loop. The field in the region of the gun and specimen chamber should be less than 3 mG for pointed filament operation. It is unusual to find a field as low as this in most buildings in which special precautions have not been taken.

Japanese workers have investigated the properties of pointed filaments extensively in recent years. The effect of gun bias on coherence for a pointed filament is considered by Ohshita, Shimoyama, and Maruse (1971), and a simple expression for the field strength in the neighbourhood of the tip has been given by Ohshita et al. (1973). Brightness data for pointed filaments are given by Hanszen and Lauer (1967).

7.6 Lanthanum hexaboride sources

These sources offer an increase in brightness of about ten times over hair-pin filaments (see Table 7.1). They have the great advantage of operating at moderate vacuum conditions (better than 10^{-6} Torr), which can be obtained in most microscopes without modifications. A marginal gun vacuum can often be improved by attending to the gun vacuum seal. Other leaks are easily traced with a partial pressure gauge or mass spectrometer (see Section 10.8). Details of one experimental system are given in Ahmed and Broers (1972), in which a rod of LaB_6 with a ground tip of about 10 μm radius is heated indirectly with a tungsten coil. Another system, available commercially uses a thin strip of carbon containing a slot to support and heat the highly reactive LaB_6 emitter. This carbon strip is supported on the pins of a conventional hair-pin base, so that the filaments can be used as a plug-in replacement for hair-pin filaments. Their heating requirements are similar to those of a hair-pin filament (about 5 W). These sources are more expensive than hair-pin filaments but give longer life so that the cost per hour may be comparable. To prevent oxidation of the tip, a minimum cooling time of 10 seconds must be allowed before admitting air to the gun after turning down the heating current. Since these filaments operate at a lower temperature than tungsten filaments, electrons are emitted within a narrower range of energy (Batson, Chen, and Silcox 1976). We can expect that high-resolution images recorded using these sources will show reduced chromatic aberration (see Sections 2.8.2 and 4.2). Because of their convenience and performance LaB_6 sources can be recommended as the most cost-effective sources currently available for HREM work, provided that an adequate gun vacuum can be obtained.

7.7 Field-emission sources

Field-emission sources are now available for several CTEM instruments operating at 100 kV. The filament consists either of a heated or an unheated single crystal of tungsten. The high electrostatic field (about 4×10^7 V cm^{-1}) in the region of the tip enables electrons to tunnel through the lowered potential barrier at the tungsten surface. The (111) or (310) tip orientations are often used. The theory of field emission has been reviewed by Dyke and Dolan (1956). The high brightness of these sources (see Table 7.1), the small illumination spot possible at the specimen, and the high spatial coherence of the radiation makes this source the best currently available for high-resolution CTEM. Fresnel fringes obtained with such a source and demonstrating this high degree of coherence are shown in Fig. 8.7 (see also Section 8.8). Since the coherence length in the plane of the final condenser aperture may be comparable with the size of this aperture, the coherence length at the specimen cannot be found using the simple formula (4.11). The large coherence length (up to 100 nm) obtainable allows the largest and smallest spatial periods to contribute to phase-contrast images and so can be expected to improve their quality (see Chapter 4).

The possibility of illuminating only a very small area (about 20 nm) with a field-emission source enables a modified focusing technique to be used with important advantages for radiation-sensitive materials. A structurally unimportant area of specimen can be illuminated and used for focusing at high magnification, while an adjacent area, unaffected by radiation damage during focusing and at the same height and focus condition, may be subsequently illuminated and used for the final image recording. The dose for the second area can be accurately controlled (independently of the coherence condition) and no refocusing is necessary. These are important advantages for high-resolution biological work.

The environmental requirements for a field-emission source are more severe than for other sources. In particular, magnetic shielding in the region of the gun and specimen is important to reduce stray a.c. magnetic fields. This is discussed in Chapter 9. A magnetically shielded stage may be necessary. The life of field-emission tips operating at 100 kV appears to be about 80 hours (Glaser, personal communication), though a wide variation is to be expected. At lower voltages, tips may last many hundreds of hours. The full width at half maximum of the energy-loss distribution for a field-emission source is about 0.3 eV. The tips must be 'flashed' before use, that is, heated briefly to remove contaminants and smooth the tip profile. For a field-emission gun the 'Wehnelt' is at positive potential with respect to the tip, but nevertheless can be used to stabilize the beam current (Cleaver and Smith 1973). The tips can be fabricated by electrolytic etch using sodium hydroxide in a similar way to that used for making thermionic pointed filaments.

Unfortunately, the high cost of field-emission guns and the down-time of about 24 hours required to re-establish ultra-high vacuum in the gun chamber (between 10^{-8} and 10^{-9} Torr) after filament change restricts the use of these instruments to a few research laboratories. The article by Crewe (1971) provides a useful introduction to the subject of field-emission sources.

The most important practical problem in producing commercial field-emission sources at 100 kV has been the control of emission stability. Before purchasing a field-emission microscope it is most important to examine the focused illumination spot at maximum imaging magnification to check for a stable intensity distribution.

References

Ahmed, H. and Broers, A. (1972). Lanthanum hexabordie electron emitter. *J. Appl. Phys.* **43**, 2185.

Anderson, W. H. J. and Mol, A. (1970). Simultaneous measurements of the brightness and the energy distribution of electrons emitted from a triode gun. *J. Phys. D* **3**, 965.

Badde, H. G., Drescher, H., Krefting, E. R., Reimer, L., Seidel, H., and Buhring, W. (1971). Use of Mott scattering cross sections for calculating backscattering of 10–100 kV electrons. *Proc. 25th Meet. EMAG*, p. 74. Institute of Physics, Bristol.

Batson, P. E., Chen, C. H., and Silcox, J. (1976). Use of LaB_6 for a high quality, small energy spread beam in an electron velocity spectrometer. *Proc. 34th Ann. EMSA Meet.* p. 534. Claitors, Baton Rouge.

Bradley, D. E. (1961). Simple methods for preparing pointed filaments for the electron microscope. *Nature (Lond.)* **189**, 298.

Broers (1976). Personal communication.

Brophy, J. J. (1972). *Basic electronics for scientists*, 2nd edn. McGraw-Hill, New York.

Cleaver, J. R. A. and Smith, K. C. A. (1973). Two lens probe forming systems employing field emission guns. *Proc. 6th Ann. SEM Symp.* Chicago, p. 49.

Crewe, A. V. (1971). High intensity electron sources and scanning electron microscopy. In *Electron microscopy in materials science* (eds U. Valdre and A. Zichichi). Academic Press, New York.

Dyke, W. P. and Dolan, W. W. (1956). Field emission. *Adv. Electron. Electron Phys.* **8**, 89.

Dyke, W. P. Trolan, J. K., Dolan, W. W., and Barnes, G. (1953). The field emitter: fabrication, electron microscopy and electric field calculations. *J. Appl. Phys.* **24**, 570.

Fernandez-Moran, P. F. (1966). Applications of improved point cathode sources to high resolution electron microscopy. *Proc. 6th Int. Congr. Electron Microsc.* Kyoto, Vol. 1 p. 27.

Grivet, P. (1965). *Electron optics*. Pergamon, London.

Haine, M. E. and Einstein, P. A. (1952). Characteristics of the hot cathode electron microscope gun. *Br. J. Appl. Phys.* **3**, 40.

Haine, M. E., Einstein, P. A., and Bochards, P. H. (1958). Resistance bias characteristics of the electron microscope gun. *Br. J. Appl. Phys.* **9**, 482.

Hanszen, K. J. and Lauer, R. (1967). Richtstrahlwertmessungen mit dem Zwei-Blendenverfahren an Electronenstrahlern mit Kugelkathoden von 250 bis 1.5 μm Durchmesser. *Z. Naturforsch.* **22a**, 238.

Hibi, T. and Takahashi, S. (1971). Relation beteen coherence of electron beam and contrast of electron image of biological substance. *J. Electron Microsc.* **20**, 17.

Kamminga, W. (1972). Numerical calculation of the effective current density of the source for triode electron guns with spherical cathodes. In *Image processing and computer aided design in electron optics* (ed. P. Hawkes). Academic Press, New York.

Lauer, R. (1968). Ein einfaches Modell fur Electronenkanonen mit gekrummter Kathodeno-berflache. *Z. Naturforsch.* **23a**, 100.

Munakata, C. and Watanabe, H. (1962). A new bias method of an electron gun. *J. Electron Microsc.* **11**, 1962.

Oshita, A., Shimoyama, H., and Maruse, S. (1971). Temperature of point cathode and fresnel fringes. *J. Electron Microsc.* **20**, 281.

Oshita, A., Shimoyama, H., and Maruse, S. (1973). Field strength at the cathode tip of the point cathode electron gun. *J. Electron Microsc.* **22,** 135.

Shimoyama, H., Ohshita, A., and Maruse, S. (1972b). Relativistic considerations on electron optical brightness. *Jap. J. Appl. Phys.* **11,** 150.

Shimoyama, H., Ohshita, A., Maruse, S., and Minamikawa, Y. (1972a). Brightness measurement of a 500 kV electron microscope gun. *J. Electron Microsc.* **21,** 119.

van der Mast, K. D., Barth, J. E., and LePoole, J. B. (1974). A continuously renewed laser heated tip in a T.F. emission gun. In *8th Int. Congr. Electron Microsc.,* Canberra, p. 120.

8

MEASUREMENT OF ELECTRON-OPTICAL PARAMETERS AFFECTING HIGH-RESOLUTION IMAGES

This chapter describes several simple techniques for measuring the important electron-optical constants which affect the quality of high-resolution images. They do not require the microscope to be dismantled. The methods described here are chosen for their simplicity and convenience and include the use of optical diffractograms for the measurement of defocus, spherical aberration astigmatism, and 'resolution' as discussed in Section 8.7 and 8.10.

8.1 Objective-lens focus increments

On a modern well-designed instrument the finest objective-lens focus step Δf should be of the same order as the variation in focal plane position due to electronic instabilities. From eqn (2.39) an upper limit is

$$\Delta f = C_c\left(\frac{\Delta V_0}{V_0} + 2\frac{\Delta I}{I}\right) \tag{8.1}$$

This gives 4.2 nm for a lens with $C_c = 1.4$ mm and $\Delta I/I = \Delta V/V\ 10^{-6}$. A typical measured value for the smallest step in the focus controls of the JEM-100B is 4.0 nm. There are several methods for measuring these steps—three are given in this book, two below and one in Section 8.7. Of these the optical bench method is probably the simplest and quickest. The first method is an adaptation of the method of Hall (1949) and uses the displacement with focus of diffraction images.

1. Prepare MgO crystals by passing either a continuous or holey carbon film through the smoke of some burning magnesium ribbon. This can be

264

ignited, with some difficulty, using a Bunsen burner. Examine the crystals in the microscope.

2. MgO has the sodium chloride structure of two displaced face-centred cube lattices. Crystals in which the beam is in the direction of the [100] or [111] zone-axis are common. Those of the first kind appear as square crystals. The presence of Pendellösung fringes indicates that the beam runs in the direction of a diagonal of the cube face. The diffraction pattern for the [100] case is shown in Fig. 8.1. Crystals in the [111] orientation appear hexagonal. Identify the reflections in the diffraction pattern of a suitably thin crystal. For a crystal near the zone-axis orientation observed with no objective aperture in place, a set of images of the crystal will be seen (one 'diffraction image' for each strong Bragg reflection) which merge near Gaussian focus and separate laterally with increasing defocus. Using the objective aperture to isolate the central beam and one other strong beam, identify the Miller indices of a particular diffraction image. By measuring this lateral image displacement for a known number of focal steps, eqn (3.17) can be used to obtain a value of Δf.

3. Align the microscope as described in Chapter 10. Align the illuminating system, obtain current centre, check cold-fingers for continuity and correct astigmatism on the carbon background with no objective aperture.

4. Without touching the objective aperature holder, record a double exposure of the selected crystal at two focus values close to Gaussian focus. The two focus values chosen should fall within the span of a single focus control (say n steps apart) and the magnification should be chosen such that some movement of the diffraction image is observed with this focus change. The full range of the finest focus control may have to be used.

5. The separation of diffraction images is next measured from the

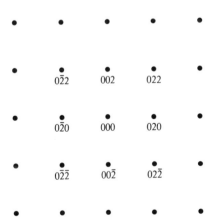

Fig. 8.1 The (100) reciprocal lattice plane for MgO. The lattice spacing is 0.21 nm for the 002 reflection. Amongst MgO smoke crystals this pattern is easily recognized from the perfect squares formed in the diffraction pattern.

micrograph using an optical microscope and graticule. The electron-optical magnification must be measured using the methods discussed in Section 8.3. For the finest focal steps one is measuring distances of 1–2 nm.

6. If the required focal plane separation in the object space for n steps is $n \, \Delta f$, an expression for this follows from eqn (3.17) and an analysis similar to that accompanying Fig. 3.5. This gives

$$\Delta f = d/(n\theta M)$$

where d is the lateral translation of the diffraction image measured from the micrograph recorded at magnification M. Here θ is the angle between the central beam and the reflection used, obtained from Bragg's law. At 100 kV, $\theta = 17.55$ mrad for the (002) reflection of MgO.

The second method is simpler than the first, although some machine-shop work may be required. Make up a specimen holder in which the specimen grid is held at a fixed angle of, say, 45° to the incident beam. Place a thin carbon film in the holder and examine it in the microscope. Select a region near the centre of the grid and record an image at the minimum contrast focus (see Section 10.4). All points on a line across the specimen at the same height will be imaged at this focus. Advance the finest focus control by several clicks, and, without touching the translation controls, record a second image. By measuring the distance between the lines of minimum contrast in the image, it is a simple matter to find their difference in height, that is, the focus change between image recordings. By experiment, you should choose a focus change for which the line of minimum contrast does not move out of the field of view. After developing and drying the plates, place one plate on top of the other over an illuminated viewing screen. The alignment of the plates is not critical if the line of minimum contrast moves by a large fraction of the plate's length. Measure the distance between the two lines of minimum contrast at right angles to the lines and divide by the magnification (see Section 8.3). For a specimen at 45° to the incident beam, the focus change is then equal to the resulting distance.

A third method for measuring focus increments is described in Section 8.7.

8.2 Spherical aberration constant

Figure 2.18 shows the effect of spherical aberration on a ray leaving the specimen at a large angle. For an axial point on a crystalline specimen, a high-order Bragg reflection is such a ray. The resulting displacement between bright- and dark-field images at Gaussian focus can be used to measure C_s (Hall 1949). Alternatively, the in-focus displacement of a bright-field image formed with no objective aperture as the illumination is tilted can be used to give a value for C_s. Equation (2.35), referred to the image plane, gives the image translation as

$$d = MC_s\theta^3 \tag{8.2}$$

Two bright-field images are recorded successively at exact focus with the central beam aligned with the optic axis and at a small known angle to it. The resulting image translation gives d and hence C_s. The method in detail is as follows.

1. Evaporate a thin layer, less than 20 nm thick, of any low-melting-point FCC metal onto a holey carbon film to form a fine-grained polycrystalline film. With the thinnest films grains will be roughly spherical. The film should be thin enough to leave exposed areas of carbon for astigmatism correction and the grains should be small enough to leave holes between some grains. More details of evaporation techniques are given in Holland (1956), Pashley (1964), and Mathews (1975). The first of these references provides all details of the experimental methods for evaporation.

2. Examine the film and align the microscope with particular attention to current-centring. Use an uncovered area of carbon to correct the astigmatism as discussed in Chapter 10. Next, take a double exposure of a hole in the film with the central beam successively on the optic axis and then tilted, so that an identified diffraction ring falls on the optic axis. The position of the axis should be marked after current-centring using the screen pointer. Two tilt conditions can usually be pre-set with the microscope bright- and dark-field channels. Use no objective aperture and take both images at exact focus, judging this from the sharpness of the image. A small error in focusing is not important if a high-order diffraction ring is chosen, thus ensuring that the phase-shift due to spherical aberration dominates that due to defocus.

3. To ensure that the central beam has been accurately aligned, repeat step 2 four times with the beam tilted into four orthogonal directions. If the microscope has been accurately aligned, and there is no astigmatism, the four image displacements will be equal. Recordings of the tilted and untilted diffraction patterns (double-exposed) are also required to measure θ. This makes a total of eight double exposures. A magnification of 25 000 is adequate for the images.

4. For FCC metals the tilt angle is obtained from

$$\theta = \frac{r}{r_0} \frac{\lambda}{a_0} (h^2 + k^2 + l^2)^{1/2}$$

where a_0 is the cubic unit-cell constant ($a_0 = 0.408$ nm for gold), r_0 is the measured distance between the central beam and the diffraction ring with indices (h, k, l), and r is the exact measured distance between the two central beams appearing on the doubly exposed diffraction pattern. A small condenser aperture should be used. Equation (8.2) can then be used to obtain a value for C_s with d the image separation under tilt θ.

An indexed ring pattern for FCC metals is given in Fig. 8.2. Magnification calibration can be obtained from an optical ruled diffraction grating. At the magnification suggested (25 000) it will be necessary to relate the size of a surface feature recorded at this magnification to that of the same feature

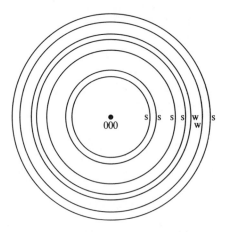

Fig. 8.2 The ring diffraction pattern from a polycrystalline face-centered cubic material used for determining the spherical aberration constant of a microscope. To assist in identifying the rings they are labelled w (weak) and s (strong). Their radii have the ratio of the square roots of 3, 4, 8, 11, 12, 16, 19, 20, 24, . . . The indices of the reflections shown, from the centre, are (000), (111), (200), (220), (311), (222), (400), (313). A beam stop is useful when recording the diffraction pattern to prevent the central beam from overexposing the emulsion.

recorded at a lower magnification where many lines of the grating can be seen.

Note that the spherical aberration constant C_s depends on the lens focal length according to Figs 2.12 and 2.14. A shorter focal length reduces spherical aberration. Thus the value of C_s obtained applies only for the particular objective-lens current and specimen height used for the calibration. Make a note of the objective lens current used during calibration—this should be close to the value found to give images of highest resolution (see Sections 1.2 and 8.4). A second method of measuring C_s is given in Section 8.7.

Where values of C_s are needed for computer-simulated images, it is important to estimate the sensitivity of the images to C_s. For a simple three-beam image, a black–white contrast reversal occurs for a change ΔC_s given by $\pi \Delta C_s \lambda^3 U_h^4 / 2 = \pi$ (see eqn (3.24)), that is, values of C_s are required to within an accuracy of $\Delta C_s = 2/(\lambda^3 U_h^4)$ for a spatial frequency U_h. At high resolution ($U_h > 3.3 \, \text{nm}^{-1}$, $\Delta C_s < 0.3$ mm at 100 kV) this accuracy is difficult to achieve; however, the method described above will be found to be more accurate than the optical diffractogram technique described in Section 8.7. Note that the error ΔC_s is independent of C_s so that the allowable fractional error is increased by reducing C_s. The major source of error in the above method (accurate to within perhaps 10 per cent) arises from the neglect of focusing effects. An improved method which is independent of Δf has been described by Budinger and Glaeser (1976) (accurate to within perhaps 3 per cent), together with a fuller discussion of sources of errors.

8.3   Magnification calibration

At high resolution, crystal lattice fringes form the most convenient length standard, but an accuracy in magnification calibration of better than 5 per cent is very difficult to achieve. The important sources of error are the accuracy with which the lens may be reset to a particular value, and changes in specimen height between the specimen of interest and the calibration specimen. This is a particularly important problem for molecular images. For many periodic specimens, a cell dimension will be known from X-ray work (Wyckoff 1963). A tilted specimen may introduce large height changes and consequent error. From eqn (2.9) the effect of a small change Δf in specimen height at large magnification is

$$\frac{\Delta f}{f_0} = -\frac{\Delta M}{M} \tag{8.3}$$

where ΔM is the magnification change and f_0 the objective-lens focal length. Thus, if a height change equal to the thickness of a grid is allowed on specimen change, the result is a 6 per cent error in magnification for $f_0 = 1.6$ mm.

Partially graphitized carbon is particularly convenient in that it can be dusted on to fragile biological specimens (Heidenrich 1968). Lattice fringes obtained from this specimen in all directions are a sensitive assurance of accurate astigmatism correction, as shown in Fig. 8.3. The spacing is 0.34 nm. The asbestos fibres crocidolite and chrysotile (Yada 1967) are also suitable standards. Using the finest tweezers, these can be teased apart and onto a grid already containing a specimen. Crocidolite has spacings of 0.903 nm along the fibre direction and 0.448 nm for fringes at 60° to the edge, while the spacings of chrysotile asbestos are 0.45 nm, 0.46 nm, and 0.73 nm for a specimen observed normal to the fibre axis. Chrysotile appears to be more radiation-sensitive than crocidolite, which is available commercially.

All these specimens have the advantages that their orientation is simple to determine and they may be added to a prepared specimen. Any of the specimens listed in Chapter 5 can be used for routine magnification calibration. The important point for accurate high-resolution length measurement is that the calibration specimen be on the same grid and as near to the unknown specimen as possible, and that lens currents are not altered between recordings of these specimens, apart from necessary refocusing. The change in height between the calibration specimen and the unknown specimen is easily measured from the calibrated focus change needed to bring each into focus.

For routine work fringes obtained from any of the test specimens available commercially will provide magnification calibration for a particular specimen holder, stage, pole-piece, magnification setting, and high voltage within about 5 per cent. Where a variable-height stage is used, a correction

Fig. 8.3 Lattice fringes suitable for magnification calibration obtained from partially graphit-ized carbon. The presence of clear fringes in all directions is an indication of accurate astigmatism correction. The fringe spacing is 0.34 nm. A specimen of partially graphitized carbon can be kept permanently on hand as a preliminary check on the microscope stability and resolution.

chart must be drawn up giving the true magnification in terms of indicated magnification and measured objective-lens current (Section 8.4) needed to focus the specimen at a particular height. From Section 2.3, the magnifica-tion change due to variation in objective-lens focal length is seen to be inversely proportional to the lens focal length or proportional to the square of the objective-lens current. Owing to magnetic hysteresis, the lens field strength (and hence the magnification) will depend on whether a particular set of lens currents are approached from a high or a lower lens current. Some instruments include an automatic lens-current cycling device which improves the resetting accuracy of magnification to about 3 per cent.

There is an increase in magnification between the viewing phosphor and the film plane on most microscopes. The ratio of magnifications on the two planes is

$$\frac{M_1}{M_2} = \frac{v_1}{v_2}$$

where M_1 and M_2 are the magnifications on the phosphor and film planes respectively and v_1 and v_2 are distances between the projector lens pole-piece and the phosphor and film plane. These distances can be

Table 8.1.   List of useful larger lattice spacings for mangification calibration. The last column shows the method, the letters referring to the imaging conditions shown in Fig. 5.1

Specimen	Symbol	Lattice (hkl)	spacing (nm)	Method
Cu-phthalocyanine	$C_{32}H_{16}N_8Cu$	001	1.26	(a)
Cu-phthalocyanine	$C_{32}H_{16}N_8Cu$	201	0.98	(a)
Crocidolite	—	along fibre	0.903	(c)
Potassium chloroplatinite	K_2PtCl_4	100	0.694	(a)
Cu-phthalocyanine	$C_{32}H_{16}N_8Cu$	001	0.63	(c)
Potassium chloroplatinate	K_2PtCl_6	111	0.563	(a)
Pyrophyllite	$Al_2O_3\cdot 4SiO_2\cdot H_2O$	020	0.457	(a)
Potassium chloroplatinite	K_2PtCl_4	001	0.412	(a)
Molybdenum tri-oxlate	MoO_3	110	0.381	(a)
Potassium chloroplatinite	K_2PtCl_4	100	0.347	(c)
Gold	Au	111	0.235	(b)

measured approximately from the microscope. A similar equation relates the magnification on the film plane to that on the plane of an energy analyser, image intensifier, or other image sensor. At high magnification Section 2.2 indicates that the depth of focus is sufficiently great for all these images to be treated as 'in focus'.

Measurements of fringe spacing should be taken over as large a number of fringes as possible to reduce errors, using either a travelling optical microscope or a transmission optical microscope with graticule. Counting a large number of fine fringes without error is tedious and is greatly facilitated by the use of a photographic enlarger whose magnification has been accurately calibrated. Measurements should be taken from the enlarged image rather than a print, which is subject to shrinkage. A list of lattice spacings useful for magnification calibration is given in Table 8.1.

8.4   Objective-lens current measurement

In high-resolution work, where specimen height changes are frequently made, an indication of objective-lens current is useful for magnification calibration (see Section 8.3). This is provided on some instruments. If not, the voltage across a low value (say, one ohm), high-stability resistor in series with the lens should be measured using a digital voltmeter. From eqns (8.3) and (2.38) we see that for 5 per cent accuracy in magnification calibration the voltmeter should have a dynamic range of at least three decades. A more accurate instrument is required for the determination of C_c (see Section 8.5). A three-digit voltmeter (not 'two and a half' digits) cannot be expected to show any variation for the three finest focus controls on most instruments. Objective-lens currents commonly lie in the approximate range 0.5–1.5 A.

8.5   Chromatic aberration constant

A simple technique for the determination of C_c follows from eqn (2.39),

$$\Delta f = C_c\left(\frac{\Delta V_0}{V_0} - 2\frac{\Delta I}{I}\right)$$

If the high voltage V_0 is kept constant and the method described in Section 8.1 is used to measure the defocus Δf for a measured change in objective-lens current (see Section 8.4), the chromatic aberration constant can be found using

$$C_c = \frac{I\,\Delta f}{2\,\Delta I}$$

where I is the mean value of lens current used and ΔI is the change in lens current for defocus Δf. A measurement of C_c using the variation of focal length with high voltage is also possible but less convient. For most high-resolution work, however, it is the quantity Δ defined in eqn (4.9) which is of interest, and this is most easily measured using optical diffractograms (see Section 8.7). The effect of changes in C_c on image quality are indicated in Fig. 4.3.

8.6   Astigmatic difference

A technique for correcting astigmatism is discussed in Chapter 10. Occasionally, however, it may be necessary to measure the astigmatic difference C_a which appears in eqn (3.23).

When astigmatism is severe, C_a can be obtained from measurements of Fresnel fringe widths. The width of the first Fresnel fringe formed at the edge of, say, a small circular hole in a thin carbon film is very approximately

$$y = (\lambda\,\Delta f)^{\frac{1}{2}} \tag{8.4}$$

for defocus Δf. From a measurement of y, the focus defect can thus be obtained. The effect of astigmatism is to separate longitudinally by an amount C_a the position of exact focus for edges at right angles. This can be seen from eqn (3.23), which contains in addition to the usual focus term an azimuthally dependent 'focus' describing astigmatism. If the maximum and minimum fringe widths measured in perpendicular directions are y_{max} and y_{min} referred to the object plane, then eqn (8.4) gives

$$C_a = \frac{y_{max}^2 - y_{min}^2}{\lambda}$$

The following practical considerations are important.

1. The clearest fringes will be obtained using well-defocused illumination ($\theta_c < 1$ mrad), and a correspondingly extended exposure time. From

Chapter 4 the coherence conditions for observation of fringes of width d is

$$\theta_c < \frac{\lambda}{2\pi d}$$

2. The micrograph should be taken close to Gaussian focus, where the fringe asymmetry is most pronounced. Over-focused fringes (first fringe dark, lens too strong) are more 'contrasty' than under-focused fringes, a subject of considerable research (see Fukushima, Kawakatsu, and Fukami (1974), which contains references to earlier work).

A more sensitive measurement of C_a can be obtained from an optical diffractogram of a thin carbon film (Section 8.7).

8.7 Optical diffractometer measurements

The optical diffractometer is undoubtedly the most useful and important tool at the disposal of the microscopist for fault diagnosis in high-resolution work and for the measurement of instrumental parameters. It was developed by Thon and co-workers (Thon 1971). The construction and use of an optical diffractometer suitable for high-resolution electron microscopy has been described elsewhere (Beeston, Horne, and Markham 1972), and recent promising developments in the design of 'on-line' diffractometers are discussed in Bonhomme and Beorchia (1978) and Spence and Bleha (1980). In high-resolution electron microscopy these optical diffraction patterns ('diffractograms') have five main uses.

1. To detect and measure astigmatism.
2. To detect specimen movement during an exposure.
3. To measure the focus defect at which a micrograph was recorded.
4. To measure the microscope's spherical abberation constant.
5. To measure the damping envelope constant Δ and θ_c (see Sections 4.2 and 8.10).

Diffractograms have also been used to provide information on the structure of specimens, since they exposure the phase of the scattered electron wave. However, the main use for an optical bench in a laboratory turning to high-resolution work will be in training microscopists to correct image astigmatism accurately. Astigmatism may be difficult to detect in an image, but shows up readily in an optical diffractogram as an elliptical distortion of the characteristic ring pattern. Equally important is the sensitivity of an optical diffractogram to image movement during an exposure (drift) which may also be difficult to detect in the image. Figure 8.4 shows a simple optical set up. This design does not require an optical bench or tedious alignment procedures, is extremely compact, and allows the viewer to see both the diffractogram and the micrograph simultaneously.

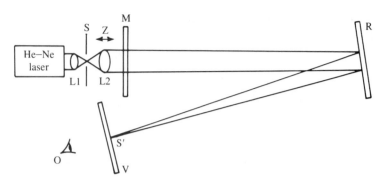

Fig. 8.4 Compact optical diffractometer for HREM. The observer, at O, sees both the diffractogram in transmission on the ground-glass screen at V, and the region from which the diffractogram (or 'optical microdiffraction' pattern) is obtained on the micrograph at M. R is a front-surfaced mirror used to fold the beam-path for compactness. Focusing is achieved by *small* motions of L2 along Z, whose focal length (a few centimetres) separates it approximately from S. Note that S is conjugate to S', as for the SAD mode in STEM. The position of the micrograph M along Z is not critical. L1 is a ×30 optical microscope objective.

Additional mirrors may be used to pass the beam up a wall and reflect it from the ceiling, thus allowing a bench-top design. The thinnest possible holey carbon films should be used for the image recording. By adding graphitized carbon to these and including in the image some carbon lattice fringes, an inbuilt angular calibration standard is obtained, thereby eliminating the need to measure the quantities in eqn (8.5). For preliminary work, take pictures at moderate electron-optical magnification (say 100 000), since the higher magnification images produce rather small, noisy optical diffraction patterns requiring substantial optical enlargement. Panchromatic film must be used with the popular red-light helium–neon lasers, and the careful use of an optical exposure meter will save expensive Polaroid film otherwise wasted on trial exposures. Polaroid film is available for producing positive prints (type 52) and for producing prints plus a negative (type 55). A well-darkened room is essential. The focal lengths of the lenses used are not critical—they can be measured by holding a lens several feet from any light source and measuring the distance between the lens and its focus. Once the bench has been set up, diffraction patterns similar to Figs 8.5 and 8.6 should be seen on the screen. A single broad ring only will be seen if the micrograph has been taken close to the optimum focus of eqn (6.16a). The radial intensity variation in the optical diffraction pattern is approximately proportional to $\sin^2 \chi(\theta)$ as outlined in Sections 3.5 and 4.2. Figure 8.5 shows the effect of astigmatism on an optical diffractogram, while the effect of drift is indicated in Fig. 8.6. These two effects should be clearly distinguished. The optical diffractogram provides an excellent differential diagnosis of these common image imperfections.

Optical diffractograms can also be used quantitatively. A simple FORTRAN program is given in Appendix 1 to perform the necessary data

Fig. 8.5 The optical diffraction pattern of an electron micrograph of thin amorphous carbon film. The electron image was recorded with an incorrectly adjusted stigmator and this produced the elliptical distortion of the optical diffraction pattern seen above. Figure A1.1 gives an example of correct stigmator adjustment, resulting in a circular ring pattern. The procedure for adjusting the stigmator is given in Section 10.1. The bar across the pattern shown above is a beam stop used to exclude the central bright spot from the image. Notice the difference between this diffraction pattern (which contains astigmatism but no drift) and that shown in Fig. 8.6, which contains drift with little astigmatism.

analysis. Given the radii of the rings appearing in an optical diffraction pattern, this program calculates both the focus setting Δf and the spherical aberration constant C_s using the method of Krivanek (1976). By measuring the ring radii in two orthogonal directions it is also possible to find the coeffiient of astimatism C_a. It is also possible to use the small programmable 'pocket' calculators for the same analysis. If this is done, it becomes possible to have on hand values of C_s, Δf, and C_a for a particular micrograph as soon as it has been photographically processed and dried. In order to use the program given in Appendix 1 it is necessary to know the radii of the rings in the optical diffraction pattern at their maximum (or minimum) intensity. These must be supplied in units of $S_n = \theta_n/\lambda$, where θ_n is the electron scattering angle which gives rise to a particular ring n. The quantities S_n are simply related to the radii r_n of the diffractogram rings as measured from a print taken from the optical bench. The formula is

$$S_n = \frac{Mr_n}{\lambda_1 f} \tag{8.5}$$

where M is the electron microscope magnification, λ_1 is the laser light wavelength and f is the focal length of the optical lens. For the helium–neon

Fig. 8.6 The optical diffraction pattern of a micrograph of an amorphous carbon film in which the specimen has moved during the exposure. The specimen movement is in the direction AA' at right angles to the direction of maximum ring contrast and results in low contrast along this line.

laser, $\lambda_1 = 632.8$ nm. The procedure for using the computer program is then as follows. (1) Set up the optical bench as indicated in Fig. 8.4 and take a positive print of the optical diffraction pattern of a micrograph of a thin carbon fim. When using a 1.5 mW laser the exposure time is likely to be several seconds for a micrograph of average optical density. (2) Fix the print and allow it to dry. Measure the radii of as many minima and maxima as possible. (3) Convert these radii to the quantities S_n using eqn (8.5.). Be sure to use the same units for λ, r_n, and f. Enter these numbers into a computer as described in Appendix 1 and run the program.

A simpler and more accurate method of finding the S_n is to include some lattice fringes in the original micrograph. These fringes produce sharp spots on the optical diffractogram which can be used to calibrate the ring patterns. If the Bragg angle at the original specimen for a spot observed on the optical diffractogram is θ_B, then the S_n for the rings are given by

$$S_n = \frac{2r_n\theta_B}{r_B\lambda_{\text{electron}}} \tag{8.6}$$

where r_B is the distance measured on the optical diffractogram from the central spot to the sharp spot arising from the Bragg reflection θ_B. If the diffractogram rings appear elliptical, the coefficient of astigmatism is easily found by measuring two sets of ring radii along the major and minor axes of the ellipse. The difference between the two focus values obtained for these two sets of data is then equal to C_a (see Fig. 2.20).

The accuracy of the diffractogram technique for measuring C_s is, unfortunately, rather poor. It increases with both the order and the total number of rings used. An error analysis of the method shows that best results are obtained at a large under-focus value where many high-order rings are seen. In small-unit-cell crystal imaging (few beams) the determination of C_s is critical to image interpretation—for example a change of 0.4 mm in C_s reverses the black–white contrast of silicon images ($\lambda = 0.037$, $\Delta f = -300$ nm, $C_s = 2.2$ mm). At the point-resolution limit (eqn (6.17)) the change in C_s needed to reverse contrast is 38%, independent of C_s and λ. Since the diffractogram analysis may give values of C_s varying by more than a millimetre within the same through-focus series, the C_s determination is then best done by the more accurate methods of Section 8.2 for a known specimen position (focusing current). In large-unit-cell crystals of known structure, C_s can be refined by quantitative image-matching, since, with many beams contributing to the image, a unique image-match is readily obtained. (In small-unit-cell crystals there are many pairs of values of C_s and Δf which give the same image, as shown in Fig. 5.1(d).)

An ingenious method has also been developed by Krakow and co-workers for obtaining the entire through-focus transfer function from a single diffractogram, using an inclined amorphous carbon specimen. A recent application of this technique can be found in Frank, McFarlane, and Downing (1978/9). One such diffractogram will frequently allow the resolution-limiting effects of chromatic aberration and beam divergence to be distinguished as a function of focus. The atlas of computed diffractograms of this type being prepared by Dr J. Frank should greatly assist the rapid characterization of electron lenses, and will allow users to identify at a glance the particular resolution-limiting factor under specified conditions.

In general, then, to ensure that a high-resolution image can be calibrated with approximate values of C_s, Δf, and C_a it is only necessary to include in the image a small region of thin amorphous carbon. A region of contaminant will also serve. Bourret and co-workers have found that the contrast of diffractogram ring patterns can be improved by making a reduced copy of the electron micrograph on Kodak electron-image plates before recording the diffractogram.

8.8 Transverse coherence width

Spatial coherence in electron microscopy is characterized by $\gamma(r, t)$, the complex degree of coherence. As discussed in Chapter 4 the width X_c of this function is directly related to the incident beam divergence θ_c if the illuminating aperture (C2 of Fig. 2.1) is incoherently filled. For hair-pin filaments this is normally so (see Chapter 4). The semi-angle subtended by the illuminating aperture can either be measured directly from the microscope or obtained from the diameter of diffraction spots. In the first case, $\theta_c = r/L$ with r the radius of the C2 aperture and L the distance between this and the specimen. The transverse (spatial) coherence width is then given by

$$X_c = \frac{\lambda}{2\pi\theta_c} \tag{8.7}$$

If the second method is used, care must be taken to avoid exceeding the linear range of the emulsion used when recording the diffraction pattern of a specimen of known structure by suitable choice of film speed and camera length. If the measured diameter of a diffraction spot is d and the measured distance between the central spot and a spot with scattering angle β is D, then

$$\theta_c = \beta\left(\frac{d}{2D}\right)$$

Here β is twice the Bragg angle. The diameter of weak high-order spots will be found equal to that of the central beam and these are less likely to saturate the emulsion. An advantage of this method over direct measurement is that full account is taken of the objective-lens pre-field (see Section 2.9). This is important for an immersion lens operated at short focal length, as used for high-resolution work.

A more direct impression of coherence effects can be obtained from recordings of Fresnel fringes. Figure 8.7 shows Fresnel fringes obtained from an instrument fitted with a field-emission source. The number of fringes observed is related to the coherence width X_c as indicated in a simplified way below. This simplified model of Fresnel diffraction allows the fringes to be thought of as arising physically from interference between an extended line source along the specimen edge (the wave scattered from the specimen edge) and the unobstructed wave. The correct detailed form of Fresnel fringes must, however, be obtained from the analysis of Chapter 3 using the Fresnel propagator. A plane wave incident on an edge can be represented by the radiation emitted by a coherent point source in the

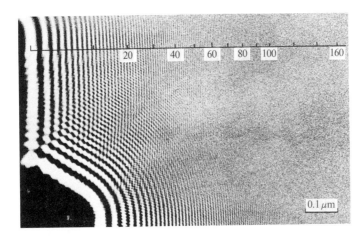

Fig. 8.7 The very large number of Fresnel fringes which can be observed using a field-emission source at 100 kV. More than 160 fringes can be counted at the edge of this molybdenum oxide crystal. The high source brightness allows a very small illuminating aperture to be used (see also Section 4.5 for a discussion of the effect of source size on coherence).

distant plane of an illuminating aperture. A translation of this point source in a plane normal to the specimen edge has the effect of shifting the fringe system as shown in Fig. 8.8. Independent point sources separated by the diameter of the aperture will establish fringe systems which are slightly out of register. The resultant sum of intensities produces a set of fringes whose contrast is thereby degraded. Sources separated by d introduce a fringe shift $\Delta z \, \theta$ where Δz is the focus defect. An approximate expression for the positions of the fringe maxima with a coherent source is

$$x_n = \sqrt{2 \, \Delta z \, \lambda} \, (n - k)^{1/2} \qquad (8.8)$$

where k is a constant $(k < 1)$ such that $2\pi k$ specifies the phase shift introduced on scattering at the specimen edge. The 'period' of the nth fringe is $(x_n - x_{n-1})$. For two sources with separation d, fringes will be more or less washed out beyond the nth fringe where

$$(x_n - x_{n-1})/2 = \Delta z \theta \qquad (8.9)$$

Equations (8.8) and (8.9) can be solved for n. Equation (8.8) then gives the distance x_0 over which fringes will be clearly seen. This is

$$x_0 = A_0 \lambda / \theta_c$$

where A_0 is a constant of the order of unity involving k. An accurate value of this contant can only be obtained from the full treatment for a continuous extended source given in Chapter 3, which unfortunately does not give a simple result. Note that through the constant k the fringe visibility depends on the scattering properties of the specimen. A comparison of this result

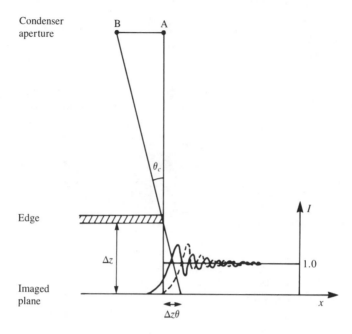

Fig. 8.8 Fresnel fringes formed by two distant separated point sources. The intensity is not zero at the specimen edge. The fringes formed by source B alone (shown dotted) would be displaced with respect to those formed by source A alone (continuous line). The fringes become finer with increasing distance from the edge, which is taken to be opaque. The more realistic case for electron microscopy of an edge with complex transmission function is treated in Chapter 3. A line of independent point sources along AB sets up many fringe systems slightly out of register and therefore blurred beyond a certain distance from the edge.

with eqn (8.7) suggests that within the limits of this simple model, the transverse coherence width X_c is proportional to the distance over which Fresnel fringes are observed measured normal to the specimen edge. A comprehensive analysis of Fresnel diffraction appears in the article by Komrska (1971).

As shown in many texts (e.g., Hirsch, Howie, Nicholson, Pashley, and Whelan 1977) the angle θ_c falls sharply as the illumination is defocused. The secret of recording many clear Fresnel fringes is to use a well-defocused second condenser, the smallest spot (maximum demagnification) on the first condenser, low magnification (about 15 000), and long exposure. So long as the final illuminating aperture remains incoherently filled (see Chapter 4 for methods of determining this), the use of a smaller, brighter source does not increase the coherence in the specimen plane, but simply produces a more intense image. The coherence depends only on the illumination angle θ_c.

Three other techniques have been used to give an indication of coherence length. The electron biprism gives a direct measurement of $\gamma(x_{1,2})$ from the contrast of the central fringe. Theoretical treatments in the English language for this instrument can be found in Gabor (1956), Hibi and

Takahashi (1969), and Komrska (1971). Details of the simple modifications necesary to operate the Siemens 101 electron microscope as an electron biprims are given in Merli, Missiroli, and Pozzi (1974). Another ingenious technique uses the translation of lattice fringes with defocus to estimate X_c (Harada, Goto, and Someya 1974). Interference fringes have also been observed within the region of overlap of two diffraction spots (discs) using crossed crystals to obtain sufficiently close orders (Dowell and Goodman 1973). These fringes arise from interference between waves emitted from separated points within the illuminating aperture and thus indicate a degree of coherence across the illuminating aperture (see Section 4.5). The determination of coherence width from optical diffractogram measurements is discussed in Section 8.10.

8.9   Electron wavelength and camera length

A method has been given which enables the electron wavelength and microscope camera length to be obtained from two measurements of Kikuchi line spacings on the diffraction pattern of a specimen of known structure. The camera length L is related to the measured distance D between two identified reflections or Kikuchi lines by

$$Dd = L\lambda \qquad (8.10)$$

where d is the crystal lattice spacing giving rise to the observed Bragg reflection. Physically, the camera length corresponds to that distance between the specimen and a film plane (without lenses) which would be needed to produce a diffraction pattern of the same size as that observed in the microscope, where lenses are used to magnify the diffraction pattern formed in the objective-lens image focal plane.

Normally eqn (8.10) is used to find d from a measurement of D with λ and L known. However, if a diffraction pattern could be recorded in the correct orientation, containing two identified zone-axes A and B a measured distance R apart, the camera length L could be found from $L = R/\theta$ where θ is the angle between A and B (assumed small). This angle is known if the crystal structure is known. Zone axes are revealed by a characteristic pattern of Kikuchi lines. Thus, in principle L can be found without knowing the electron wavelength. This value of L enables the wavelength λ to be found subsequently from eqn (8.10).

In practice it is not necessary to record two zone-axes on one micrograph, since the problem can be recast in the form of known angles and two measured distances between Kikuchi lines by selecting a favourable orientation. Simple expressions for λ and L in terms of these measurements are given together with a worked example in the paper by Uyeda, Nonoyama, and Kogiso (1965). This is probably the simplest and most accurate wavelength determination method available for electron micros-copy. The error $\Delta\lambda/\lambda$ is typically only about 5 per cent of the error in the ratio of Kikuchi line separations. The overall accuracy may be considerably less that 0.5 percent. Somewhat simpler variations of this method are described by Høier (1969) and Fitzgerald and Johnson (1984), together with a worked example. The method allows the ratio d/λ to be determined to about 0.1 per cent.

8.10   Resolution

There is no simple way to define, let alone measure, the resolution of an electron microscope under the most general conditions. The essential difficulty arises from the fact that, under coherent conditions, the ability of a microscope to distinguish or resolve neighbouring object points depends on the scattering properties of those object points. Thus, resolution depends on the choice of specimen (in particular, on the magnitude of any phase-shift introduced on scattering) and it is strictly necessary, even for weakly scattering objects, to specify the nature of the specimen when specifying resolution. The ability to resolve the individual atoms of a particular molecule is thus no guarantee that those of a different atomic species will be resolved, even given similar inter-atomic spacings and instrumental conditions. Since the resolution of a microscope should be a property of the instrument alone (not the sample), a simple definition of resolution cannot be given. Under the normal conditions of strong multiple scattering in electron microscopy, even greater difficulties arise, as briefly discussed below.

The Rayleigh criterion cannot be applied in high-resolution electron microscopy, since it was devised for incoherent imaging conditions, where it is possible to define point-resolution in a way which is independent of the object. The situation in high-resolution electron microscopy is more akin to coherent optical imaging, in which it is customary to give the transfer function of a lens rather than to specify a single number for its resolution. Only by assuming a phase shift on scattering of exactly $-\pi/2$ from an idealized weakly scattering point object is it possible to plot out an idealized impulse response for the microscope, using the Fourier transform of the transfer function, as has been done in Fig. 3.6. This 'impulse response' is the image of an object chosen to be sufficiently small to make the image independent of small changes in the size of the object. Since the 'impulse response' of a modern medium voltage (300–400 kV) electron microscope is about 0.17 nm, a point test object much smaller than this, i.e. of subatomic dimensions, would be required in practice to observe the impulse response directly. Thus direct observation of the impulse response is clearly impractical, quite apart from the problems of finding a suitable invisible substrate. Several more practical proposals for characterizing the resolution performance of electron microscopes have, however, been put forward, both for strongly and weakly scattering speciments, and these are briefly discussed in turn below.

For the strongly scattering majority of electron microscope specimens it is generally useful to think in terms of the specimen lower 'exit'-face wavefunction as the 'object', and to consider the resolution-limiting effects of the electron lenses which image this wavefunction. Both theory and experiment support the idea (essentially the column approximation of diffraction contrast theory) that this exit-face wavefunction is locally determined by the specimen potential within a narrow cylinder less than half a nanometre in diameter for thin specimens. However, there is no guarantee that the dynamical exit-face wavefunction will contain simple peaked functions in one-to-one correspondence with object point scatterers. Unfortunately, this exit-face wavefunction is not observable, so that a comparison of this function with the final image is not possible. All that is known *a priori* about the dynamical exit-face wavefunction is that it is locally determined and preserves the point-group and any translational symmetry of the specimen potential projected in the beam direction, under most common operating conditions (the image may not preserve symmetry, if astigmatism is present and resolution limited). However, reliable methods do exist for computing the exit-face wavefunction (see Section 5.5) for specimens of known structure. The most general methods, therefore, for specifying the resolution of an electron microscope would be based on a comparison of computed and experimental images of a specimen of known structure, thickness, and orientation. This comparison could be based on a 'goodness-of-fit' index or R factor similar to that used in X-ray diffraction and would allow the extraction of structural detail out to (or beyond) the information-resolution limit of the microscope. The limits to the accuracy with which atomic positions in crystals of known structure can then be determined by this method are set ultimately by noise (see Section 5.13). The variation of this R factor between the computed and experimental images of a known structure with small changes in the atom positions of the computed images would then give an impression of instrumental resolution. If the specimen were crystalline, the choice of a large unit-cell ensures that the transfer function is sampled at a large number of points.

The situation for weakly scattering objects is generally a great deal simpler and the definition and measurement of resolution under these conditions has received a great deal of attention in the literature. Before discussing these methods, it is worthwhile considering how the microscopist can be assured that his micrographs were recorded from a specimen region satisfying the weak-phase object approximation. At least three tests are possible. (1) In an instrument fitted with convergent-beam or STEM facilities, the specimen thickness can be measured from the microdiffraction pattern (see section 11.4). (2) If wedge-shaped specimens are used, the kinematic scattering region can be selected at will (see Fig. 10.4(a)). (3) For crystals, one can be reasonably sure that weak-scattering conditions obtain if the measured Bragg electron diffraction intensities accurately predict the corresponding optical diffractogram intensities (see eqn (3.36)). The damping envelope constants must also be known (see Section 4.2).

These damping envelope constants θ_c and Δ in practice exercise the dominant influence on the form of the microscope impulse response under weak-scattering conditions. Thus, rather than citing a single number for the point- or information-resolution limit of his instrument, the microscopist interested in high-resolution work will be chiefly concerned with determining the constants C_s, Δf, θ_c and Δ for each image recorded. All of these depend, through the specimen position ('height'), on the objective lens current (see Section 2.7, and Appendix 3). As briefly mentioned in Section 1.2, an early task of the microscopist is to determine the specimen position which gives the most favourable combination of these constants. The measurement of C_s, Δf, and C_a has already been discussed, and an ingenious method of measuring Δ and θ_c using two micrographs recorded under identical conditions has been described by Frank (1976) (see also Saxton 1977). This relies on the fact that the Young's fringes seen in the optical diffraction pattern from a similar pair of superimposed, slightly displaced micrographs extend only to the band-limit of information common to both micrographs. The fringe contrast thus falls to zero at the onset of electron noise (represented by high-order Fourier components) which is in principle the only differing contribution to the two micrographs. By plotting this band-limit of common information against defocus, it is possible to determine both Δ and θ_c. Since $\Delta = C_c Q$ (eqn (4.9)), where Q involves the quadrative addition of instabilities due to high-voltage and lens-current fluctuations, together with the effect of the thermal spread of electron energies, any one of these quantities can also be determined if all the others are known. As discussed in Appendix 3, the band-limit due to partial spatial coherence moves to higher spatial frequencies with increasing defocus. A practical difficulty with the Frank method is that vibration from the plate transport mechanism between recordings may alter the specimen height (and thus the focus setting). Experiments show that this change in Δf varies between an amount which is less than that imposed by lens current instabilities up to about 10 nm, depending on the specimen. The Frank method, therefore, may give a slightly pessimistic estimate of the highest-resolution information which can be obtained. Once all these constants have been determined, the transfer function can be drawn out using eqn (4.8).

The problem of measuring the point-resolution of an instrument without using *a priori* information about either the object or the instrument has also been discussed by Frank (1974). His suggestion is that the half-width of the cross-correlation function formed between two successive recordings of the same amorphous carbon film be used to define the point-resolution of an electron microscope. The method is limited to weak-phase objects or incoherent imaging, and requires image digitization. Nevertheless, *this appears to be the best practical proposal for a standardized point-resolution measurement yet put forward,* since all instrumental aberrations and instabilities are included, while the effects of electron noise are minimized through cross-correlation. An asymmetrical correlation function indicates astigmatism (Frank 1975). The formation of a cross-correlation function is greatly facilitated in STEM (Cowley, personal communication). These functions are widely used in geophysics (Smylie 1973), where numerous applications of similar techniques can be found.

The method commonly used by manufacturers of seeking the smallest point-separation in the image of a thin carbon film is unsatisfactory, owing to the effects of electron and Fresnel 'noise'. At high resolution the correlation found between point-separations measured on successive plates is found to be poor (Dowell, Farrant, and Williams 1966). For point-separations greater than 0.5 nm, atomic clusters formed by evaporation of PtIr alloy are useful. These specimens are available commercially. A light evaporation of gold on carbon is also easy to prepare (see Section 1.3) and forms similar small atomic groups. Both these specimens are useful for practising alighment and astigmatism correction and to enable the user to become familiar with through-focus contrast effects on small particles, as discussed in Chapter 6. The most convincing point-resolution test is a recording of a molecule containing heavy atoms in a known arrangement.

Lattice images are also widely used as a resolution test (Heidenreich, Hess and Ban 1968). This subject is fully discussed in Chapters 5 and 10. It is also possible to estimate instrumental resolution by comparing computer-simulated images with experimental many-beam structure images recorded without an objective aperture. If the specimen structure and thickness are known, together with the microscope focus setting, illumination aperture, and spherical aberration constant, images can be computed for a range of values of Δ until a good match is obtained. This will then determine the resolution limit due to incoherent instabilities.

The use to which an electron microscope is to be put is an important consideration in determining the 'kind' of resolution required. Two resolution limits have already been defined (see Section 4.2). For studies based on image processing, it is the information-resolution limit which is all-important (eqn (4.11)). On the other hand, if the microscope is to be used for the study of defects or other non-periodic specimens whose images one wishes to interpret in a straightforward way, then the point-resolution (eqn (6.17)) is of paramount importance. At the present state of the art, optical diffractogram analysis offers the best method of measuring both these resolution limits. The development of an on-line diffractometer would greatly facilitate the control of conditions needed for optimum resolution and would offer particular advantages for high-voltage and biological applications. Such a device would also allow on-line spatial filtering of the optical representation of the electron image on a convenient scale (Bonhomme and Beorchia 1978; Cowley and Spence 1979).

A cautionary comment must be added on the use of axial three-beam fringes and tilted-illumination two-beam fringes for resolution measurement. Expressions for the image intensity for both these cases have been given in Section 5.1. While in principle they may offer a method for measuring the information-resolution limit, there are a number of practical difficulties. Firstly, for axial three-beam fringes and the focus settings of eqn

(5.10), half-period fringes may appear and give a spurious impression of very high-resolution detail, beyond the true band-limit set by incoherent instabilities (see eqn (4.11)). Secondly, it is important to stress that the analysis of Section 4.2 (eqn (4.8) in particular) assumes either kinematic scattering or, more generally, a strong zero-order diffracted beam (which may also occur as a result of Pendellösung oscillations). Unless a wedge-shaped crystal is used, or microdiffraction facilities are available, it is not usually possible to be certain that this condition has been satisfied for a particular image region and therefore that the 'damping envelope' concept can be applied. These very fine fringes are unlikely to be recorded at the Scherzer focus since the microscopist can maximize their contrast by selecting a focus setting which makes the slope of $\chi(u)$ zero in the neighbourhood of the important inner Bragg reflections (see Section 5.8 for more details). By equating this condition (eqn (5.76)) to the focus condition for obtaining half-period fringes (eqn (5.10)), we find

$$a_0^4 = C_s \lambda^3 / n = u_0^{-4}$$

as the crystal periodicity which will produce axial half-period fringes of highest contrast in an instrument operating at particular values of C_s and λ. Fringes obtained under these conditions allow the largest illumination aperture to be used (producing the most intense final image) for the smallest contrast and 'resolution' penalty, and it is likely that many of the sub-Ångstrom fringe spacings reported in the literature were recorded under these conditions. *Such fringes do not measure instrumental resolution in any useful sense,* and must be thought of as, at best, a technique for measuring the instability limit due to mechanical vibration. Note that for half-period, axial three-beam fringes of this type, the damping envelope concept described in Section 4.2 does not apply (despite the presence of a strong central beam), since it neglects terms such as $\Phi_h \Phi_{-h}$ of eqn (5.9).

The comments in Section 5.13 on the accuracy with which atomic positions can be determined for crystals of known structure should be read in conjunction with this section.

References

Beeston, B. E. P., Horne, R. W., and Markham, R. (1972). Electron diffraction and optical diffraction techniques. In *Practical methods in electron microscopy* (ed. A. M. Glauert) Vol. 1. North-Holland, Amsterdam.

Bonhomme, P. and Beorchia, A. (1978). A light optical diffractometer for electron microscopical images operating in line. In *Proc. 9th Int. Congr. Electron Microsc.*, Toronto, Vol. I, p. 86.

Budinger, T. F. and Glaeser, R. M. (1976). Measurement of fucus and spherical aberration of an electron microscope objective lens. *Ultramicroscopy* **2**, 31.

Cowley, J. M. and Spence, J. C. H. (1979). Innovative imaging and microdiffraction in STEM. *Ultramicroscopy* **3**, 433.

Dowell, W. C. T. and Goodman, P. (1973). Image formation and contrast from the convergent electron beam. *Phil. Mag.* **28**, 471.

Dowell, W. C. T. Farrant, J. L., and Williams, R. C. (1966). The attainment of high resolution. In *Proc. 6th Int. Congr. Electron Micscrosc.*, Kyoto, p. 635.

Fitzgerald, J. D. and Johnson, A. W. S. (1984). A simplified method of electron microscope voltage measurement. *Ultramicroscopy* **12**, 231.

Frank, J. (1974). A practical resolution criterion in optics and electron microscopy. *Optik* **43**, 25.

Frank, J. (1975). Controlled focusing and stigmating in the conventional and scanning transmission electron microscope. *J. Phys. E.* **8**, 582.

Frank (1976). Determination of source size and energy spread from electron micrographs using the method of Young's fringes. *Optik* **44**, 379.

Frank, J., McFarlane, S. C., and Downing, K. H. (1978/9). A note on the effect of illumination aperture and defocus spread in bright field electron microscopy. *Optik* **52**, 49.

Fukushima, K., Kawakatsu, H., and Fukami, A. (1974). Fresnel fringes in electron microscope images. *J. Phys. D* **7**, 257.

Gabor, D. (1956). Theory of electron interference experiments. *Rev. Mod. Phys.* **28**, 260.

Hall, C. E. (1949) Method of measuring spherical aberration of an electron microscope objective. *J. Appl. Phys.* **20**, 631.

Harada, Y., Goto, T., and Someya, T. (1974). Coherence of field emission electron beam. In *Proc. 8th Int. Congr. Electron Microsc.*, Canberra, Vol. 1, p. 110.

Heidenreich, R. D., Hess, W. and Ban L. L. (1968). A test object and criteria for high resolution electron microscopy. *J. Appl. Crystallogr.* **1**, 1.

Hibi, T. and Takahashi, S. (1969). Relation between coherence of electron beam and contrast of electron image. *Z. Angew. Phys.* **27**, 132.

Hirsch, P. B., Howie, A., Nicholson, R. B., Pashley, D. W., and Whelan, M. J. (1977). *Electron microscopy of thin crystals.* Krieger, New York.

Høier, R. (1969). A method to determine the ratio between lattice parameter and electron wavelength from Kikuchi line intersections. *Acta Crystallogr.* **A25**, 516.

Holland, L. (1956). *The vacuum deposition of thin films.* Wiley, New York.

Komrska, J. (1971). Scalar diffraction theory in electron optics. *Adv. Electron. Electron. Phys.* **30**, 139.

Komrska, J. (1973). Intensity distributions in electron interference phenomean produced by an electrostatic bi-prism. *Opt. Acta* **20**, 207.

Krivanek, O. L. (1976). A method for determining the coefficient of spherical aberration from a single electron micrograph. *Optik* **45**, 97.

Mathews, J. W. (1975). *Epitaxial growth.* Academic Press, New York.

Merli, P. G., Missiroli, G. F., and Pozzi, G. (1974). Electron interferometry with the Elmiskop 101 electron microscope. *J. Phys. E* **7**, 729.

Pashley, D. W. (1964). Thin metal specimens. In *Modern developments in electron microscopy* (ed. B. M. Siegel), p. 83. Academic Press, New York.

Saxton, W. O. (1977). Spatial coherence in axial high resolution conventional electron microscopy. *Optik* **49**, 51.

Smylie, D. E. (1973). Analysis of irregularities in the earth's rotation. In *Methods in computational physics* Vol. 13, p. 391. Academic Press, New York.

Spence, J. C. H. and Bleha, W. (1980) A real time optical image projection system for electron microscopy. *J. Microsc.* **120**, 121.

Stroke, G. W. and Halioua, M. (1973). Image improvement in high resolution electron microscopy with coherent illumination (low contrast objects) using holographic de-blurring deconvolution, III, Part A, Theory, *Optik* **37**, 192.

Thon, F. (1971). Phase contrast electron microscopy. In *Electron microscopy in materials science* (eds. U. Valdre and A. Zichichi), p. 570. Academic Press, New York.

Uyeda, R., Nonoyama, M. and Kogiso, M. (1965). Determination of the wavelength of electrons from a Kikuchi pattern. *J. Electron Microsc.* **14**, 296.

Wyckoff, R. (1963). *Crystal structures,* 2nd edn. Wiley, New York.

Yada, K. (1967). Study of chrysotile asbestos by a high resolution electron microscope. *Acta Crystallogr.* **23**, 704.

9

INSTABILITIES

The reduction of mechanical and electrical instabilities in commercial CTEM instruments over the last ten years has made high-resolution electron microscopy possible. These instabilities produce an incoherent broadening of the microscope point-spread function (point image) which limits the resolution and can be likened to the imposition of a limiting objective aperture. The measurement of the instabilities described in this chapter is an essential first step in setting up a high-resolution laboratory, and a great deal of time will be saved if measurement facilities are installed at the outset so that when a more subtle fault develops it becomes a simple matter to rule out the easily measured instabilities such as magnetic-field interference and vibration. The optical diffractogram should also not be overlooked as a diagnostic aid for incoherent instabilities.

The book by Alderson (1975) contains a great deal of useful information which complements the material of this section.

9.1 Magnetic fields

An impression of the effect of a.c. magnetic fields on high-resolution images can be obtained by applying a known field while observing lattice fringes. A small transformer is a suitable source of 50 Hz magnetic field which will produce image blurring when held near the specimen chamber. A field applied near the illuminating system results in a loss of intensity owing to the resulting oscillatory beam deflection. Some manufacturers quote a sensitivity for their machines to a.c. fields; a figure of 0.2 nm μT^{-1} (horizontal component at 50 Hz. 1 $\mu T = 0.01$ gauss) is typical for the effect of a field on image resolution. A typical figure for the sensitivity of the illumination spot to deflection is $1\,\mu\text{m}\,\mu\text{T}^{-1}$. Note that the effect of the stray field is greatest where the electron velocity is lowest—for example, near the filament. The microscope lenses themselves have a shielding effect and leakage is most likely at the junction between lens casings. A small homogeneous constant magnetic field such as that due to the Earth is of no consequence in CTEM; however, an inhomogeneous constant field may introduce astigmatism. Permanent magnets should consequently be kept well clear of the microscope.

For high-resolution work an a.c. field of less than $0.3\,\mu\text{T}$ (horizontal component at 50 Hz) is required in the vicinity of the specimen chamber This may be difficult to achieve since the fields commonly found in laboratories are often greater than $0.5\,\mu\text{T}$ owing to mains electrical wiring. Stray fields can be reduced by ensuring that only one ground return for the microscope exists, by moving ancillary equipment, and, if necessary, by re-routing the mains wiring. A common mistake is to place, say, an electrically operated vacuum gauge on the microscope cabinet connected to a different socket and ground return from that used by the microscope, which may be a special high-quality ground. If the microscope and vacuum gauge chassis are in contact a ground-loop can result producing a large a.c. field. Similarly, all additional safety equipment, such as electrical water-valve solenoids (which will be grounded through the water plumbing) should be checked for ground-loops by disconnecting the ground and observing the effect on the focused illumination spot and measured field strength near the specimen chamber. A complete analysis of ground-loops and shielding techniques can be found in the text by Morrison (1967).

Pointed filaments are more sensitive to a.c. fields than hair-pin filaments and the sensitivity increases further for a field-emission source. The simplest test for 50 Hz a.c. fields is as follows. Form a small focused illumination spot at a magnification of about 15 000 times. Run the spot quickly across the viewing phosphor using the beam-translate controls. A zig-zag after-image on the phosphor indicates stray field interference with the illumination system. This effect will be found in many laboratories, where the use of pointed filaments would be of limited value without some prior reduction in the ambient a.c. field.

While the microscope itself used as an oscilloscope in this way is probably the most sensitive indicator of stray fields, a field can easily be measured with a coil connected to an oscilloscope. This will also indicate the important frequency components of the interference. For maximum sensitivity, use the largest possible number of turns of very fine wire—the windings of a mains transformer have been used. It is important to use a long, twin-core shielded lead between the coil and oscilloscope to ensure that the field set up by the transformer in the oscilloscope is not included in the measurements. The oscilloscope should be used in the balanced input mode with the braid and the user grounded. For calibration, the Biot–Savart law gives the field directed concentrically around a long straight wire as

$$B = 20 \frac{I}{D} \ \mu T$$

where I is the current in amps passing through a wire distance D cm from the coil. This gives the sensitivity of the coil in, say, millivolts per microtesla. The field used for calibration should be much larger than any electronic noise.

Severe a.c. field interference causes image distortion (see Fig. 9.1) and a unidirectional blur on Fresnel fringes. If the interfering field is homogeneous and cannot be eliminated at its source, a magnetic field compensator is available which applies a field of opposite phase to the interfering field. A sensor coil is used to measure the ambient field which is amplified and applied, with reverse phase, near the specimen chamber.

A description of this equipment can be found in Gemperle, Novak, and Kaczer (1974). These devices are most effective against interference by homogeneous fields.

The microscope itself should not contribute more than about 0.05 μT to the ambient a.c. field. While ground loops are a common cause of a.c. field interference, other causes include electric motors (in air conditioners,

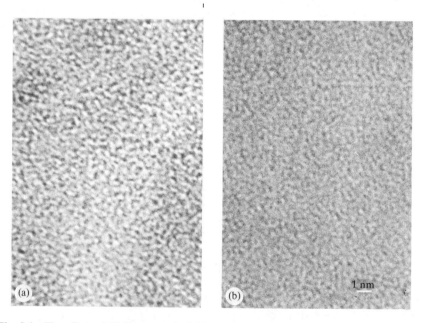

Fig. 9.1 The effect of 50 Hz magnetic field interference on the high-resolution image of an amorphous carbon film. The field strength is 3 μT (30 mG) measured outside the specimen chamber in the direction of maximum strength and would be considerably reduced at the specimen itself. This field is due to the current in the goniometer motor drives, which must be switched off before recording images on some electron microscope models. The field is off in (a) and on in (b).

elevators, rotary pumps, etc.) and transformers. These can often be rotated to a position of minimum interference. The undesirability of siting a microscope near an elevator motor, autotransformer, fluorescent light, electric welding unit, railway track, tram track, or NMR unit is discussed further in Alderson (1975), which should be read with care before choosing a site for a new instrument.

Since the amplitude of image blurring due to a.c. magnetic field interference depends on the accelerating voltage, it is easily distinguished from mechanical vibration using image recordings taken at, say, 50 and 100 kV. If the source of the interference is mechanical, it will not change with voltage.

9.2 High-voltage instability

The incoherent image broadening due to high-voltage fluctuations, resulting in variations in all the lens focal lengths, must be reduced to well below the spatial coherence width for it to be possible to obtain a high-resolution phase-contrast image. A high-voltage (h.v.) stability of a few parts in 10^6 is just sufficient, as discussed in Section 2.6.2. The important sources of instability are:

1. noise in the reference voltage;
2. the stability of the high-voltage divider resistor chain;
3. dirt in the oil tank;
4. the cleanliness of the gun chamber;
5. the vacuum in the gun chamber.

It is unfortunate that microdischarge reaches a local maximum at about 10^{-4} Torr, so that a stable gun must be operated well above or below this pressure. A vacuum of 5×10^{-6} Torr (measured in the rear pumping column) is generally adequate for high-resolution work, and for instruments fitted with one diffusion pump this will only be obtained with the microscope in excellent condition.

High-voltage instability is revealed by flickering on the beam current meter and by fluctuations in the size of the caustic image (see Section 10.2). If either of these are encountered, dirt in the gun chamber is the most likely cause. Small amounts can be removed by conditioning, that is, by operating the gun for a short period above the voltage at which it is intended to be used. Major cleaning must be done using fine metal polish and alcohol and finished with a burst of dry nitrogen gas under pressure. The Wehnelt hole must be examined after cleaning every time it is replaced, particularly if a small hole is used with a pointed filament (see Section 7.5). The Wehnelt should be cleaned by ultrasonic agitation. Examine the main gun chamber O-ring for hair-line cracks—a plentiful supply of spares is needed.

A good deal of the labour of high-resolution electron microscopy consists of time spent cleaning the microscope gun chamber. Instability which persists after cleaning the gun is likely to be due to poor vacuum. This may be caused by leaks elsewhere in the column, by an ageing diffusion pump heating element or oil supply, or, more likely, by outgassing from photographic plates. Outgassing may also be due to an old specimen fragment lodged in the column, or to the microscope interior itself. Instabilities orginating in the microscope high-voltage supply can be isolated by disconnecting the high-voltage cable from the oil tank, but this job is best left to an engineer. A method for distinguishing high-voltage instabilities from those due to objective-lens current fluctuations is described in Agar, Alderson, and Chescoe (1974). This is based on the fact that, while a change in objective-lens current tends to rotate the image, the image blurring due to high-voltage fluctuations acts in a radial direction.

The most sensitive permanent monitor of high-voltage stability is an oscilloscope. For high-resolution work it is strongly recommended that a suitable oscilloscope be kept permanently on hand near the microscope. On most instruments a connection is available (usually on the oil tank) from which a signal of a few hundred volts may be taken off through shielded leads to an oscilloscope. This voltage is proportional to the microscope accelerating voltage. The type of connector and the magnitude of the voltage vary from microscope to microscope and the installation of a permanent oscilloscope high-voltage stability monitor is best discussed with the microscope company engineers. In choosing an oscilloscope for this application, the input sensitivity is important; for most electron microscopes, an oscilloscope with an input sensitivity of better than 10 mV per division will be needed. Since, together with specimen drift, high-voltage instability is perhaps the most common cause of spoilt image recordings at high resolution, *the installation of a permanent oscilloscope high-voltage monitor is strongly recommended.* However, in order to minimize stray magnetic fields from the oscilloscope power transformer (and a possible ground-loop) it is important to remember to turn the oscilloscope off, and to disconnect it from the wall socket after checking the stability, before using the microscope. A ground-loop is possible if the metal case of the oscilloscope is in contact with the metal of the microscope cabinet and if the wall socket used for the oscilloscope has a separate ground return from that used by the microscope.

9.3 Vibration

Mechanical vibration is rarely a problem on well-sited modern instruments, and specimen movement during an exposure is more likely to be due to thermal expansion of the stage or specimen (see Section 9.4). Alderson (1975) contains a full discussion of vibration effects and the choice of site for a microscope. Seismologists are the experts in this field and usually have access to vibration-measuring equipment. A useful booklet (*An introduction to vibration measurement*) is available from Bruel and Kjaer which explains the operation of accelerometers and the use of spectrum analysers.

The microscope and its supports can be thought of as a mechanical oscillator. The resonant frequency of a mechanical oscillator is inversely proportional to the square root of the product of mass and compliance. The aim in minimizing electron microscope vibration is to keep the resonant frequency of the microscope system lower than any likely exciting vibrations, say less than 20 Hz. This is done by making the mass and compliance large and the microscope structure very rigid to minimize the number of coupled oscillators. The mass can be made large by siting the microscope in a basement on a massive concrete block set into the ground and insulated from the building structure. Alternatively, the compliance may be increased by supporting the microscope on air bags or car tyres.

Roughing pumps, water pumps for building heating and cooling, and turbulent cooling water flow in the microscope are common causes of vibration. A change in the water pressure can be used to diagnose this latter cause. The optical diffractogram of a thin carbon film (see Section 8.7) will indicate the presence of drift or vibration with the characteristic rings washed out in the direction of any vibration or drift. This is a useful test if a double image exposure does not indicate the presence of systematic drift, yet vibration is suspected. It enables the effects of astigmatism and vibration to be distinguished, as shown in Fig. 8.6. The diffractogram must be obtained from a bright-field image. A useful diagnostic technique is to attempt to exaggerate the fault observed by applying known fields or vibrations. If the microscope is fitted with an electron detector, a sensitive indication of vibration effects or magnetic fields is obtained by feeding the output from this detector into an oscilloscope or spectrum analyser. A small focused spot is formed across a fixed specimen edge, so that movement of the spot appears as a fluctuation in the output of the detector.

A laser light-beam reflected onto a wall from the surface of a dish of mercury can be used to give a rough measure of vibration amplitudes, however only differences between the displacement of the dish and that of the laser can be detected in this way. The construction of an inexpensive vibration monitor using a crystal microphone is described by Nicholson (1973). These vibration measurements frequently reveal an appreciable difference between vibration levels during the day and those recorded at night. The human hand is also an extremely sensitive vibration detector which should not be overlooked.

9.4 Specimen movement

Specimen movement during an exposure (drift) is the most common cause of spoiled image recordings at high resolution. The amount of drift is referred to the object plane and so depends only on the exposure time and not on the magnification. In high-resolution STEM the effect of drift is to cause a distortion of the image, less serious than the incoherent blurring caused in CTEM. The common causes of drift are discussed below.

1. The specimen grid must make good electrical contact with the specimen holder and any retaining screw or clip must be tight. The specimen itself must either be a conductor or coated with conducting material such as carbon to prevent it from accumulating charge under the action of the beam. A charging specimen is easily identified by the erratic movement or jumping which occurs. A specimen picked up from paper with the finest tweezers will occasionally include a paper fibre collected by the tweezers. If dust particles are frequently found with the specimen a fan should be installed in the microscope room, drawing air in through a dust filter and thereby maintaining the microscope room at a slight positive pressure. Dust is carried out under doors and cracks by the continuous draught established by the fan. Smoking should not be allowed in the microscope room. Always examine a specimen first at low magnification to check for the presence of charging dust particles or fibres which may be out of the field of view at high magnification. These may introduce a variable amount of antigmatism.

2. It is important on top-entry tilting holders that there is good electrical contact between the portion of holder which is free to tilt and that which is not. Since jewelled bearings are often used, a metal spring must be used to provide electrical contact and this must be effective. Do not lubricate any parts of the specimen holder.

3. There must be no contact between the specimen holder and the anti-contamination cold-finger. A cold-finger whose height is adjustable is an advantage for high-resolution work since one may want to use the shortest possible focal length, thereby crowding the specimen, cold-finger, and objective aperture into a very small gap a few millimetres wide. An electrical continuity test is sometimes useful. There should be an open circuit between the frame of the microscope (connected to the specimen) and the external extension of the cold-finger into the nitrogen reservoir. The continuity should be checked through the full range of specimen tilt positions.

4. A clogged lens cooling-water filter is a common cause of drift. The restricted water flow causes the lenses to overheat, which in addition to causing drift may also cause the illumination spot to wander as the lens winding impedance varies with temperature. The external lens casing should not feel hot—if it does, a blocked water filter is a likely cause. Microscopes using internal recirculated water supplies generally do not have this problem. The duty cycle of the thermoregulator should be checked every week as this will indicate any reduction in cooling-water pressure. A refrigeration unit is avoided in some microscope designs by running the entire system somewhat above room temperature.

5. The translation mechanism must be operating properly. The O-rings surrounding the push rods to the specimen stage must contain the correct amount of vacuum grease. Too little or too much can cause drift. Any pull-off springs must supply the correct tension over their full range. The mechanism can be tested by operating one translation control (say the X direction) several turns clockwise, then the same number of turns anticlockwise to bring the specimen to its original position. If the Y control has not been touched, there should be no movement in the Y direction.

For bright-field images, where exposure times are short (a few seconds) specimen drift is not generally a problem. For high-resolution dark-field images, exposures of up to 30 seconds are common and for these, if the conditions mentioned above are met, the main requirement is complete thermal equilibrium of the specimen and stage. This may take several hours to achieve if the lenses have been turned off. The longer the microscope can be left running, with all liquid-nitrogen reservoirs full, the better for long-exposure high-resolution work. Before taking such a recording, examine the specimen at the highest magnification available for signs of drift using an area of specimen adjacent to that of interest. Use the translate controls to 'home in' on the area of interest, using small oscillatory X and Y excursions of the translation controls to relieve any strain in the mechanism. A modest jolt to the microscope bench will be seen (at high magnification) to further settle the specimen.

For even longer exposures, a procedure which has been found successful is to allow the microscope to stabilize, set up all the experimental conditions exactly, then turn down the filament and leave the microscope for several hours with an adequate supply of liquid nitrogen. On returning, it is only necessary to turn up the filament, focus, and record the image without touching the specimen-translate controls. The thermal expansion of the specimen holder rod used on side-entry stages generally makes these less suitable for long-exposure dark-field high-resolution work.

Specimen movement is easily measured by taking a double exposure with several seconds separation. It also appears in an optical diffractogram of a thin carbon film as a fading of the ring pattern in one direction, thereby distinguishing the effect of drift from that of astigmatism (see Section 8.7). A slow unidirectional drift in focus is usually due to thermal gradients in the microscope column. These can be reduced by eliminating draughts, by ensuring that the lens-cooling-water temperature is close to room temperature and is supplied at the correct flow rate, and by leaving the magnification setting near maximum when the microscope is not in use.

9.5 Contamination and the vacuum system

Contamination rates well below 0.1 nm min^{-1} are frequently quoted for modern instruments. While this figure depends on a variety of experimental conditions such as the specimen thickness and temperature, no serious difficulty should be experienced with contamination on an instrument fitted with a modern design of anti-contamination cold-finger held at liquid-nitrogen temperature. Contamination refers to the build-up of decomposed carbon on a specimen. There is some evidence (Knox 1974) that this forms in a crown or annular mound with a diameter equal to that of the illumination spot if this is small, leaving the central illuminated region relatively free from contaminant. In a clean instrument, the main sources of contaminant are the specimen, the diffusion pump vapour, photographic plates, and O-ring grease. Mass spectrometer analysis shows a marked increase in vacuum components likely to cause contamination when a specimen holder which has been touched is introduced into the microscope owing to finger grease. For this reason, the specimen holder should always be handled with disposable or nylon gloves. It can be cleaned using pure alcohol and ultrasonic agitation. Many O-rings will not require any grease, on those that do (movable and frequently changed parts) use a minimum and carefully check for dust particles trapped in the grease on the O-ring. With side-entry holders the vacuum-seal O-ring should be examined for dust under an optical microscope every time before the holder is inserted into the microscope. Once an O-ring has established a good seal it should not be interfered with.

The contamination rate can be reduced by heating the specimen with a large illumination spot, or by improving the overall microscope vacuum. The rate is often measured by noting the reduction in the size of small holes in a carbon film after several minutes. A great deal of research has been done on contamination, particularly in scanning microscopy, and the article by Hart, Kassner, and Maurin (1970) contains references to much earlier work. To obtain a sufficiently low contamination rate for convergent-beam or STEM work, it is usually necessary to pre-heat or 'bake-out' the specimen under vacuum. Thus, specimen 'bake-out' facilities would be useful if, for example, the convergent-beam mode were to be used on the Philips EM 400 to seek specimen areas in the correct orientation for lattice imaging. This would also greatly increase the time during which high-contrast images could be obtained.

Specimen etching is occasionally experienced if oxygen or water vapour are present near the specimen. If this is observed, the most likely cause is a vacuum leak in the specimen exchange mechanism. Some specimens themselves release oxygen under the action of the beam. Water vapour is well condensed by the anti-contamination device if this is at liquid-nitrogen temperature (where the water vapour pressure is 10^{-9} Torr); however, it quickly vaporizes as the temperature is allowed to rise (water vapour pressure is 1 Torr at $-20\,°C$).

A microscope vacuum of at least 10^{-6} Torr is required for high-resolution work, measured in the pumping column at the rear of the microscope. The vacuum affects both the contamination rate and the high-voltage stability. The vacuum will be inferior in the region of the photographic plates or film (unless they have been very well dried) and may be greatly improved within the specimen anti-contaminator owing to the cryopump action of these cold surfaces. A good vacuum in the gun chamber is important, though this is difficult to measure directly. Apart from leaks, the main limitation on the microscope vacuum is adsorbed molecules on the internal walls of the instrument. On an instrument with a low leak rate, an order of magnitude improvement in vacuum can generally be achieved by heating the rear pumping line with electrically heated tapes for several hours. The pumping lines must not be heated beyond a temperature which will damage any components. The lens casings cannot be baked-out in this way, but by adjusting the thermoregulator they can safely be run rather warm for some hours with a similar effect.

A useful simple test for the microscope leak rate can be made by timing the decline in measured vacuum to a particular point after the instrument has been closed down—that is, after closing the main diffusion pump valve to the microscope and all roughing lines. A weekly check will indicate the development of any new leaks which are easily traced using a mass spectrometer or partial pressure gauge as a leak detector (see Section 10.9). If one is not fitted, a large leak can sometimes be found by squirting small quantities of acetone around the suspected area and watching for a change in the vacuum gauge. A log book kept of the number of hours between filament changes will also serve to indicate the quality of the vacuum. An hour meter, actuated by the high-voltage relay, is probably the cheapest vacuum gauge, if used to supply readings for a filament-life log book.

References

Agar, A. W., Alderson, R. H., and Chescoe, D. (1974). Principles and practice of electron microscope operation. In *Practical methods in electron microscopy* (ed. A. M. Glauert). North-Holland, Amsterdam.

Alderson, R. H. (1975). The design of the electron microscope laboratory, In *Practical methods in electron microscopy* (ed. A. M. Glauert). North-Holland, Amsterdam.

Gemperle, A., Novak, J., and Kaczer, J. (1974). Resolving power of an electron microscope equipped with automatic compensation of transverse A.C. magnetic field. *J. Phys. E* **7**, 518.

Hart, R. K., Kassner, T. F., and Maurin, J. K. (1970). The contamination of surfaces during high energy electron irradiation. *Phil. Mag.* **21**, 453.

Knox, W. A. (1974). Contamination formed around a very narrow electron beam. *Ultramicroscopy* **1**, 175.

Morrison, R. (1967). *Grounding and shielding techniques in instrumentation.* Wiley, New York.

Nicholson, P. W. (1973). A vibration monitor for ultramicrotomy laboratories. *J. Microsc.* **102**, 107.

10

EXPERIMENTAL METHODS

This chapter describes some of the practical problems found in recording high-resolution images and the procedures necessary to overcome them. The best way to learn to record and interpret good high resolution images is to take a lot of micrographs of thin carbon films, and analyse these using diffractograms as described in Section 8.7. Then proceed with the project described in Appendix 5 ("The challenge of HREM").

To obtain point detail consistently on a scale less than 0.5 nm requires that the microscope be kept in excellent condition, and a specialized instrument reserved for high-resolution work is essential. Objective-lens pole-piece and specimen stage changes must be kept to a minimum. The user must have confidence that there are no old specimen particles lodged in the objective pole-piece, left there by previous users. Apertures must be clean. The microscope site must be satisfactory with respect to vibration, magnetic fields, humidity, temperature, and dust (see Chapter 9 and Alderson 1975). The spherical aberration constant, measured under the proposed experimental conditions, must be less than 2 mm (see Section 8.2). *However, it is the electronic stability and reliability of the microscope which are most important for high-resolution work.* High-voltage stability must be close to $\Delta V/V = 3 \times 10^{-6}$ per minute or better, with a similar figure for the objective-lens current stability. A full discussion of the influence of electronic instability can be found in Section 4.2, Appendix 3, and Section 2.8.1. A contamination rate better than 0.1 nm min^{-1} is required, and this is of particular importance for high-resolution dark-field work (see Section 9.5). Thermal drift of the specimen must be minimized (see Section 9.4). A vacuum of at least 6×10^{-6} Torr, measured at the top of the rear pumping line, is required to minimize contamination and high-voltage instability. The interior of the microscope, in particular the gun chamber, must be scrupulously clean. The lattice fringes seen in small particles of partially graphitized carbon form a useful test specimen (see Section 8.3) and these specimens are available commercially. In many laboratories such a specimen is kept permanently on hand, and, before starting work each day on any high-resolution project, clear images of the 0.34 nm lattice fringes in this specimen are obtained using a non-tilting specimen holder to confirm that the microscope is operating correctly. This practice is strongly recommended.

A large investment of time and energy is required to keep an instrument in top condition. A high-resolution image project should not be tackled unless the facilities are available for this level of maintenance and, more important, unless the microscopist is fairly certain that the results obtained from the project will justify the time and effort involved.

10.1 Astigmatism correction

The setting of the objective stigmator is critical at high resolution. Some machines provide direction (azimuth) and strength controls for the stigmator, while others allow the strength to be varied in orthogonal directions (see Section 2.8.3). The astigmatism should be approximately corrected by adjusting the controls in a systematic way for a symmetrical Fresnel fringe (see Agar, Alderson, and Chescoe 1974). The more accurate correction needed at high resolution can then be obtained by observing the structure of a thin amorphous carbon film at high magnification. An electron-optical magnification of at least 400 000 is needed to correct astigmatism accurately in 0.3 nm image detail using ×10 viewing binoculars. Adjust both the focus controls and the stigmator for minimum contrast, which should occur at a sharply defined setting. Under-focusing the objective by a few tens of nanometres should then show the grainy carbon film structure with no preferential direction evident in the pattern. This is a difficult procedure requiring considerable practice. An optical diffractogram is the quickest check on astigmatism correction (see Section 8.7) and is essential when learning to correct astigmatism for ultimate performance. The following points are important.

1. The eyes must be fully dark-adapted. Allow at least 20 minutes in the darkened microscope room.

2. Since both are well within the lens field, the objective aperture and lower cold-finger position and state of cleanliness critically affect the required stigmator setting. Never touch either after astigmatism correction and before recording an image. The important thing about astigmatism is that it does not vary in time and that it is correctable. For this reason a small amount of contamination on the objective aperture may not matter so long as the field set up by it can be corrected magnetically using the stigmator. A dirty aperture which sets up a fluctuating field is more serious.

Test for the amount of contamination on the objective aperture by moving the aperture while viewing the image. There should not be a violent degradation in astigmatism with small aperture movements. The hole in the cold-finger can be inspected for cleanliness by finding a magnification range at which it limits the field of view. Test this for cleanliness by moving the cold-finger slightly, if it is adjustable, while observing the low-magnification image.

3. The astigmatism correction needed varies with the type of holder used, the high voltage, and the objective lens strength. It appears to depend weakly on the magnification, possibly owing to flux linkage between the lenses.

4. A common cause of serious astigmatism is a piece of dislodged or evaporated specimen in the objective pole-piece. Examine the pole-piece regularly with a dentist's mirror, but do not remove it unnecessarily. The soft iron of the pole-piece is easily scratched.

5. Unless the cold-finger reservoir is continuously topped-up with liquid nitrogen by a pump, the image will rapidly become astigmatic when the reservoir boils empty, which causes the cold-finger to move as it warms up. There is thus no point in correcting the astigmatism until the cold-finger reaches liquid-nitrogen temperature. This may take at least half an hour after first filling the reservoir.

6. Astigmatism can result if an intense Bragg reflection is caught on the objective aperture edge. This situation should be avoided.

The sensitivity of high-resolution images to astigmatism can be judged by observing the lattice fringes of graphitized carbon, in bright field with untilted illumination, while adjusting the stigmator. To obtain fringes running in all directions (Fig. 8.3), both stigmators must be set very accurately. One-dimensional fringes are more easily obtained, but these do not ensure accurate astigmatism correction.

Finally, it is worth repeating that the astigmatism must be corrected each time the objective aperture is moved.

10.2 Taking the picture

It has been said that the skill of high-resolution electron microscopy lies in knowing when to take the picture. High-voltage instability and specimen drift are to some extent unpredictable effects, so that a certain amount of luck is required for the longer-exposure recordings. The following procedure should minimize the number of spoiled image recordings—these precautions become increasingly important as the exposure time is increased. A list of conditions which should be checked before recording high-resolution images is given in Section 10.10.

1. At least 20 minutes in a darkened room is required for full dark-adaptation of the eyes. There are two processes involved—the rapid dilation of the pupil and a slower but important chemical process.

2. The binocular focusing is important. Focus each eyepiece separately (closing one eye), then adjust one eyepiece with both eyes open until a sharp image of the phosphor grain is seen. On medium- and high-voltage machines a video system will normally be fitted (see Section 11.3). This should be carefully adjusted for high contrast.

3. The anti-contamination liquid-nitrogen dewar should have been full for at least half an hour. Astigmatism is introduced if the temperature of the lower anti-contamination cold-finger is allowed to alter (owing to thermal expansion), so that this must be allowed adequate time to reach thermal equilibrium. The specimen must have been out of contact with the lower cold-finger for several minutes—this is important if a tilting stage is used with an extended top-entry holder operating at short focal length to reduce spherical aberration (see Sections 1.2 and 2.7). At the extremes of tilt, the specimen may touch the lower cold-finger, resulting in specimen cooling and drift. The thermal stability and cleanliness (to prevent charging effects) of the three components immersed in the objective-lens field (cold-finger, objective aperture, and specimen) are of the utmost importance for high-resolution work (see Fig. 2.9). By switching to a low-magnification range where the field of view is limited by the lower cold-finger, the cleanliness of this aperture can be examined.

4. Check for high-voltage stability both by observing the electron beam current meter, which should not flicker, and by examining the caustic image. On sitting down at any microscope on which it is intended to do high-resolution work, the first thing to do is to watch the beam current meter pointer closely for at least 20 seconds. Intermittent flickering of the meter during this time indicates that the instrument is unsuitable for high-resolution work. Using this test, most instruments in routine use will be found to be unsuitable without some maintenance (usually gun cleaning and leak repair). An occasional flicker (say, once every 20 seconds) may not be important, since these voltage spikes are often very short compared to both the exposure time and the persistence time of the human eye. The oscilloscope monitor (Section 9.2) can be used to distinguish fast spikes from an increase in the continuous background fluctuation which determines the resolution limiting factor Δ in eqn (4.9). The caustic image is obtained by switching to diffraction mode without a specimen or imaging apertures in place. Adjust the diffraction focus for a disc with a bright outer ring. This ring contracts or expands with changing high voltage and can be examined through the binocular.

5. The microscope must be correctly aligned as described in the instruction manual. Check the illuminating system, in particular gun tilt, condenser astigmatism, and condenser aperture centring. The gun tilt will need frequent readjustment during the first hour of operation. This is important, since for the highest resolution one is generally using a small illuminating aperture and so requires the maximum possible source brightness. Check the gun-bias and make a note of the illumination conditions. These will be needed if a comparison with computed images is intended. If the microscope is not fitted with a high-voltage 'wobbler', obtain current centre for the objective lens by adjusting the beam tilt until no image translation is observed at the centre of the screen with change in focus (see Agar *et al.* (1974)). A difference at the object plane of about 0.5 μm is common between current centre (the point about which the image rotates with change in lens current) and voltage centre (the point about which the image expands with change in high voltage). If a high-voltage wobbler is fitted, alignment by voltage centre is preferred since, for an axial image point, this confers some immunity to high-voltage instability due to micro-discharge during long exposures. This, and the energy spread of electrons leaving the filament are the main sources of chromatic aberration for thin high-resolution specimens where energy losses in the specimen itself are generally negligible. The use of voltage-centre alignment minimizes chromatic aberration for axial image points due to these effects at the expense of image distortion due to fluctuation in lens current (which would be minimized by using current-centre alignment). Once aligned, modern instruments should not require any other major adjustments other than those mentioned above. An excellent review of the practical procedures used to correct severe misalignment and the geometric and chromatic field aberrations can be found in the article by Riecke (1975).

6. The diffraction conditions required must be set up and recorded. For many-beam lattice images of small crystals, a motor goniometer is required since top-entry stages, particularly if operated at short focal length, are not eucentric. It is necessary to operate both the translate and tilt controls simultaneously to keep the same region of specimen in view while tilting. This is difficult for the smallest crystals, which are usually also the thinnest, and therefore most suitable. The method for doing this is described in Section 10.4. A zone-axis orientation is usually required and it is important to ensure that the region of crystal which is 'in focus' is that from which the recorded diffraction pattern is obtained, since this may form the basis of subsequent image simulation. After obtaining the required tilt condition, make a final check for the clearance between the specimen and the lower cold-finger using an electrical continuity device if this is fitted. The current centre obtained in (5) above ensure that the illumination direction coincides with the optic axis, so that the central diffraction spot can now be used to locate the objective aperture (if one is to be used) with respect to the optic axis (see Fig. 5.1).

7. Check the specimen for drift by observing the image at the highest magnification available for several seconds. Use a region of specimen near that to be subsequently recorded.

8. Once a suitable region of specimen is found, 'home in' with the translation controls (see Section 9.4) and give a modest jolt to the microscope bench to settle the specimen. The final adjustments will be to the focus and stigmator controls (see Sections 10.5 and 10.1). To record the image, operate the plate advance mechanism then wait for a few seconds to allow mechanical vibrations to decay. The final focus adjustment, in which a setting is established a certain number of 'clicks' from the minimum contrast condition, must be made *after* operating the plate advance mechanism. Vibration from this mechanism may alter the specimen height by up to 10 nm. Open the shutter carefully without otherwise touching the microscope. Don't talk or move any metal objects during the exposure. Magnetic field variations of up to 0.4 μT have been recorded as a result of the movement of a wristwatch. The effects of loud speech on lattice images can easily be observed through the binocular. Finally, close the shutter. The disturbance caused by the plate (or film) transport mechanism and the time taken for this to decay can be seen by watching the lattice image after closing the shutter. Fringes generally remain obscured for several seconds after the transport has stopped, giving an indication of the sensitivity of the lattic fringe image to mechanical vibration.

10.3   Finding and recording lattice fringes—an example

The main requirement for finding lattice fringe images is patience. The following is a step-by-step outline of the procedure used to find and record the image shown in Fig. 10.4, which is used as an example of the many-beam lattice imaging method. Alternatively, crystals of MgO may be examined, as suggested in 'The challenge of high resolution electron microscopy' (see Appendix 5 and Fig. 8.1). They have the advantages of requiring very simple specimen preparation, while providing an internal thickness calibration. However small crystals are more difficult to align than a large single crystal. The silicon image was obtained from a chemically-thinned wedge-shaped specimen of silicon with the incident electron beam travelling in the [110] crystal direction. Approximately seven Bragg reflections contribute to the image; however, this number is not sufficient to allow the individual silicon atom columns to be resolved. It is assumed that a suitably thin silicon wedge has been prepared using either commercially available jet thinning equipment or laboratory-built thinning equipment (see Section 5.14). We further assume that the optimum specimen position in the objective-lens pole-piece has been found (see Sections 1.2 and 2.7). This is specified by the objective-lens current shown on the digital read-out current meter (see Section 8.4) needed to focus a specimen at the correct height. When using a top-entry stage with specimens surrounded by a thick rim such as chemically thinned silicon and germanium, spacers or washers placed above the specimen may be needed to obtain the correct focusing current. Consistent results between specimens can be obtained by thinning these specimens most of the way from one side, as shown in Fig. 10.1, so that only one special washer need then be kept for these specimens. No such problems arise with side-entry stages in which the specimen height is continuously adjustable. However side-entry specimen holders should be examined under a low-power optical microscope to ensure that the specimen-retaining clip is accurately positioned and making firm contact

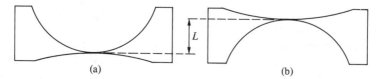

Fig. 10.1 For chemically thinned specimens, the specimen height in the microscope (and therefore the lens current needed to focus) may depend on which way up the specimen is when inserted into the instrument. Inverting the specimen may alter the height by an amount L which is comparable with the lens focal length. In order to work at small focal length, and so reduce aberrations, one usually wants the specimen very close to the lower anti-contamination cold-finger. This can be achieved by thinning most of the way from one side, and consistently inserting specimens in the holder in orientation (a) above.

with the specimen. For both types of holders the specimen must make firm mechanical and electrical contact with the specimen holder.

1. The preliminary checks given in Section 10.10 should be conducted. These procedures must be followed every day before commencing work if long fruitless searches for lattice fringes are to be avoided. For example, high-voltage instability which may be difficult to detect on the beam current meter may develop owing to a small leak in the gun chamber. Such a fault may prevent fine lattice fringes from being seen, and would only be revealed by a vacuum measurement (see Sections 9.2 and 9.5) or oscilloscope instability test (see Section 9.2).

2. Load the standard specimen (partially graphitized carbon—see Section 8.3) into the microscope and confirm that the 0.34 nm fringes can be seen. A non-tilting specimen holder can be used.

3. Fit the silicon specimen to the tilting specimen holder, being careful not to damage it. Wear gloves at all times when handling the specimen holder. Check the position of the retaining clip on side-entry holders. On top-entry holders in which the specimen is retained by a threaded ring the tightening torque applied to this ring is critical, and brittle specimens are easily broken. Treat silicon and germanium as you would a small piece of glass.

4. Load the silicon specimen into the microscope and bring the specimen to a horizontal position. Find the hole in the centre of the specimen. The illumination system must now be adjusted. Where a pointed filament is used (see Section 7.5), the filament image should be examined at a magnification of about 50 000 and made symmetrical by adjusting the gun tilt controls. Figure 10.2(a) shows an image of a correctly adjusted pointed filament. In Fig. 10.2(b) we see the effect of introducing a stray a.c. magnetic field (see Section 9.1), in this case caused by switching on the goniometer motor power supplies. The central tip of the filament image provides a sensitive test of the presence of stray fields. (The images shown in later figures were obtained with the goniometer power supply switched off.) Setting up the pointed filament involves obtaining the correct combination of filament

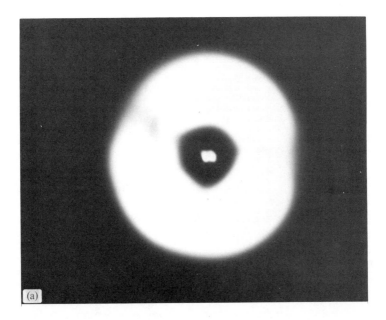

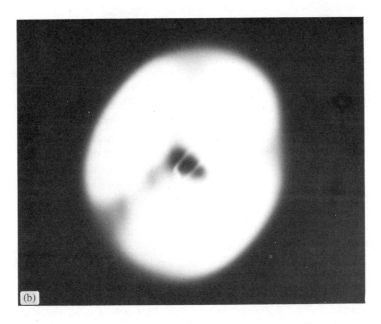

Fig. 10.2 (a) Focused pointed-filament image obtained from the filament type described in Section 7.5 and used to form the images shown in Fig. 10.4. The form of this image depends chiefly on the gun-bias setting, the Wehnelt hole size, and the filament tip height. At the high magnification used to form lattice images, only a small portion of the bright ring falls over the plate camera. A great deal of time can be wasted looking for lattice fringes on an instrument in which the filament has not been set up correctly. (b) The effect of 50 Hz stray a.c. magnetic field on the filament image shown in (a). Note the alteration in the appearance of the filament tip. This field was due to a piece of ancillary equipment and washed out all fine lattice fringes. A simple test for these fields, in which the spot is run across the screen using the beam deflection coils, is described in Section 9.1.

height, gun-bias, and filament-heating current (see Section 7.5). The gun-bias setting corresponding to the lowest beam current should be used, to minimize chromatic aberration (see Fig. 4.3 and Sections 5.19 and 2.8.2). A clean condenser aperture must be selected (examine it by switching the microscope to lowest magnification) and centred. The images shown in Fig. 10.4 were obtained with a condenser aperture which subtends a semi-angle of 1.16 mrad at the specimen, as can be measured from a recording of the diffraction pattern taken with focused illumination (see Fig. 10.7). A diameter of 200 μm is a typical condenser aperture size. Factors affecting the choice of condenser aperture size are discussed in Sections 5.19, 4.2, and 4.6. The condenser aperture is centred by swinging the condenser focus controls through the exact focus position and adjusting the aperture position to obtain an image which expands and contracts symmetrically. The condenser astigmatism controls must also be set correctly, and the maximum strength for the first condenser lens must be selected (sometimes called 'spot size') which will subsequently provide sufficient intensity for focusing at high magnification (see Sections 4.5, 9.5, and 7.1 for other factors affecting the choice of first condenser lens excitation). If the filament is being used for the first time it will be necessary to check the gun-tilt controls every 15 minutes or so for the first hour of operation.

5. The specimen must now be accurately tilted into the zone-axis or symmetrical orientation. Figure 10.4 was obtained from a silicon crystal in the [110] zone-axis orientation, that is, with the electron beam travelling in the [110] crystal direction. A method for doing this is described in Section 10.4 for top-entry stages. Once the specimen is approximately oriented, check that it is not touching the lower or upper cold-fingers (see Section 10.9) and that the image is in focus at the correct (optimum) objective-lens current (see Section 1.2). When using tilted specimens there may be an appreciable difference between the lens current needed to bring regions at the edge of the specimen into focus and that needed to focus the central specimen region.

6. A preliminary astigmatism, drift, and focusing check should now be made. Examine a thin edge of the specimen at a magnification of about 600 000 and find the minimum-contrast setting of the stigmator and focus controls without an objective aperture, as described in Section 10.5. Once this has been found, weaken the objective lens by about 90 nm (see Section 8.1) to increase the image contrast and watch the image carefully for about five seconds. It should appear completely motionless, and the background graininess of the image should appear sharp. With experience the appearance of this background becomes a useful rapid guide to image quality. If the image is not sharp or is unstable check the following.

1. The anti-contaminator has reached thermal equilibrium.
2. The high voltage is stable (see Section 9.2).
3. The specimen is not touching either cold-finger.
4. A dirty objective aperture has not accidentally been left inserted into the microscope.

5. Intermittent erratic jumping of the image commonly indicates that it is charging up. Silicon and germanium are sufficently good conductors to prevent this from happening so long as an earth return-path exists for charge to leak away. If charge appears to be accumulating on the specimen, check that it is making good electrical contact with the specimen-holder ring, and that an electrical connection exists between this ring and the specimen-holder body across the jewelled bearings. Special spring-metal washers are provided for this purpose (see Section 9.4). It is a waste of time to look for lattice fringes unless a stable, sharp image of the background in a very thin specimen region can be obtained. Check that sufficient intensity is available to enable the stigmator and focus controls to be set correctly (see Sections 10.1 and 10.5) at a magnification of about 600 000.

7. A search must now be made for a suitable area of specimen. A uniform wedge is required if the change in image appearance with thickness is of interest. Only by comparing computer-simulated images for various thicknesses with experimental images of a wedge-shaped specimen can one be certain of the image interpretation and so select regions of crystal thickness in which a simple image interpretation is possible. A convenient range of thickness would include two bright-field Pendellösung thickness periods. The separation of these 'equal thickness' fringes in the image depends both on the wedge angle and the electron-optical magnification. To obtain these fringes, insert the smallest objective aperture to include only the central diffracted beam (without altering the specimen orientation) and examine the image at a magnification of about 500 000. Search the wedge edge for a region in which approximately two dark Pendellösung bands fill the portion of the viewing screen lying above the photographic plates. The bands should be even and the wedge edge smooth. Figure 10.3 shows such a region from which the lattice image in Fig. 10.4 was obtained. Note that the image extends slightly beyond the edge of the wedge. It is a mistake initially to seek a wedge with the smallest possible angle (Pendellösung bands widely separated), since specimens are invariably bent in these regions. Remove the objective aperture after 'homing in' with the translation controls (see Section 9.4) once a suitable region of specimen has been found. It may be necessary to examine several specimens in order to find a suitable specimen area.

8. If a stable image can now be seen showing sharp background contrast, make small adjustments to the stigmator orientation and fine-focus controls until a lattice image is seen. For Fig. 10.4, final adjustments to these controls were made at an electron-optical magnification of 640 000. Owing to the use of a non-standard objective-lens current, a correction factor has been applied to the indicated magnification to obtain this true magnification—see Section 8.3. The viewing binoculars usually provide an additional magnification of 10. Even without an objective aperture in place, a faint image of the Pendellösung fringes should be seen, whose contrast depends on the focus setting. Look for fringes initially along the thinner side of the first Pendellösung fringe. Check that both binocular eyepieces are

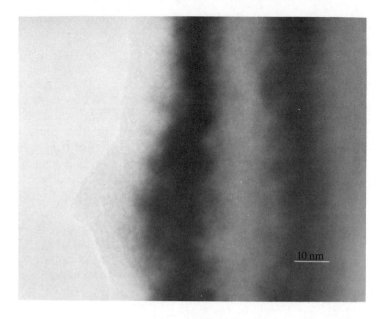

Fig. 10.3 Pendellösung fringes (the dark vertical bands) formed in silicon in the symmetrical [110] zone-axis orientation (see Fig. 5.1(c)), formed using a small central objective aperture which excludes all but the central beam. Calculations give the increment in thickness between these bands as 23 nm; they therefore provide a useful method of calibrating the thickness scale for lattice images obtained from wedges. They also indicate the shape of the wedge—the thickness is constant along lines of constant intensity. Kinematic or single-scattering images are obtained on the thinner side of the first dark fringe, so that images of defects in this thickness range may be interpreted by the theory of Sections 6.1 and 6.2 (see also Section 5.3.1). The lattice image is obtained by removing the objective aperture, given the necessary instrumental stability, and is shown in Fig. 10.4(a) for this same specimen region.

focused (see Section 10.5) and that the image remains stable at least for the length of time needed for an exposure. If the image is too dim, switch to lower magnification and check the form of the filament image. Check also that the most intense part of the filament image is being used. Look 'into' the image rather than 'at' it. Knowing what you expect to see will soon lead to more rapid identification of a fringe image—this can only come with experience. Remember that the photographic emulsion will record far more detail than you can see. If a stable image is seen showing lattice fringes of moderate contrast over part of the viewing area you can be confident that the developed plates or film will show fringes of strong contrast in most areas. The stigmator adjustment is critical. On some machines where orientation and amplitude controls are used, the more sensitive orientation control is a wire-wound potentiometer. On these microscopes, as you look into the binocular at the electron image you should feel the wiper of the orientation stigmator potentiometer move from one turn on the resistance winding to the next. This is sufficient to convert a lattice image in silicon observed under these conditions from one showing sharp 'tunnels' (that is, showing several sets of crossed lattice fringes) to a pattern of fringes all of which run in the same direction. Making the smallest adjustments to the stigmator control, select the image showing sharp tunnels. Adjust the fine-focus control for maximum contrast.

9. Operate the plate-advance mechanism, make final focus adjustments, pause, and record the image (see Section 10.2). *Write down the objective-lens current and focus defect* (see Section 10.5). A very dim desk lamp is useful here.

For silicon images observed under the above conditions you will find several focus settings which give sharp, apparently identical images of high

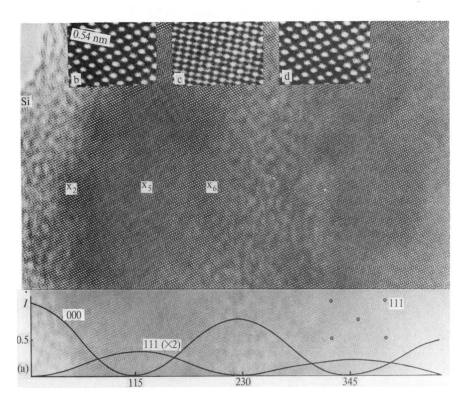

Fig. 10.4   (a) Lattice image of silicon in the [110] symmetrical orientation taken at 100 kV accelerating voltage. The edge of the specimen can be seen at top left, showing Fresnel edge fringes. Enlargements of regions x_2, x_5, and x_6 are shown at insets (b), (c), and (d). The image at x_5 gives a false impression of the structure (shown in Fig. 10.4(b)), while that at x_2 and x_6 faithfully represents this projection of the diamond structure to the resolution available. The intensity of the four (111) type Bragg beams contributing to the image, together with that of the central beam, is shown below in approximate registry with the image using a horizontal thickness scale in Ångstroms. The interpretation of these images is discussed further in Section 5.14 (see also Spence, O'Keefe, and Kolar 1977, and R. Glaisher and A. Spargo, Ultramicroscopy **18,** 323 (1985)).

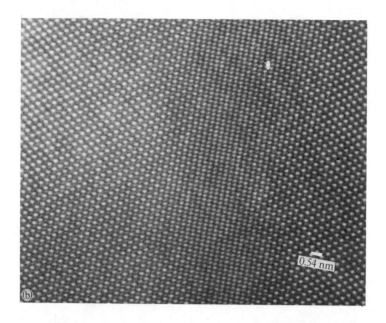

0.54 nm

(b)

Fig. 10.4 (b) Portion of the image shown in Fig. 10.4(a) around x_5 at higher magnification. The change in image appearance with thickness (which increases from left to right across the picture) is clearly shown. Near the centre of the image at a thickness of approximately 11.5 nm the central beam has been extinguished (see lower half of Fig. 10.4(a)) by Pendellösung, resulting in a 'half-period' image (see Section 5.1) which does not reveal the crystal structure. This image results solely from interference between the four (111) type reflections. The amplitudes *and phases* of all Bragg beams contributing to the image (see Fig. 10.8) must be studied in order to understand the thickness dependence of these images (see Section 5.14).

contrast. These are Fourier images (see Section 5.14). Between these focus settings the lattice image seems to disappear. The focus increment between 'identical' images (images which appear identical through the viewing binocular) for silicon viewed in the [110] direction at 100 kV is $\Delta f_0 = 86$ nm. Apart from an unimportant half-unit-cell translation, these Fourier images are in fact identical in regions of perfect crystal. Only one image, however, will include a high-contrast image of any crystal defects which may be present—this is the 'optimum focus' image (see Section 6.2).

As discussed in Section 5.19, for small-unit-cell crystals, the perfect crystal lattice image may show very low contrast (or be absent entirely) at this optimum focus if, as is common with older microscopes and metallurgical or semiconductor problems, the fundamental lattice spacing exceeds the point-resolution of the instrument (see Sections 4.2 and 5.19). The highest-contrast lattice images may then occur at very large under-focus (~300 nm). The focus increment between Fourier images can be used to calibrate the focal steps of the microscope focus controls (see Section 8.1).

If a through-focus series is required it is important to bear three points in mind.

1. Vibration from the plate or film-advance mechanism following each image recording may alter the height, and thus the focus condition, of the specimen by an amount greater than the electronic focal increment selected.

2. Magnetic hysteresis may make it necessary to avoid reversing the direction of the lens current change.

3. It may not be possible to return reproducibly to an earlier focus setting if this would involve changing ranges between the various focus knobs. This problem is eliminated on the newer digital machines, which use a system of 'continuous' focus adjustment. On older machines where the span of the finest focus control covers a range of, say, 94.5 nm, divided into perhaps 21 4.5 nm steps, a problem arises if a high-contrast image is observed in the middle of this range and one wishes to record the next-but-one Fourier image. To do this it would be necessary (for silicon) to increase the focus setting by 86 nm, which is only possible by incrementing the second-finest focus control. In practice the focus calibration does not appear to be preserved between ranges.

The exact crystal orientation used for the final image recording is difficult to determine since the smallest selected area aperture will include a much larger region of specimen than that included in the lattice image. Problems associated with the design and construction of specimen-tilting devices which will allow specimens to be tilted accurately and reproducibly to the same orientation provide an important limit to the development of the methods of high-resolution electron microscopy. The need for accurate orientation setting increases with specimen thickness. Having set the crystal as close as possible to the zone-axis orientation using the method of Section 10.4, the best practical guide to orientation is probably the symmetry of the image. Since this depends on the stigmator setting, the possibility of inadvertently compensating for a small orientation error arises by using the stigmator. This 'cylindrical lens' (see Section 2.8.3) can be used to 'defocus' a strong pair of fringes and so reduce their contrast relative to a weaker pair of fringes running at right angles to these. The result may be a symmetrical two-dimensional image obtained from a specimen in which the five Bragg reflections contributing to the image do not have equal strength owing to an orientation error. A check on this unwanted effect is given by the optical diffractogram method described in Section 8.7.

For the finer fringes it may be necessary to search specimens for several days before a good image is obtained. It is difficult, however, to concentrate for more than about two hours at a time. The anti-contaminator must not be allowed to warm up during the day (see Section 9.5), as this would result in a lengthy delay while the specimen area regains thermal equilibrium.

The material of Section 5.19 is intended to complement this section.

10.4 Adjusting the crystal orientation using non-eucentric specimen holders

Much of the time spent on high-resolution work is devoted to specimen-orientation corrections. The tedious alignment process (which becomes increasingly critical in thicker crystals) is greatly facilitated on an instrument fitted with microdiffraction facilities. Here the symmetry properties of the convergent-beam pattern from bent specimens can be used to seek crystalline regions in the exact, required orientation.

A systematic method of bringing a small crystal into the zone-axis or symmetrical 'cross grating' orientation is outlined below for instruments not fitted with microdiffraction facilities. It applies to top-entry stages fitted with a motor-driven goniometer fitted with a pair of foot pedals to control the orientation and tilt of the specimen. The general principle of the method can also be applied to double-tilt side-entry stages if it is found that the specimen position which produces the highest-quality images (see Section 1.2) does not coincide with the specimen height needed for eucentric operation.

1. Insert the specimen into the microscope and set the goniometer to the neutral position so that the specimen disc is approximately horizontal. We shall assume that the required zone-axis orientation is close to this position. Switch the microscope to diffraction mode and observe the focused diffraction pattern of a small crystal or thin region of specimen edge.

2. A pattern similar to that shown in Fig. 10.5 should be seen. The arc of intense spots is a portion of the Laue circle which traces the intersection of the Ewald sphere (see Fig. 10.5) with the reciprocal lattice plane of interest. We wish to alter the specimen orientation with respect to the incident beam so that this Laue circle contracts to a point at the centre of the pattern. The incident beam is taken to be aligned with the microscope optic axis using the method of voltage or current centring described in Section 10.2. The difficulty is that any changes in specimen orientation will be accommpanied by a translation of the specimen, thus bringing a new region of specimen into the area which contributes to the diffraction pattern. This difficulty is overcome by obtaining an *image* of the specimen within each diffraction spot. Do this by adjusting the 'diffraction focus' control (taking the diffraction pattern out of focus) until a poor-quality image of the specimen region of interest is seen within the central diffraction disc, as shown in Fig. 10.6. The 'rule of thumb' mentioned in (3) below requires that the diffraction lens is always weakened, not strengthened, to take it out of focus.

3. Now adjust the specimen orientation using the foot controls while simultaneously adjusting the main specimen-translate controls in such a way that the same region of specimen remains imaged within the central diffraction disc. This procedure requires practice and patience. A useful rule of thumb is that in order to contract the Laue circle, the tilt and orientation

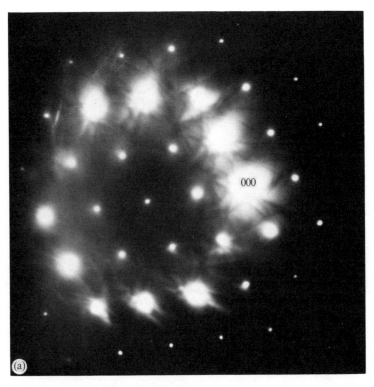

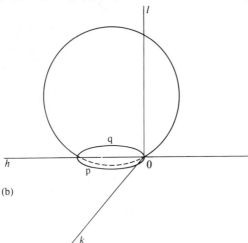

Fig. 10.5 The Laue circle. The central beam is labelled (000). This circle, the bright ring of spots in (a), is formed by reflections which occur near the intersection of the Ewald sphere (shown in (b)) with the plane hk, whose normal is $0l$. To bring the crystal into the orientation needed for lattice imaging, the sphere (which always passes through the origin) must be rolled across to the right so that $0l$ forms a diameter and the Laue circle opq contracts to a point. The pattern in (a) was obtained from a silicon crystal near [110] orientation at 100 kV accelerating voltage.

Fig. 10.6 Defocused central diffraction spot from a small triangular-shaped silicon crystal. The out-of-focus diffraction pattern makes each diffraction spot into a low-magnification image of the specimen, allowing simultaneous observation of the crystal orientation and position. As the specimen is tilted, corrections can be made to the translation controls to keep the same region of specimen in view.

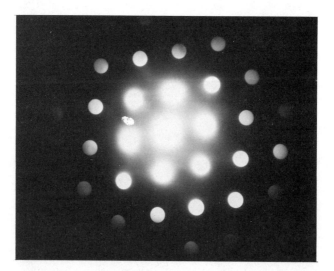

Fig. 10.7 Diffraction pattern from the silicon crystal used to form the images in Fig. 10.4. The condenser lens is fully focused, showing the illumination conditions used when recording the images. The 'beam divergence' can therefore be obtained from a picture such as this (see Section 4.6). The crystal is close to the symmetrical [110] zone-axis orientation (see also Section 5.6).

controls should be altered in such a way that the specimen tends to move toward the centre of the Laue circle. Figure 10.5 shows the diffraction pattern as it appears for a specimen close to the zone-axis orientation, while Fig. 10.7 shows the focused zone-axis diffraction pattern. There is little point in setting the orientation more accurately than this for ductile materials, since local bending takes the specimen through a range of orientations.

10.5 Focusing techniques

Focusing techniques differ according to the diffraction conditions used and the image interpretation required. The wobbler focusing aid cannot be used at high resolution, nor does the disappearance of the Fresnel fringe coincide, either with Gaussian focus or a useful reference focus (Cockayne 1970). The complexity of these fringes for a phase edge, for which fringes occur both inside and outside the specimen edge, makes them of little use at high resolution (see Section 3.1).

1. For dark-field images recorded using tilted illumination and a central aperture (Fig. 6.12), the optimum focus is best judged by eye, taking into account all the precautions mentioned in Section 10.2. Only by close optical examination of a lot of micrographs will the microscopist learn to recognize a 'sharp' image, free from drift, astigmatism, and inaccurate focusing. At a magnification of 300 000 using a ×10 binocular, the eye can resolve detail at the resolution limit of the microscope, but the contrast is generally too low to make this useful and one usually focuses on detail about the size of a small molecule. This will ensure that the transfer function is 'flat' to a resolution of, say, 0.6 nm (Section 6.5). Obtaining point-detail dark-field images on a conventional instrument showing structure much finer than this is largely a matter of trial and error at the present stage of development of CTEM. Nevertheless, the change in image appearance with defocus in dark field is far less dramatic than in bright field, so that focusing may not be as critical. In practice it does appear possible to correct for the image distortion introduced by the asymmetric diffraction conditions by using the stigmator (see Fig. 6.8 and Section 6.5). A dark-field reference focus condition, based on the appearance of small holes, has been mentioned in Section 6.5. Bear in mind that detail smaller than the viewing phosphor grain size cannot be seen. A typical phosphor grain size is 15 μm; however, this varies from instrument to instrument and is easily measured using an optical microscope and a graticule. The resolution of a phosphor screen is limited by multiple scattering of the incident beam, and lies in the range 40–100 μm.

2. For bright-field phase-contrast images of small molecules on ultra-thin carbon supports (see Section 6.2) where the objective aperture has either been removed or chosen according to eqn (6.16b), the optimum focus can be obtained by counting focus steps in the under-focus direction (weaker lens) from a particular reference focus until the value of eqn (6.16a) is obtained. Several images are recorded in the neighbourhood of this focus. The focal steps are calibrated as described in Section 8.1. The reference focus is the focus condition at which a thin amorphous carbon film shows a pronounced contrast minimum when viewed at high magnification. This arises in the following way. If the transfer function (eqn (6.11)) is drawn out for increasing under-focus (Δf negative), it will be seen that the band of well-transmitted spatial frequencies (Fourier components which suffer a 90° phase shift) becomes narrower and moves beyond the range of object detail which can easily be seen through the binocular. Thus the contrast decreases initially with increasing under-focus from the Gaussian focus ($\Delta f = 0$). At the minimum-contrast focus a second loop develops in the intermediate spatial frequency range ($d = 0.6$ nm for $C_s = 1.4$ mm) which increases the image contrast as it grows into the form of Fig. 6.1. The minimum contrast occurs at about 30 nm under-focus for $C_s = 1.4$ mm and can be found for other values of C_s by drawing out a series of transfer functions. Alternatively, it can be taken approximately equal to the optimum dark-field focus given by eqn (6.26), since this makes $\chi(u) \approx 0$ to the highest resolution possible. (Note that the PCD approximation would give minimum contrast for $\Delta f = 0$ from eqn (5.26).)

Subject to the limit set by radiation damage, several images are usually recorded close to the required focus, which can be chosen to suit either the projected charge density approximation or the projected potential image interpretation, depending on the specimen thickness and the resolution required (see Sections 3.4 and 5.3).

All these focusing techniques may result in severe radiation damage. For many biological specimens one is focusing on the 'ashes' of the destroyed specimen. There is nevertheless some evidence that the damaged remnants of a large molecule will preserve the overall shape of the molecule to some resolution in projection. The possibility of focusing on a region of specimen adjacent to that subsequently used for recording the image has been mentioned in Section 6.9. The newer image-intensifier devices have been found to assist greatly in focusing radiation-sensitive materials (see Section 11.3).

3. The focusing technique used when recording many-beam lattice images of large unit-cell crystals (structure images) has been outlined in Section 10.3. In practice the microscopist will adjust the focus control to produce the image of highest contrast. The selection of the Scherzer (optimum) focus can then be confirmed using an optical diffractogram (Section 8.7). Image calculations have, in most cases, confirmed that, in thin regions of large unit-cell crystals (those whose unit-cell dimensions exceed the point-resolution of the microscope), these high-contrast images are those which may be interpreted straightforwardly in terms of the specimen's structure. For metals and semiconductors, however, where the unit-cell dimensions are frequently smaller than the microscope point-resolution (eqn (6.17)), the lattice images of highest contrast will occur at the focus setting given by eqn (5.76) which is much larger than the Scherzer optimum setting (eqn (6.16)). Optimum-focus images for the analysis of defects in these crystals can be obtained by counting calibrated focus steps from the minimum-contrast condition and confirming the choice of focus using an optical diffractogram. Regions of perfect crystal in these images, however, may show rather poor (or non-existent) contrast; nevertheless, any defects present will be imaged faithfully to the point-resolution limit of the instrument. Advances in the structural analysis of defects in these small-unit-cell crystals clearly depends on either the reduction of the point-resolution limit or advances in image processing techniques (see Appendix 3).

In a well-designed microscope, all images of interest should fall within the span of a single focus-control knob. The focal increments should be comparable with the focusing stability limit set by high-voltage instabilities and lens-current instabilities (see Section 8.1). To achieve both these conditions on some instruments it may be necessary to consult with the manufacturing company's engineers about modifications to the focus-control assembly. At the present state of development, the stability limit usually suggests a smallest focus increment of about 4 nm. To cover a reasonable range of images one therefore requires perhaps 40 steps on the finest control. The range of focus settings over which images of good contrast can be seen depends on the size of the illuminating aperture used (and so, indirectly, on the brightness of the electron source) as discussed in Section 5.2.

It sometimes happens that the highest-contrast image occurs near the end of the range of the finest focus control, and that images are required at focus settings on either side of this optimum focus. On some instruments it will be found possible to move the position of the optimum focus image nearer to the middle of the finest focus control range by switching one of the coarse focus controls to a lower setting and then returning it to its original setting. Alternatively, try using a different combination of coarse focus controls to obtain approximately the same focus current.

10.6 Substrate films

In this section some of the methods which have been used to obtain an 'invisible' specimen support are briefly reviewed. The discussion is mainly concerned with attempts to image small molecules.

For high-resolution work, the thinnest possible substrate is required. Ideally the substrate used to support a molecule of interest should be invisible. No truly satisfactory substrate has yet been found and all substrates introduce unwanted contrast effects of some kind. In dark-field experiments it is the elastic and inelastic scattering from the substrate which limits the image contrast (see Section 6.6), since it provides image background. Substrate thickness fluctuations on an atomic scale, and Fresnel 'noise' are an even more important problem when using a thin amorphous substrate. Fresnel diffraction effects can be minimized by reducing the coherence of the illumination, which unfortunately also reduces any pure phase contrast from the atomic structure of interest. The possibility of obtaining high-resolution images under conditions where the illumination semi-angle θ_c equals the semi-angle subtended by the objective aperture has been investigated by Japanese workers (Nagata, Matsuda, Komoda, and Hama 1976). Their work shows a clear reduction in background substrate noise as the illumination angle is increased; however, the contrast of heavy atoms imaged at high resolution has yet to be determined by this method.

Coarser thickness fluctuations in the substrate can be eliminated by photographic processing (Whiting and Ottensmeyer 1972). A third unwanted contrast effect complicating the interpretation of small molecule images is the presence of unwanted atomic species, arising, say, from impure chemicals used in specimen preparation.

Amorphous carbon films a few nanometres thick can be prepared by indirect evaporation. Pure carbon is evaporated onto freshly cleaved mica or clean glass with a glass slide interposed between the carbon source and the mica as described in Section 1.3. Evaporation techniques are described in detail in Holland (1956) and Chopra (1969). Films can be floated off onto the surface of distilled water with care. Since a film of the correct thickness will be invisible in light, a minute drop of ink can be added to the mica after evaporation which will enable the film to be seen on the water surface. The films are picked up onto holey carbon films drawn up from below the surface of the water and allowed to dry. On examination in the microscope, the holes in the carbon film should be covered with an ultra-thin layer of amorphous carbon.

The thickness of substrate films is difficult to measure, but a vibrating-crystal film-thickness monitor will give an estimate of the thickness of evaporated films. The important sources of error with this method are differences in temperature between the reference crystal and the evaporation crystal, and the assumed density of amorphous carbon. A figure of 1850 kg m^{-3} is commonly used, though this is likely to be in error for the thinnest films. By evaporating a large enough area to weigh directly, an estimate of the thickness can also be obtained. A third method of thickness determination uses electron energy-loss spectra (Egerton 1976). Pendellösung fringes have also been used to estimate the thickness of crystalline supports (Spence and Spargo 1970).

Thin flakes of graphite and magnesium oxide have also been used to support small molecules (Hashimoto, Kumao, Hino, Endoh, Yotsumoto, and Ono 1973; Moodie and Warble 1976). The thinnest flakes of graphite are obtained by crushing the flakes under alcohol in a small mortar, followed by ultrasonic agitation. The flakes are picked up onto a holey carbon film which should provide many single thin flakes lying across holes in the carbon. In this technique the objective aperture is placed centrally about the optic axis and the illumination is tilted so that the aperture excludes all substrate Bragg reflections (Fig. 6.8). The objective aperture used must be matched in size to the diffraction pattern for a particular objective-lens focal length. There is usually enough variation in the sizes of commercial apertures to allow this. Scattering occurs into the aperture from any non-periodic scattering such as surface heavy atoms or contaminant. The effectiveness of this technique can probably be improved by working at ultra-high vacuum, since the advantage of a periodic substrate is lost if it contains, say, 1 nm thickness of contaminant on top and bottom surfaces.

Magnesium oxide crystals are easily prepared by passing a holey carbon film through the smoke of a burning magnesium ribbon. It has been suggested that foreign molecules attach less easily to clean surfaces (Uyeda 1974) and the author's experience with graphite substrates has been that heavy atoms and molecules added to these adhere in rows along surface atomic steps on the graphite. This behaviour is also seen in Iijima's work (Fig. 6.4). Groups of heavy-atom molecules can be seen directly through the binocular in this way at high magnification and their presence can be verified by examining graphite flakes untreated with the heavy-atom compound (see Section 6.2). Rhenium trichloride, which is soluble in the alcohol used for crushing the graphite flakes, is a useful heavy-atom test specimen. Reproducible results from studies of adatoms on surfaces requires the sophisticated techniques of ultra-high vacuum electron microscopy (see Section 12.8).

Other substrates which have been tried with varying degrees of success include BeO (Mihama and Tanaka 1976), Al_2O_3 (Formanek, Muller, Hahn, and Koller 1971), beryllium (Komoda, Nishida, and Kimoto 1965), boron (Dorignac, Maclachlan and Jouffrey 1979), and the silicon Pendellösung minimum (Spence 1975). Attempts have also been made to image unsupported molecules across small holes in carbon films (Tanaka, Higashi-Fujime, and Uyeda 1975). The Chicago group have been experimenting with the growth of graphite films using carbon monoxide passed over heated nickel films (Mlinarsky 1977). So far the thinnest and most successful substrates used have been those prepared by *in situ* electron beam thinning (see Section 6.2).

Techniques for manufacturing holey carbon films are described in Kay (1965) and Dowell (1970). The difficulty with all holey carbon film preparation techniques is reproducibility. The Dowell technique, while more elaborate, appears to give more consistent results. A micro-syringe is useful to measure the quantity of paraffin used which, together with the vigor with which the weighing bottle is shaken, determines the size of the holes. Holes about 20 nm in diameter are often obtained, but the results tend to vary from batch to batch. Holey carbon films are also available commercially.

Lattice images can be obtained from crystals supported by either continuous or holey carbon films, though the contrast is higher in the latter case if a crystal is found extending beyond the edge of the substrate.

10.7 Photographic techniques and micrograph examination

Clear accounts of the response of photographic emulsions to electrons are given in Valentine (1966) and Agar *et al.* (1974). This section contains brief notes on some aspects of photography relevant to high-resolution work.

Briefly, the optical density D of an electron micrograph is proportional to electron exposure E up to an optical density of between 1 and 2. This contrasts with the situation in light photography where the response is more closely logarithmic. For electrons, then,

$$D = kE$$

where the optical density D is defined as

$$D = \log_{10}(I_0/I)$$

with I_0/I the ratio of incident to transmitted light intensity for the developed micrograph. Thus a micrograph of density 2 transmits 1 per cent of the light incident on it and so appears dense black in transmission. It is the reflected light which is important when viewing prints. The electron speed k is listed in Agar *et al.* (1974) for most modern emulsions. With E in electrons per square micrometre, the speed k has units electron^{-1} μm^2. The reciprocal of the speed is the emulsion sensitivity and gives the number of electrons per square micrometre necessary to expose an emulsion to unit density. The sensitivity is typically about unity.

For high-resolution dark-field images recorded at high magnification using a pointed filament or LaB_6 source, a fast film with $k = 0.5$ at 100 kV is required with high detective quantum efficiency (DQE). DQE has been discussed in Section 6.8, and suitable emulsions are given in Section 10.8. For minimum exposure methods or other techniques for high resolution which use moderate or low magnification and trial-and-error focusing, it may be important to know the spatial resolution of the emulsion to ensure that this does not limit the image resolution. The resolution r in line pairs per millimetre is *approximately* $r = \frac{1}{2}d$, where d is the width of the emulsion point-spread function. This width is approximately equal to the electron penetration depth in the emulsion, and is typically between 5 and 10 μm for 100 kV electrons. It is thus commonly supposed to be independent of the film grain size (which increases with film speed) for all but the fastest speeds, and sets a limit to the maximum useful optical enlargement of about 30 times. In practice, for short electron exposures, statistical electron noise may set a lower limit. Recent work, however, indicates that a substantial improvement in emulsion resolution *for electrons* can be obtained by using

NTB 3   NTB 2   SO 163   Emulsion
                                              exposure

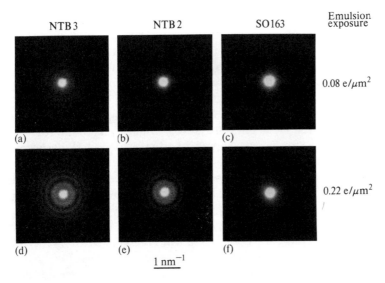

0.08 e/μm^2

(a)   (b)   (c)

0.22 e/μm^2

(d)   (e)   (f)

1 nm^{-1}

Fig. 10.8 Diffractograms obtained from low-exposure electron images recorded on three different emulsions but otherwise identical conditions. Two electron exposure levels are shown. All images were recorded at an electron–optical magnification of 50 000 giving electron exposures at the specimen of 200 electrons nm^{-2} (upper series) and 550 electrons nm^{-2} (lower series). The limiting spatial frequency observed in these diffractograms is 1/65 μm^{-1} (upper series) and 1/40 μm^{-1} (lower series), corresponding to 1.3 nm and 0.8 nm detail at the specimen, respectively. The differing performance of the emulsion is thought to be due to differences in their electron modulation transfer functions and possibly to the effects of additive noise in the emulsions. (Diffractograms courtesy of Dr W. Chiu.)

nuclear-track emulsions for minimum exposure work, as shown in Fig. 10.8 (Chiu and Glaeser 1978).

At the magnifications commonly used for high-resolution images in CTEM (300 000–700 000), the recorded image will require further optical enlargement of between 15 and 30 times if 1 nm detail is to occupy approximately 1 cm on the final print. A search for promising areas on the micrograph is most easily undertaken using an optical microscope fitted with zoom magnification control. Alternatively micrographs recorded on film can be examined on a microfilm reader which provides a very large field of view and about the right magnification. A preliminary examination at low magnification, say ×10, will quickly reveal those micrographs which must be discarded because of drift or astigmatism.

Some special precautions are needed when searching for images of isolated small molecules. It is difficult to avoid scratching the lower side of the plate when operating the optical microscope specimen stage. These scratches on the lower face of the glass will appear as out-of-focus detail which can be misleading. Always refocus on the lower side of the micrograph once a promising area is found to check for false detail. A calibrated eye-piece graticule is essential to ensure that one is concentrating on detail in the correct size-range. A recording of lattice fringes at the same electron-optical magnification will be required to calibrate the graticule (see Section 8.3). An optical magnification must be chosen at which the film grain itself is not distracting and it is generally better to err on the side of too low magnification initially, since a large field of view is required to gain an overall impression of the average substrate noise, against which the particular molecular shape is sought.

A Polaroid print taken from the optical microscope will show regions of high specimen potential as dark areas on a bright background for dark-field phase-contrast electron images and as bright areas on a dark background for bright-field electron images. The process of black and white reversal in making prints is known as phase reversal. If the preferred phase for the final print of small non-periodic features is black detail on a white background, it will therefore be necessary to use a Polaroid negative film.

No particular difficulty should be experienced in enlarging micrographs of bright-field lattice images where an optical enlargement of twenty times is often sufficient. A full discussion is given in Agar *et al.* (1974). Enlargements of bright- and dark-field images of small molecules at atomic resolution require slightly more elaborate precautions. For high-contrast a dense micrograph is required with correspondingly low transmission, since the contrast is proportional to the optical density (Valentine 1966). At an enlargement of thirty times, a 200 W light source, possibly forced-air cooled, will then be needed to obtain sufficient image intensity to allow the enlarger to be focused accurately. The quartz–iodine projection lamps are ideal for this purpose and their small source size further increases the contrast of the print by about one grade. The position of the bulb must be adjustable to obtain uniform image illumination and the enlarger should be used with its optic axis horizontal. A screen is placed about 2 m away, to which the printing paper is fixed. Lattice fringes are again needed for magnification calibration.

It is sometimes possible for light from the electron filament to reach the electron emulsion. On very long exposures the result is a blurred over-exposed patch which cannot easily be avoided but is rarely troublesome. Some practical considerations when handling plates for high-resolution work are indicated below.

1. When loading plates into the microscope cassette, each plate should be tapped on its edge against the bench to dislodge small glass particles. The perfection of the emulsion is important it it is to be examined subsequently at an optical magnification of thirty times or more.

2. Use a water rinse between developer and fixer and fix for twice the time taken for the emulsion to clear. Wash for a full 30 minutes in running water. Drying marks can be a serious problem and are avoided by giving the plates a final rinse for about 10 minutes in a large distilled water bath containing a magnetic stirrer and a fewdrops of photographic wetting agent. Dry the plates in a dust-free cabinet at about 35 °C. Do not attempt to speed up the drying process in any way. The safest storage for plates is their original boxes—molecular detail is easily lost if plates are stored in paper or plastic bags. They should not be stored back to back.

3. Drying marks remaining on the back of a glass plate can be removed by polishing the glass side of the plate lightly with television antistatic cream. The emulsion side should not be touched, but a camel-hair brush may be used to remove dust.

4. A fine corrugation or reticulation pattern on a scale not much larger than the film grain is sometimes seen. This is caused by too great a temperature difference between the photographic solutions and the water used for washing. Tap water is generally well below room temperature. Ideally the wash water should be at the same temperature as the chemicals used.

5. Over-exposed films or plates can frequently be salvaged by using Farmer's reducer (available from photographic suppliers), which slowly and uniformly reduces the optical density of a developed micrograph. This reduction in optical density results in an unavoidable loss of information since, for electron exposures, the contrast and signal-to-noise ratio both increase with exposure (Valentine 1966).

10.8 Emulsions for high-resolution CTEM

The photographic emulsions presently available for high-resolution electron microscopy are discussed below.

Ilford have ceased production of all glass plates for electron microscopy. Their EM4 series was well suited to both routine bright-field microscopy and long-exposure dark-field work, with a speed $k = 0.6$. Ilford's replacement film SP332 will resemble the slower EM5 emulsion which, while satisfactory for bright-field recordings, is too slow for dark-field work. Ilford are looking into the possibility of producing a faster film for electron microscopy.

Kodak have no plans to end production of their Electron Image plates. These give a maximum speed of $k = 2$ if developed with D-19 (full strength) for 12 minutes at 20 °C with nitrogen-burst agitation. The speed can also be reduced to as low as $k = 0.2$ using full strength D-19 plus 1 gram per litre Kodak anti-fog (powder) and 2 minutes development. Kodak's 4489 film is rather too slow for dark-field work if developed as recommended, but individual workers have reported success in speeding this film up by extended development time. A new Kodak product, the SO-163 film gives a similar speed range to their Electron Image plates. The maximum speed for this film is $k = 2.7$ when developed with D-19 (full strength) for 12 minutes at 20 °C with nitrogen-burst agitation. Since the speed can also be reduced to $k = 0.2$ this is a useful general purpose emulsion. An intermediate speed for bright-field work is obtained with the SO-163 film using D-19 diluted one part developer to two parts water and 4 minutes development time (agitation, 20 °C).

For those inorganic specimens for which radiation damage is an unimportant consideration, the choice of gun bias, first condenser excitation, and illuminating aperture size will be fixed by considerations of chromatic aberration and coherence (see Chapters 4 and 5). Then the object current density j_0 is fixed for a particular electron source at focused illumination. The lowest magnification which will allow accurate focusing should then be selected. Next, the maximum exposure time must be determined as limited by specimen movement and the microscope site. A film speed is finally selected which will give an optical density of about 2, through choice of emulsion and development time, under these conditions of magnification, exposure time, and illumination intensity. This procedure ensures that the maximum number of electrons are used to form the image, thereby minimizing the electron noise granularity. An exposure of 30 seconds showing 0.5 nm detail should be possible on a well-sited instrument in thermal equilibrium. The granularity of bright-field images can be reduced by selecting a slow film and using the maximum exposure time.

In summary, the Kodak type SO-163 emulsion is probably the most useful emulsion currently available for conventional bright-field high-resolution work at 100 kV. Emulsions with very similar properties to those of the Kodak products are also available form the Fuji Photo Film Company, Ltd. of Tokyo.

The development of new emulsions for minimum-exposure biological work continues to be an active field of current research. As shown in Fig. 10.8, recent work indicates that the Kodak NTB2 nuclear-track emulsion has a greatly improved ability to record high spatial frequencies at low exposures owing to its dense, fine grain, high speed and low fog level. It thus appears to be the best currently available emulsion for recording images of radiation-sensitive materials (Chiu and Glaeser 1978). It does, however, require specialized preparation techniques. Another Kodak X-ray film. 'Industrex AA' has been found to give equally good results at greatly reduced cost.

10.9   Ancillary instrumentation for high-resolution CTEM work

In addition to the microscope itself, various pieces of ancillary equipment have been referred to throughout this book. Considerations regarding these are collected together in this section for reference.

1. A mass spectrometer or partial-pressure gauge is an extremely useful adjunct to an electron microscope. The downtime spent tracing leaks which occur with increasing frequency as the machine ages can be greatly reduced by using the mass spectrometer as a leak detector, tuned to helium, argon, or a proprietry leak-detection gas. In addition to also functioning as an accurate vacuum gauge, a mass spectrometer gives a full analysis of contamination products in the microscope (Echlin 1975). An inexpensive unit will soon repay its cost through rapid vacuum-fault diagnosis. Some units contain magnets, others operate on the principle of the radiofrequency quadrupole. If magnets are present, these must be carefully positioned in order not to disturb the electron image (see Section 9.1).

2. A supply of bottled dry nitrogen gas is useful for high-resolution work. The microscope should be filled with dry-nitrogen gas whenever internal maintenance is necessary, to reduce the amount of water vapour allowed into the column. Water vapour (present in air) is usually the largest gas component in a vacuum system and is an important source of specimen etching. The pump downtime for a microscope filled with dry nitrogen is also less because water vapour is pumped rather slowly.

3. For magnification calibration on an instrument used at various objective-lens focal lengths, a digital objective-lens current meter is useful (see Section 8.4).

4. A direct-reading image current density electrometer is useful both as an accurate exposure meter and for quantitative experiments. The dynamic range of these instruments greatly exceeds that of film, which makes them useful for obtaining the ratio of the intensity of a particular Bragg reflection to that of the incident beam. Similarly, only by taking two accurately timed exposures (one of which must be very short) can the ratio of the background intensity from a dark-field substrate to that of the incident beam be measured photographically. This is easily done using an electrometer. These instruments also simplify gun-tilt alignment (the focused spot is adjusted for maximum intensity) and are invaluable when setting up a pointed filament (see Section 7.5) or for making brightness measurements. On many instruments an electrometer is fitted by the manufacturers; a simple design is described by Hills and Garner (1973) for instruments not fitted with an electrometer. For absolute measurements of electron flux a correction may be needed for electron backscattering, particularly at lower voltages. On most instruments the phosphor itself is used to collect electrons for current measurement, on others a separate screen must be fitted, which should be removable and can be fitted, for example, in place of the movable beam stop.

5. When using pointed filaments, a meter to indicate the filament heating current is useful to enable all filaments to be run at the same temperature and so give the longest life.

6. In view of the importance of thermal stability for long-exposure dark-field recordings, it is an advantage to arrange for a liquid-nitrogen pump to keep the anti-contamination reservoir (and diffusion pump, if an oil-filled pump is used) continuously filled.

7. A rubber bulb fitted with a nozzle is always useful for blowing away dust from a specimen, or the traces of cleaning agents after cleaning the microscope gun chamber. Never blow into the internal parts of a microscope—the human breath contains a great deal of water vapour.

8. The use of an oscilloscope for high-voltage stability testing has been mentioned in Section 9.2. The microscope manufacturers can give details of the input sensitivity required of such an instrument.

9. On some microscopes a simple electrical continuity check can be installed, connected between the interior of the cold-finger liquid-nitrogen reservoir and the microscope frame. Since the specimen is also electrically connected to the microscope frame, an open circuit between frame and cold-finger reservoir indicates that the specimen is not in contact with the anti-contamination cold-fingers above and below the specimen.

10.10   A checklist for high-resolution work

The following checks are listed for rapid reference. They should be performed before commencing any high-resolution work.

1. LENS COOLING-WATER THERMOREGULATOR. DUTY CYCLE CORRECT?
2. NEW OR RECENTLY INSTALLED SOURCES OF VIBRATION?
3. METER CHECK-POINTS ON MICROSCOPE ELECTRONICS.
4. READ VACUUM
5. PLATES/FILM. OUTGASSED? LOADED?
6. LIQUID NITROGEN. COLD-FINGER IN THERMAL EQUILIBRIUM?
7. OSCILLOSCOPE HIGH-VOLTAGE STABILITY CHECK. GUN CLEANING?
8. EXAMINE FILAMENT IMAGE—STRAY MAGNETIC FIELDS?
9. ALIGN ILLUMINATION SYSTEM
10. ALIGN IMAGING LENSES
11. SPECIMEN–COLD-FINGER CONTACT?

References

Agar, A. W., Alderson, R. H., and Chescoe, D. (1974). Principles and practice of electron microscope operation. In *Practical methods in electron microscopy* (ed. A. M. Glauert) Vol. 2. North-Holland, Amsterdam.

Alderson, R. H. (1975). The design of the electron microscope laboratory. In *Practical methods in electron microscopy* (ed. A. M. Glauert). North-Holland, Amsterdam.

Chiu, W. and Glaeser, R. M. (1978). Electron exposure-dependent contrast transfer. In *Proc. 9th Int. Congr. Electron Microsc.*, Toronto, Vol. 1, p. 94.

Chopra, K. L. (1969). *Thin film phenomena*. McGraw-Hill, New York.

Cockayne, D. J. H. (1970). Electron microscope images of defects in crystal lattices. D.Phil. Thesis, University of Oxford.

Dorignac, D., Maclachlan, M. E. C., and Jouffrey, B. (1979). Low noise boron supports for high resolution electron microscopy. *Ultramicroscopy* **4**, 85.

Dowell, W. C. T. (1970). The rapid production of holey carbon formvar supporting films. In *Proc. 7th Int. Congr. Electron. Microsc.*, Grenoble, p. 321.

Echlin, P. (1975). Contamination in the scanning microscope. In *Scanning electron microscopy/1975* p. 679. I. I. Research Institute, Chicago.

Egerton, R. F. (1976). Foil thickness determination by electron spectroscopy, *J. Phys. D* **9**, 659.

Formanek, H., Muller, M., Hahn, M. H., and Koller, T. (1971). Visualization of single heavy atoms with the electron microscope. *Naturwissenschften* **58**, 339.

Hashimoto, H., Kumao, A., Hino, K., Endoh, H., Yotsumoto, H., and Ono, A. (1973).

Visualization of single atoms in molecules and crystals by dark field electron microscopy. *J. Electron Miscrosc.* **22**, 123.

Hills, G. J. and Garner, R. T. (1973). A design for a simple electronic exposure meter for use with an electron microscope. *J. Microsc.* **98**, 105.

Holland, L. (1956). *Vacuum deposition of thin films.* Chapman and Hall, London.

Kay, D. H. (1965). *Techniques for electron microscopy.* Blackwell, Oxford.

Komoda, T., Nishida, I., and Kimoto, K. (1965). Beryllium single crystal flakes as substrates for high resolution electron microscopy. *Japan J. Appl Phys.* **8**, 1164.

Mihama, K. and Tanaka, N. (1976). Beryllium oxide specimen supporting films for high resolution electron microscopy. *J. Electron Microsc.* **25**, 65.

Mlinarsky, L. (1977). Experiments with thin single crystal graphite films. In *Proc. 35th Ann. EMSA Meet.* p. 160. Claitor, Baton Rouge.

Moodie, A. F. and Warble, C. E. (1967). The observation of step growth in magnesium oxide by direct transmission electron microscopy. *Phil. Mag.* **16**, 891.

Nagata, F., Matsuda, T., Komoda, T., and Hama, K. (1976). High resolution observation of biological specimens by an incoherent illumination method. *J. Electron Microc.* **25**, 237.

Riecke, W. D. (1975). Instrument operation for microscopy and microdiffraction. In *Electron microscopy in materials science* (eds. U. Valdre and E. Rueld) Part I. Commission of the European Communities, Directorate General 'Scientific and Technical Information and Information Management', Luxembourg.

Spence, J. C. H. (1975). Single atom contrast. In *Proc. EMAG, 1975,* p. 257. Academic Press, New York.

Spence, J. C. H. and Spargo, A. E. C. (1970). Plasmon mean free path in aluminum. *Phys. Lett.* **33A**, 116.

Spence, J. C. H., O'Keefe, M., and Kolar, H. (1977). High resolution image interpretation in crystalline germanium. *Optik* **49**, 307.

Tanaka, M., Higashi-Fujime, S., and Uyeda, R. (1975). Electron microscope images of mercury atoms bound to DNA filament. *Ultramicroscopy* **1**, 7.

Uyeda, R. (1974). Innovations in specimen supporting media for high resolution. In *Proc. 8th Int. Congr. Electron Microsc.,* Canberra, p. 246.

Valentine, R. C. (1966). Response of photographic emulsions to electrons. In *Advances in optical and electron microscopy* (eds. R. Barer and V. E. Cosslett) Vol. 1. Academic Press, New York.

Whiting, R. F. and Ottensmeyer, F. P. (1972). Heavy atoms in model compounds and nucleic acids imaged by dark field transmission electron microscopy. *J. Mol. Biol.* **67**, 173.

11

ASSOCIATED TECHNIQUES

For each new detector that is fitted to an electron microscope, a new subdiscipline of electron microscopy is created. This chapter provides a very brief survey of some recently developed techniques which are compatible with high-resolution electron microscopy (HREM), and which provide complementary information. For while HREM is a powerful method for the study of the defect structure of crystals, it is subject to several important limitations. The most important of these are as follows. (1) The method provides little information on the atomic number of the elements present. (2) HREM images reveal only a projection of the crystal structure, giving no indication of atomic displacements in the beam direction (but see Goodman and Warble (1987)). (3) Film recording does not normally allow a record of dynamic events to be obtained in real time. (4) Electronic structure information is often desirable for the defects seen in HREM images. (5) Elastic relaxation effects in the very thin specimens used for HREM work may render the results unrepresentative of bulk material.

We therefore provide in this chapter a brief bibliography of references to the various techniques which address these problems, namely energy-dispersive X-ray spectroscopy (EDS), electron energy-loss spectroscopy (EELS), microdiffraction or convergent-beam electron diffraction (CBED), cathodoluminescence (CL) in STEM, and video recording of HREM images.

These new subdisciplines are closely related to certain existing well-established fields. Thus, for example, the theory of cathodoluminescence closely parallels that of photoluminescence (PL), while energy-dispersive X-ray spectroscopy has close similarities with X-ray fluorescence spectroscopy. Much electron energy-loss spectroscopy has a good deal in common with X-ray absorption spectroscopy, both with X-ray absorption near-edge structure (XANES) work and with the extended X-ray absorption fine structure (EXAFS). Both of these in turn have much in common with photoelectron spectroscopy, using either incident X-rays (XPS, or X-ray photoelectron spectroscopy) or ultraviolet light (UPS, or ultraviolet photo-electron spectroscopy). Thus, the student or research worker interested in any of the growing number of new spectroscopies found on modern electron microscopes would be well advised to consult the literature of the parent subject, always bearing in mind any differences which may be important. For example, the books by Cardona and Ley (1978), Teo and Joy (1981), Joy, Romig, and Goldstein (1986), and Egerton (1986) contain much relevant background material for this chapter.

11.1 Energy-dispersive X-ray spectroscopy and 'alchemi'

The modern energy-dispersive X-ray spectrometer, when fitted to a transmission electron microscope, is capable of detecting the characteristic X-rays of all the elements present in the sample which are of greater atomic number than sodium. Lighter elements may be detected using special 'windowless' detectors. The energy resolution of the X-ray spectrum produced is typically only about 150 eV; however, the overall efficiency is high, owing to the large collection solid-angle and 'parallel' processing method used. Incoming X-ray photons produce pulses in a lithium-doped silicon detector, the heights of which are proportional to the X-ray energy. These pulses are sorted in a multichannel analyser to form a histogram or X-ray spectrum. Unlike a grating spectrometer, which disregards photons whose energy falls outside the current slit setting, every photon arriving at the EDS detector is counted, regardless of its energy, so that this serial device has many of the advantages of a parallel detection system. The performance is limited, however, by the dead-time of the detector and by the maximum count rate which can be processed. This limit may be reached in a channel other than that of interest, thereby limiting the count rate in the channel of interest. Excitations in the silicon detector itself may also give rise to spurious peaks. An electron-probe-forming lens is required for high spatial resolution. This requirement, together with the side-entry stage needed for EDS (and possible mechanical instabilities produced by the X-ray detector and its liquid-nitrogen dewar) may conflict with the requirements for HREM work. Through the use of the symmetrical condenser–objective lens, together with a side-entry stage, however, some impressive design compromises have been achieved.

The reduction of the X-ray background is a primary concern for quantitative EDS work. Background X-rays may originate either in the illumination system, and so appear in a 'hole count' spectrum (in which the electron beam is passed through a hole in the sample), or from post-specimen interactions. These may be due to high-energy backscattered electrons, to fluorescence of the general specimen surroundings, or to transmitted high-energy electrons scattered through large angles. All the evidence suggests that one must think of the pole-piece region as filled by a 'plasma' of kilovolt electrons travelling in all directions.

The spatial resolution of the EDS technique has been the subject of a considerable literature (see, for example, Joy *et al.* (1986) for a review). Many authors have reported the results of detailed incoherent multiple-scattering calculations which follow the progress of inelastically scattered secondary electrons through a thin foil using 'Monte-Carlo' computational techniques. An exact solution to the transport equation describing this process exists, however, for thin foils (Fathers and Rez 1979). A resolution of a few nanometers is commonly achieved; however, under favourable conditions in STEM instruments a monolayer of atoms (lying in the plane of the beam) may be detected. For example, using beam-stabilization techniques and a subnanometer probe size with many hours of recording, the segregation of arsenic atoms to a silicon grain boundary has been observed on a nanometer scale (Grovenor, Batson, Smith, and Wong 1984). As a rough rule of thumb, multiple-scattering calculations and experiment suggest that the spatial resolution b in EDS is given very approximately for thin samples by

$$b = (1760/V)t^{3/2}$$

where b and t, the specimen thickness, are expressed in Ångstroms, and V is the accelerating voltage in electron volts. In practice, however, it is usually the ability to distinguish a certain species in a given matrix which is important in EDS, rather than spatial resolution. A typical limit on the sensitivity of the method is about 0.05 atomic per cent; however, this depends on many experimental parameters, including the shape of the sample (whether particles or bulk material are used) and the atomic number of other elements present.

Windowless detectors and ultra-thin window detectors have recently been developed for the detection of elements as light as boron (see, for example, Williams and Joy (1984) for a review). The determination of the necessary scaling factors for quantitative work is under way in this promising area of research. Current problems include incomplete charge-collection, peak overlap, and the contamination of windows. In windowless systems, icing of the detector must be prevented, and pumps must be used which are capable of pumping water for its removal. The method invites comparison with EELS for the microanalysis of light elements. The EELS spectrometer need not influence HREM performance, unlike the EDS detector. The background in an energy-loss spectrum is much greater than that in an EDS spectrum owing to the unfavourable shape of the EELS edges. Light-element lines in EDS may also be confused by the presence of low-energy, high-order emission lines from heavy elements present in the matrix. With the development of parallel-recording EELS systems (Schuman 1981; Egerton 1984; Krivanek, Ahn, and Keeney 1986) a need has arisen for a careful comparison of the two systems. General reviews of the EDS method can be found in Joy *et al.* (1986).

It is interesting to note that much of the background in EDS spectra from crystalline samples may arise from coherent Bremsstrahlung (CB) (Spence, Titchmarsh, and Long 1985; Korobochko, Kosmach and Mineev 1965). In passing down along a string of atoms of separation L, an electron travelling with velocity v emits monochromatic, plane-polarized X-rays of frequency $\omega = 2\pi v/L$ (in the electron rest-frame), and energy $\hbar\omega$. In the laboratory frame, taking account of the relativistic Doppler effect, the emission energy is

$$\varepsilon = \frac{hc\beta}{L(1 + \beta \sin \theta)} = \frac{12.4\beta}{L(1 + \beta \sin \theta)}$$

where ε is in keV, L is in Ångstroms, $\beta = v/c$, and θ is the take-off angle of the X-ray detector measured from the horizontal plane. This effect has been studied in detail, and a full Bloch-wave quantum-mechanical treatment has been given (Reese, Spence, and Yamamoto 1984). The atom spacings L are further related to crystal reciprocal lattice vectors $\mathbf{g}$ by $L = (\mathbf{g} \cdot \mathbf{n})^{-1}$, where $\mathbf{n}$ is a unit vector in the beam direction. Thus the effect of crystallinity in a sample is to concentrate the otherwise continuous distribution of Bremsstrahlung into a set of discrete peaks, one for each reciprocal lattice vector. The important vectors are those lying approximately antiparallel to the beam direction. Classically, the radiation results from the acceleration of the beam electron ('braking radiation'). The continuous background results from accelerations in all directions. The sharp CB peaks result from accelerations in the beam direction alone, which correspond quantum-mechanically to the umklapp process in which the crystal as a whole takes up momentum in units of $\hbar g$ (Spence and Reese, 1986). Coherent Bremsstrahlung radiation may therefore be thought of loosely as dipole radiation which emerges (non-relativistically) at 90° to the beam direction, since the axis of the dipole lies parallel to the beam and $\mathbf{g}$. Radiation in the forward direction due to transverse acceleration (known as channelling radiation, CR) has also been observed in electron microscopy (Fujimoto and Komaki 1986, personal communication). In the absence of inelastic phonon scattering, these discrete CB and CR peaks (and combinations thereof) would be the only background seen from crystalline samples. It is likely therefore that in the past there have been mistaken identifications of lines due to this mechanism (Vechio and Williams 1986). Figure 11.1 shows a typical CB spectrum obtained by EDS on a TEM instrument from a diamond sample. Since carbon X-rays are too soft to be detected by this system, no characteristic X-rays were expected from this sample. The effect is a useful one, since accelerating voltages may be chosen which position these peaks away from those sought in microanalysis, and therefore minimize the background contribution.

In addition to supplying microanalytical information, the EDS method can be used quantitatively to determine the fraction of substitutional

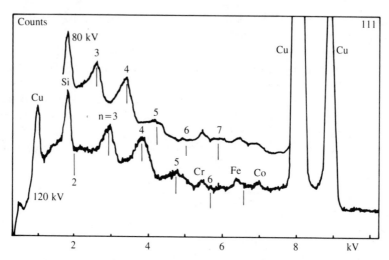

Fig. 11.1 Coherent Bremsstrahlung emission from a thin diamond crystal recorded by EDS on the Philips EM400 transmission electron microscope. Take-off angle $\theta = 20°$, accelerating voltage 80 kV and 120 kV as shown. Each numbered peak n results from a layer n of reciprocal lattice points $\mathbf{g}$ normal to the beam such that $\mathbf{g} \cdot \mathbf{H} = n$, where $H = [111]$ is the beam direction. Owing to the symmetry of the diamond lattice all structure factors in the $n = 2$ and $n = 6$ HOLZ layers are zero. The corresponding CB peaks are therefore absent.

impurity atoms which lie on particular sites in a crystal lattice. If characteristic X-rays from the trace elements of interest in crystalline regions of a TEM sample can be detected using the EDS system, then in most cases it is possible to determine the crystallographic site of the impurity. This method, known as atom location by channelling enhanced microanalysis (or ALCHEMI), involves no adjustable parameters, the fractional occupancies of the substitutional impurities of interest being given in terms of measured X-ray counts alone. Since TEM or STEM is used, the method can be applied to areas as small as a few hundred Ångstroms in diameter, while the detection sensitivity is limited by that of the EDS system to about 0.1 atomic per cent. Elements which are neighbours in the periodic table can normally be readily distinguished. The method uses the dependence on the orientation of the incident electron beam of characteristic X-ray emission, and does not require either the specimen thickness or the precise beam orientation to be known. In conventional microanalysis by EDS, a large illumination aperture is used. To obtain an orientation-dependent effect, a beam divergence smaller than the Bragg angle must be used. No dynamical electron diffraction calculations are required for the interpretation of this quantitative method. The classical problems of cation ordering in spinels, feldspars, and olivine minerals have all been studied by this method. The first application of the method was described by Taftø and Liliental (1982), while a tutorial review of the principles and applications of the technique can be found in Spence and Taftø (1983), which also contains a brief historical summary of the related effects (such as the Borrman effect in X-ray diffraction) on which it is based.

The technique depends on the fact that the characteristic X-ray emission from crystals is in fact modulated by the transverse electron wavefield in the crystal, as shown in Fig. 11.2. This intensity distribution is given by

$$I(r, t_0) = \int_0^{t_0} I(r, t) \, dt \qquad (11.1)$$

where $I(r)$ is given by eqn (5.47) with $\chi(\mathbf{g}, \Delta f) = 0$ and t_0 is now the crystal thickness. Equation (5.47) has been integrated throughout the depth of the crystal to obtain a quantity proportional to the X-ray emission yield, in the approximation of perfect localization (see Cherns, Howie, and Jacobs (1973) for a detailed discussion). Figure 11.3 shows the experimental continuous variation of X-ray emission with incident beam direction for GaAlAs (Christensen and Eades 1986). For ALCHEMI, the characteristic X-ray emission intensities from both the impurity atom and the atoms of the host crystal need be measured for only two or three crystallographic orientations of the collimated incident electron beam. A specimen thickness of about 1000 Å is sought, but the thickness is not critical and the illumination 'beam divergence' should be a fraction of the Bragg angle for a first-order reflection. In many problems it is possible to limit the substitutional sites of the impurity atom to a few likely possibilities. The crystalline specimen is then oriented in the 'systematics' or 'planar channelling'

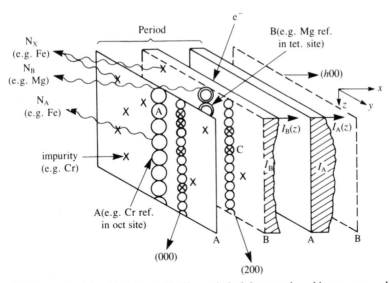

Fig. 11.2 The principle of 'ALCHEMI'. The period of the crystal stacking sequence along the systematics direction [$h00$] is indicated. The wavefield is constant in the y direction, and varies with depth along z as shown. The characteristic X-ray production is proportional to the shaded areas. Note that there is no channelling effect for small thickness. The case of a spinel crystal containing iron impurities at X is shown.

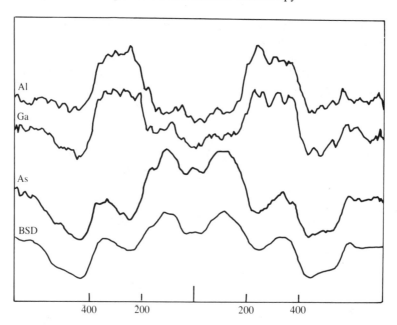

Fig. 11.3 Variation in characteristic X-ray emission as a function of angle for a low-divergence electron beam rocked across the (200) planes of GaAlAs. For the three X-ray signals, equal changes in the height of the curve represent equal percentage changes in the number of counts. The backscattered electron signal is included for reference. (From Christensen and Eades (1986).)

orientation, so that crystal planes with reciprocal lattice vector normal to the beam contain alternating candidate sites for the impurity and alternating species of the host crystal. For example, in the zinc blende structure, the [00h] systematics orientation would contain all the zinc atoms on the A planes of Fig. 11.2, while the B planes would contain only sulphur atoms. By measuring the Zn, S, and impurity X-ray emission for two incident beam directions (both approximately normal to [00h]) it is then possible to determine the fractions of the impurity which lie on the A and B planes respectively. By repeating this experiment for other sets of planes, the crystal site can be determined.

For each of these two incident beam orientations, the fast electron sets up a standing wave in the crystal whose intensity variation has the period of the lattice (e.g., from A to A', as shown in Fig. 11.2). For ideally localized X-ray emission, the characteristic X-ray emission intensity due to the ionization of the crystal atoms is proportional to the height of this standing wave at the atom concerned. The total X-ray emission from, say, the species on the A planes of Fig. 11.2 is then proportional to the thickness-integrated electron intensity I_A on the A planes, shown shaded in the figure. This area changes with changes in the incident beam direction. Because a systematics orientation has been chosen, the electron intensity is constant along the A and B planes. We let C_X be the concentration of impurities on the A planes, and $(1 - C_X)$ be the concentration on the B planes. The X-ray emission from the host atoms in known sites on the A and B planes will be used to provide an independent monitor of the electron intensities I_A and I_B, which also excite X-rays from the impurities. Thus the principle of ALCHEMI is to use the host atoms as reference atoms or 'detectors', to 'measure' the thickness-integrated dynamical electron intensity distribution. We let $N_A^{(1,2)}$, $N_B^{(1,2)}$ and $N_X^{(1,2)}$ be the six X-ray counts from elements A, B and X for a 'channelling' orientation (1) and a 'non-channelling' orientation (2) in which the electron intensities on the A and B planes are equal. Then we have the six relationships

$$N_B^{(1,2)} = K_B I_B^{(1,2)} \tag{11.2}$$

$$N_A^{(1,2)} = K_A I_A^{(1,2)} \tag{11.3}$$

$$N_X^{(1,2)} = K_X C_X I_B^{(1,2)} + K_X (1 - C_X) I_A^{(1,2)} \tag{11.4}$$

where the superscripts refer to X-ray counts obtained in two successive orientations. In addition, for the non-channelling orientation, $I_A^{(2)} = I_B^{(2)}$. Here K_A, K_B, and K_X are constants which take account of differences in fluorescent yield, and other scaling factors. These equations can be solved for C_X in a way which eliminates I_A and I_B. If we define the ratio of counts for two orientations as

$$R(A/X) = \frac{N_A^{(1)}/N_X^{(1)}}{N_A^{(2)}/N_X^{(2)}} \tag{11.5}$$

then

$$C_X = \frac{R(A/X) - 1}{R(A/X)(1 - \beta)}$$

where

$$\beta = \frac{N_B^{(1)} N_A^{(2)}}{N_A^{(1)} N_B^{(2)}}. \tag{11.6}$$

Equation (11.6) can thus be used to find the concentration of species X on the A planes (see Fig. 11.2) in terms of the measured X-ray counts $N_A^{(1,2)}$, $N_B^{(1,2)}$, and $N_X^{(1,2)}$ alone. In practice, orientation (1) is usually chosen as slightly greater than the Bragg angle, and orientation (2) is one which avoids the excitation of Bragg beams.

This technique has been applied to determine the occupancies of Cr, Al, Fe and Mg in samples of chromite spinel (Taftø 1982) and to the occupancies of Fe and trace elements in an Mg–Fe olivine (Taftø and Spence 1982). Here, results in agreement with those obtained by the Mössbauer method were found. A second independent check on ALCHEMI is described in the work of Taftø and Buseck (1982), who compared results for the occupancy of Al on the Tl site in orthoclase feldspar with X-ray diffraction results from the same sample. The occupancies derived by the two methods agreed to within less than 2 per cent. The method has also been applied to uranium atoms in nuclear waste materials (Taftø, Clarke, and Spence 1983). More recently, a generalization of the method to a larger number of sites and orientations has been described, and the effects of variations in accelerating voltage have been studied (Krishnan *et al.* 1986). Work on the calcium site occupancy in barium titanate has also been reported (Chan 1984). In semiconductor applications, a variant of this method which depends on the possibility of comparing 'absolute' X-ray intensities for two different incident beam orientations has been reported for the case of arsenic in silicon (Taftø and Spence 1983). The results of this work are shown in Fig. 11.4. Since the As emission follows the Si emission, it is concluded that both lie in the same crystal planes. By repeating this experiment for different sets of planes, a substitutional site can be localized.

The method is subject to all the limitations of conventional EDS. Thus it cannot easily be applied to elements of lower atomic number than sodium (unless a 'windowless' detector is used), and the sensitivity is restricted. However, it has some important advantages in comparison with techniques such as EXAFS, Rutherford ion backscattering (RBS) and X-ray diffraction; in particular, neighbouring species in the periodic table may be distinguished, and the method may be applied to small areas, in combination with TEM imaging. Thus real crystals (such as meteorite fragments or fine-grained polycrystalline nuclear waste materials) may be analysed, rather than crystals of specially grown synthetic analogues.

A crucial assumption of the method is that the X-ray emission process is highly localized. (The meaning of localisation is discussed in Section 11.2, below). This assumption has been tested by Self and Buseck (1983), Bourdillon, Self and Stobbs (1981), Rossouw and Maslen (1987) and Pennycook (1987). A summary of work on this problem can be found in Spence, Kuwabara and Kim (1988). Briefly, while the literature contains contradictory results, it appears that axial Alchemi may give erroneous results for X-ray energies less than about 2 kV, while the systematics method extends to some undetermined lower energy.

Since crystal planes can be found which contain interstitial sites, the method may in principle be extended to include interstitial site-occupancy determination. In addition, under the column approximation, any local change in diffraction conditions due to crystal bending or thickness changes under the probe will not affect the ALCHEMI results, since these affect both the impurity emission and that from the reference host atoms in a similar way (Spence and Taftø 1983). A final important assumption of the method, however, is that the impurities are uniformly distributed in depth throughout the crystal. Thus, the method fails for a segregated sheet of impurities lying normal to the beam at a particular depth in the crystal.

In some cases, the 'known' atomic species may segregate into distinct columns. Then the much stronger axial channelling effect may be used in a similar way (Otten and Buseck 1987). The crystallographic constraints are,

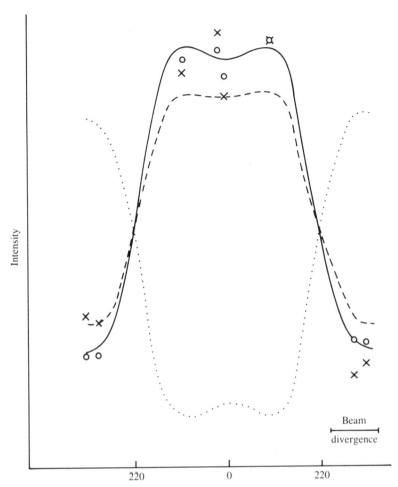

Fig. 11.4 Variation in characteristic X-ray emission from arsenic dopant atoms (crosses) in a silicon crystal with incident 100 kV beam direction, compared with that of the silicon atoms (circles). The beam divergence used is indicated to scale. Continuous line is theory for silicon; dotted line is theory for interstitial arsenic; broken line is theory for 80% substitutional arsenic (remainder incoherently positioned). (From Taftø and Spence (1983).)

however, more severe in this case, and, owing to the finer transverse oscillations of the wavefunction, unwanted localization effects become more important (Buseck, Cowley, and Eyring 1988).

By repeating ALCHEMI experiments for many orientations along the same systematics row it is found that, whereas the simple ratio of peak heights changes greatly between pairs of orientations, all these ratios reduce to the same occupancy factor C_X given by eqn (11.5) (see also Taftø and Spence (1982)). A list of references to applications of ALCHEMI, and measurements and analysis of the effect of temperature reduction on

ALCHEMI can be found in Spence and Graham (1986). The subject has also been reviewed by Krishnan (1988). These papers include references to applications to problems of dopant site location in semiconductors (Taftø and Spence 1983), rare-earth additions to thin-film garnets (Krishnan, Rez, and Thomas 1985), Al and Si in feldspars (Taftø and Buseck 1983), Ca in barium titanate (Chan 1984), Ti and Ta in the superconductor Nb_3Sn (Taftø, Suenaga, and Welch 1984) and Co in the new oxide superconductor $YBa_2Cu_3O_7$ (Shindo, Hiraga, Hirabayashi, Tokiwa, Kikuchi, Syono, Nakatsu, Kobayashi, Muto and Aoyagi (1987) amongst other work.

In studying applications of the ALCHEMI technique, it has been found that, contrary to expectations, the most parallel illumination conditions are obtained on 'twin-lens' instruments when a fully focused spot is used (Christensen and Eades 1986). This arises because of the spiral electron trajectory in the pole-piece, which introduces an inclination proportional to the size of the focused probe. Channelling effects have also been observed to influence characteristic energy-loss peaks (Spence, Krivanek, Taftø, and Disko 1981). Applications of this effect can be found in Taftø, Krivanek, Spence, and Honig (1982), where it is used to distinguish normal and inverse spinels, and in Taftø (1984) (see below). The additional complications introduced by multiple scattering of the beam electron following the energy-loss event make this a less generally useful technique.

11.2 Electron energy loss spectroscopy (EELS)

The extensive literature on this large subject has recently been reviewed by Egerton (1986). Here we briefly review those aspects of EELS most relevant to high-resolution electron microscopy. Many high-resolution electron microscopes are now fitted with either commercial or laboratory-built electron energy-loss spectrometers. These record the number $F(E)$ of transmitted electrons which, in passing through the sample, give up an amount of energy in the range E to $E + \Delta E$ to the various elementary excitation levels of the sample. $F(E)$ is related to the dielectric function of the sample, and contains discrete peaks at the energies of all the important excitations, such as inner-shell atomic ionization edges and plasma excitations.

Image intensifiers and electron energy-loss spectrometers are well removed from the sensitive pole-piece region of an electron microscope and therefore need not degrade the ultimate performance of an atomic-resolution instrument, provided that stray fields and ground-loops are avoided. Thus, it is now possible to obtain both atomic-resolution images and electron energy-loss spectra from the same identifiable submicron region of specimen. The size of the region from which spectra can be obtained will depend on the probe-forming capabilities of the instrument, and, in particular on whether this region can be identified while in the lattice-imaging mode. If mechanical adjustments to the specimen height are required when switching to the probe-forming mode, it may not be possible to locate the same region of specimen in both instrumental modes. These difficulties are avoided in dedicated STEM instruments, in which energy-loss spectra may be obtained from a sub-nanometer region at any identified image pixel at which the electron probe has been stopped.

Electron energy-loss spectra have been used chiefly to provide micro-analytical information, and, to a more limited extent, structural information from small sample regions. We first briefly discuss microanalysis by EELS. Since the primary electron (following energy loss) is detected in EELS rather than a secondary emission product, as in EDS or cathodolluminesce (which may result from a de-excitation event some distance from the primary excitation), this method has the great advantage of providing chemical information at the highest possible spatial resolution.

The resolution is limited only by the maximum size of the electron probe within the sample, and by the inelastic 'localization' (see below). Dynamical calculations for coherent probe spreading can be found in Humphreys and Spence (1979). The resolution in EELS does not depend on the range of secondary electrons or X-rays as is the case for EDS, CL and Auger spectroscopy. As examples of 'microanalysis' by EELS in STEM at the very highest spatial resolution, we cite the work of Berger and Pennycook (1982), who have detected nitrogen at a single nitrogen platelet in a diamond crystal; the work of Bourret and Colliex (1982), who have detected oxygen at the cores of dislocations in germanium; and the work of Spence and Lynch (1982), who detected a variation in elemental concentration within a single unit-cell of a barium aluminate crystal. More detailed electronic structure information has been obtained by spatially resolved energy-loss spectroscopy using a sub-nanometer probe from a single GaAs/GaInAs interface misfit dislocation by Batson, Kavanagh, Woodall, and Mayer (1986). Equally high resolution may be obtained with a parallel detection spectrometer in the TEM mode (Krivanek, Ahn and Keeney 1986). This high spatial resolution is the most important advantage of the EELS method over all other microanalytical techniques. Its chief disadvantages include the high background generally present, and the need to remove the thickness-dependent contributions of multiple inelastic scattering from the spectrum if quantitative results are to be expected from any but the very thinnest samples. This may be done by the convenient "logarithmic deconvolution" method (Johnson and Spence 1974), which solves the inversion problem of multiple inelastic scattering exactly. A full FORTRAN computer program listing for this method is given in Egerton (1986). This text also provides a very complete and pedagogically sound review of all the microanalytical (and other) applications of EELS, together with all the necessary background theoretical material.

The theoretical basis of the EELS technique has its historical origins in the general theory of the problem of stopping power for charged particles, which arises in many fields of physics. Perhaps the most important early papers are those of Bethe (1930) and Kainuma (1955). The second of these addresses the very important problem of combined multiple elastic and multiple inelastic scattering, a topic which has since received much attention (see, for example Howie (1963), Maslen and Roussouw (1984), Saldin and Rez (1986), and Craven and Colliex (1977) amongst many others). Since electrons which have been inelastically scattered may also be elastically Bragg-scattered, it is possible to form lattice images in energy-filtering instruments which select for the image only those electrons which have lost a particular amount of energy in passing through the sample (Craven and Colliex 1977). The energy-filtered images may appear almost identical to the elastic lattice image. In images filtered for a particular inner-shell excitation, the 'wrong' atomic species may therefore appear (Spence and Lynch 1982). Since the probability of single inelastic scattering is proportional to (t/λ) for a process with mean free path λ, while that for Bragg scattering (at the Bragg angle) is proportional to t^2/ξ_g^2 (see eqn (5.20)), either process may dominate, depending on the experimental conditions and material.

Since image-forming electron energy-loss spectrometers have recently been developed which may be fitted to HREM instruments for the purposes of 'chemical mapping' at atomic resolution, a comment on the interpretation of these images is appropriate. This will be complicated both by the unavoidable elastic scattering mentioned above, and by the effects of inelastic localization, which sets a weak limit on the spatial resolution of energy-filtered images.

If the inelastic electron scattering is taken to be confined to a cone of semi-angle $\theta_E = \Delta E/2E_0$ (ΔE is the energy loss), then this radiation must originate from a specimen volume whose transverse dimension is $L = \lambda/\theta_E$, where L is known as the inelastic localization, and can be taken as a measure of the size of the region in which the inelastic scattering is coherently generated. It is therefore also the size of the image element which would be formed if all this scattering were passed through a perfect lens and focused. This simple result for L is consistent with the results of more elaborate quantum-mechanical calculations (Craven, Gibson, Howie, and Spalding 1978) and with an argument based on the time-and-energy uncertainty principle (Howie 1979).

This can also be understood by considering the classical impact parameter b. A beam electron travelling at velocity v spends an amount of time $\Delta T = b/v$ in the neighbourhood of the target atom. If we think of the pulse of electromagnetic energy associated with the beam electron's passage evaluated at this atom, then this will contain Fourier components over a range $\Delta\omega = 2\pi/\Delta T$, for which the range of associated quantum-mechanical excitation energies (the beam electron energy loss) is $\Delta E = h\,\Delta\omega = hv/b$. Using $v = hk/m$ and $E_i = mv^2/2$ we then find that $L = \lambda/\theta_E = 2\pi b$, so that the localization L is approximately equal to the classical impact parameter b. Alternatively, the inelastically scattered amplitude can be written as the Fourier transform of an inelastic scattering potential in Born's approximation. The forces due to this potential affect the passing electron only when it is within a distance b of the atom, and for times of the order of ΔT. The theory of the photoelectric effect, on which the analysis of X-ray absorption near-edge structure (XANES) and much ELNES (electron energy loss near-edge spectroscopy) are also based assumes, however, almost 'instantaneous' excitation of an atom (Feuerbacher, Fitton, and Willis 1978). Here the abrupt 'switching on' of the core hole (within less than 10^{-17} s) is known as the 'sudden approximation'. No evidence for the failure of this approximation has yet been produced in inelastic electron scattering.

High-resolution electron microscopists are most likely to be interested in those aspects of EELS which can provide chemical and structural information which complements that obtainable in HREM images. For these purposes, the most promising recent area of research appears to be the study of the electron energy-loss near-edge structure (ELNES). This topic has close similarities with the corresponding X-ray technique known either as X-ray absorption near-edge structure (XANES) or near-edge X-ray absorption fine structure (NEXAFS) (see Bianconi, Incoccia, and Stipcich (1983) for a review). Figure 11.5 shows the ELNES K-shell ionization edges for silicon, aluminium and magnesium atoms in different crystal structures (Taftø and Zhu 1982). It is clear that the shape of these spectra from the same atom in different crystals does depend on the local coordination of the atom, and not on purely atomic properties. In addition, the type of atom present can be determined by comparing the energy of ionization onset with tabulated values. While some limited success can be expected in relating characteristic ELNES spectral shapes to local atomic structure by these 'fingerprinting' methods, a full understanding of this effect requires large numerical computations and a sophisticated theory. This theory extends well beyond the compass of this book, but can be found reviewed in articles by Colliex, Manoubi, Gasgnier, and Brown (1985), by Leapman, Grunes, and Fejes (1982), and by Spence (1987). There, the important concepts of secondary electron diffraction and range (on which the ELNES depends) are discussed in detail, as is the relationship of the ELNES to the density of empty states in the crystal band-structure and the role of core excitons. Some success has been achieved in matching theoretical ELNES calculations to experimental results—Fig. 11.6 shows, for example, the results of two calculations for the beryllium K edge in Be_2C (antiflourite structure) compared with experimental ELNES data. Two possible ionic states for the beryllium have been considered (Disko, Spence, Sankey, and Saldin 1986). These calculations are based on the Green's function multiple-scattering method (Durham, Pendry, and Hodges 1982) and do not incorporate core-hole effects. However, the sensitivity of ELNES to both local atomic structure (Spence 1985) and to ionicity or charge-transfer effects is evident. As a second example of this multiple-scattering approach to the analysis of ELNES, we cite the work of Lindner, Sauer, Engel, and Kambe (1986) on MgO.

It should not be thought, however, that all oscillations in ELNES spectra are due to 'solid-state' effects. For example, sharp peaks, known as 'white lines' have been known for many years in X-ray absorption spectroscopy

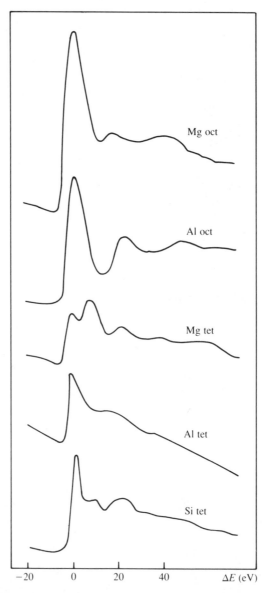

Fig. 11.5 Electron-loss near-edge structures of Si, Al and Mg atoms in different local crystallographic environments. The K edges of magnesium in olivine (octahedral), aluminium in spinel (octahedral), magnesium in spinel (tetrahedral), aluminium in orthoclase (tetrahedral), and silicon in olivine (tetrahedral) are shown. (From Taftø and Zhu (1982).)

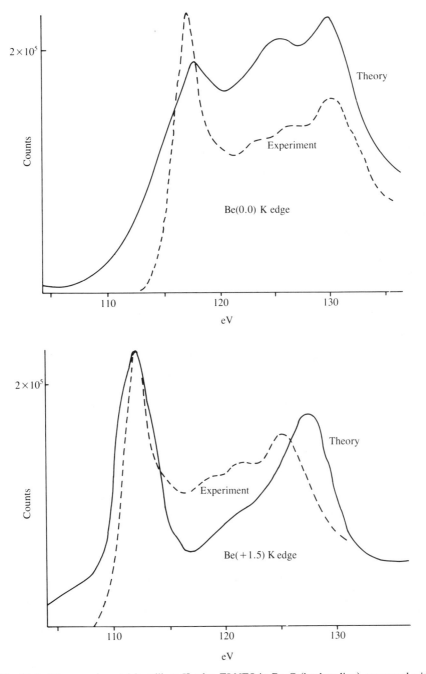

Fig. 11.6 The experimental beryllium K edge ELNES in Be_2C (broken line) compared with the results of multiple-scattering calculations for Be_2C (upper panel) and $Be_2^{1.5+}C^{3.0-}$ (lower panel). The improved fit due to the inclusion of charge-transfer effects between the Be and C atoms is clear. (A charge of three electrons has been transferred to the C atom in the lower panel. See Disko *et al.* (1986) for details.)

and these are of purely atomic origin. Detailed atomic calculations giving good agreement with ELNES and X-ray experimental data for the 'anomolous' $L_{2,3}$ white line intensity ratios in 3d transition metals are reported in Waddington, Rez, Grant, and Humphreys (1986), and give a good impression of the state of the art in this area.

Several attempts have also been made to apply the methods of extended X-ray absorption fine structure (EXAFS) to EELS spectra. While some success has been obtained (see Teo and Joy (1981) for a review), the limited energy range of EELS spectrometers (0–2.5 kV) has frequently meant that in complex crystals one is dealing with many overlapping edges in a crowded spectrum. It is suggested that ELNES (which deals with the first 40 eV or so beyond an edge) be distinguished from extended energy-loss fine structure (EXELFS) (which deals with the remainder of the edge) at that energy at which the ejected secondary electron's wavelength is approximately equal to the nearest-neighbour interatomic distance. Physically, these regimes are distinguished by the importance of multiple scattering for the secondary electron in the ELNES range.

Most of the remaining topics in EELS research lie outside the scope of this book; however, interested readers are particularly referred to work on the momentum transfer or orientation dependence of ELNES by Leapman, Fejes, and Silcox (1983), and on the ALCHEMI-like channelling effects on ELNES reported by Taftø and Lehmfuhl (1982) (see also Spence et al. (1981)). For example, Taftø (1984) has succeeded in separating the near-edge structures of the same atom in two different (non-equivalent) sites in the unit cell through the use of the channelling effect. The considerable body of literature on the study of plasmon and surface plasmon excitations by EELS should also be consulted (for a review, see Raether (1980)). An atlas of EELS spectra covering all the stable elements of the periodic table has been published (Ahn and Krivanek 1982). The text by Egerton (1986) and the EELS bibliography indexed by topic (Egerton and Egerton (1983) are particularly recommended for newcomers to the subject.

In summary, there is now some modest hope that in the future it will become possible in favourable cases to determine interatomic distances, atomic types, and possibly also interatomic angles around identified species, by the analysis of ELNES spectra taken from regions imaged by HREM. The most promising recent development has been the advent of commercial parallel-detection EELS systems (Krivanek et al. (1986)). Larger and faster small computers and some theoretical advances are still required, however, to make this goal achievable.

11.3 Image intensifiers and video systems

Since the first real-time video recordings of individual atom movements in STEM (Crewe, Langmore, and Isaacson 1975), many groups around the world have come to realize the importance of this technique for the study of atom motion in and on crystals. The spectacular results which this technique has provided on surface phase transitions in semiconductors are described in Section 12.8 (see also frontispiece). Video recording of HREM images in TEM has recently gained popularity for several reasons. (1) the falling cost of TV-rate digital image acquisition systems (particularly for small computers) makes many image-processing techniques available on-line to the TEM user. These include on-line drift correction and on-line fast Fourier transform analysis for diffractogram observation, to assist with focusing, astigmatism correction, and contrast enhancement. An array processor is essential for 'real-time' analysis. (2) By feeding back signals to the microscope, automated focusing, astigmatism correction, and alignment become possible (Saxton, Smith, and Erasmus 1983; Krivanek, Stadelman, Higgs, Chen, and Disko (1983; Koster, Van den Bos and Van der Mast 1987). (3) Weak-beam and atomic-resolution 'movies' have revealed new aspects of phase transformation and defect interactions in solids, such as the finding that dislocations in semiconductors generally remain dissociated while in motion (Cockayne, Hons, and Spence (1981)). (4) Radiation-damage effects can be minimized by examining the first few frames of a recording of a new specimen area, before serious damage occurs. (5) For teaching and demonstration purposes, video display is a great asset.

A typical image intensifier and video recording system for HREM with computer interface is described in Spence, Disko, Higgs, Wheatly, and Hashimoto (1982), and commercial systems are also available (Swann 1986, personal communication (Gatan Corp.)). In this and similar systems, the aim is to detect the arrival of every beam electron, corresponding to a detective quantum efficiency of unity. Each beam electron generates several thousand photons in the scintillator, many of which are lost through various inefficiencies. In this particular system, a single-crystal yttrium–aluminate garnet (YAG) detector screen has been use to provide higher spatial resolution, speed, and uniformity than that possible with powder-phosphor screens. A general review of the principles of computer interfacing to electron microscopes is given in Rez and Williams (1982), and an analysis of the detective quantum efficiency of electronic image-recording systems can be found in Herrmann and Krahl (1982). Using similar systems, recent work has revealed the growth of small crystals, which can be followed in real time, row by row, from the atomic resolution images (Wallenberg, Bovin, and Smith 1985). Iijima and Ichihashi (1985) have also observed 'atomic clouds' outside gold-atom clusters (see Section 12.11). Surface-profile imaging at atomic resolution has also revealed fascinating new dynamic effects (Sinclair, Yamashita, and Ponce 1981; Smith and Marks 1985).

There seems little doubt that the direct digital acquisition of images and full computer control of electron microscopes will become the standard practice in the near future. Indeed the accurate alignment, focusing, and stigmation of HREM images on the newest machines is barely possible without these aids. The main reason for this at present is that a video display allows convenient contrast enhancement of the low-contrast HREM image of amorphous material used for focusing and alignment. This will be an important step towards truly quantitative electron microscopy. The problems remain, however, of the large dynamic range of diffraction patterns, of finding a satisfactory image-output device for 'hardcopy', and of image-detector design for high-voltage machines. For quantitative still image recording and convergent-beam work, a system based on the YAG single-crystal screen coupled by fibre-optics to a liquid-nitrogen-cooled 512×512-pixel $\times$ 14-bit charge coupled device (CCD) would appear to offer advantages (Spence and Zuo 1988). Here a 16.4 mm diameter YAG crystal was used, ground sufficiently thin to give a spatial resolution of about 25 line-pairs per mm. The CCD pixel size is $25 \times 25 \, \mu$m, and the background noise level was measured to be 7 counts per second at $-130\,°$C. The advantages of CCD cameras for the recording of convergent-beam patterns from sub-nanometer regions with large dynamic range is now well established (Mochel and Mochel 1986). However, the limited frame speed (about four per second), and large data storage requirements of these systems limits their usefulness for real-time imaging (i.e. for recording 'movies'). A CCD system allows long integration times without the introduction of the 'read-out' noise common to integrating systems operating at TV speeds, which limits dynamic range. With this system a dynamic range of $2^{14} = 16384$ is obtained. Each image occupies about 0.5 Megabyte (1 byte = 8 bits) and, with high speed data transfer, takes 1.5 minutes to transfer to a storage disk. The CCD/YAG combination appears to be more sensitive than Kodak type SO-163 film, requiring shorter exposure times. It has the great advantage of allowing images and diffraction patterns to be viewed and manipulated (e.g. contrast enhancement, Fourier transform, background subtraction) immediately after they are collected at the microscope.

It seems likely that the study of time-dependent phenomena at atomic resolution will be one of the major growth areas of HREM; however, the observation of reproducible effects will require much more highly controlled vacuum conditions than those present on current machines. A review of computer-based image-aquisition schemes can be found in *J. Microsc.* **127,** Part 1 (1982).

11.4   Electron microdiffraction

A clear trend has emerged for the provision of microdiffraction or convergent-beam electron diffraction (CBED) facilities on modern HREM instruments. To some extent the information obtainable from microdiffraction patterns complements that available from HREM images. Traditionally, convergent-beam patterns have been used for the following purposes.

1. For the determination of the crystal periodicity in the beam direction. This is given by

$$C_0 = 2/(\lambda U_0^2) \qquad (11.7)$$

where λ is the electron wavelength and U_0 is the radius of the first-order Laue zone (FOLZ) ring seen in the microdiffraction pattern. A unit cell's dimensions, and its angle in the plane normal to the beam can also be obtainable from the zero-order Laue zone pattern, but accuracy is limited.

2. To align the crystal zone axis with the electron beam. Modern HREM instruments may provide probe sizes as small as about 30 Å, so that the alignment of the region of interest for HREM becomes possible if no mechanical alteration to the specimen height is required in changing from the microdiffraction to the HREM mode.

3. In thicker specimens of small-unit-cell crystals, the two-beam CBED method of specimen thickness determination may be applied (Blake, Jostsons, Kelly, and Napier 1978).

4. In favourable cases, it may be possible to determine the space-group of the crystal (Steeds and Vincent 1983; Eades, Shannon, and Buxton 1983; Goodman 1975; Tanaka, Sekii, and Nagasawa 1983). However, a thicker region of crystal will generally be required for the CBED pattern than for the HREM image, since CBED patterns from very thin crystals show little contrast. A possible solution to this problem lies in the use of the Tanaka wide-angle CBED method (Eades 1984). This method allows convergent-beam patterns to be obtained over an angular range larger than the Bragg angle, without overlap of orders. It may therefore be used to reveal contrast at higher angles in CBED patterns from very thin crystals, if sufficiently large areas of uniform thickness are available. The ability of this technique to isolate orders without overlap is particularly valuable for the thin specimens of large-unit-cell crystals (with correspondingly small Bragg angles) commonly used for HREM.

5. For the refinement of low order structure factors in crystals by comparison of experimental and computed CBED rocking curves. An example of this technique, applied to a study of bonding in GaAs, can be found in Zuo, Spence and O'Keefe (1988), which contains references to earlier work.

In general, the HREM method is most useful for crystals with a short period in the beam direction and large cell dimensions normal to the beam, since these give the maximum information in this projection. Such crystals are the least favourable for CBED work (since orders are likely to overlap), for which the opposite relative dimensions are preferable, unless a wide-angle technique can be used. The HREM method is also preferable for the study of the defect structure of crystals and for the observation of superlattices, which give rise to very weak reflections in a CBED pattern but to intense points in an HREM image, particularly in thicker regions. The great power of the CBED method lies in its ability to provide symmetry information on the specimen structure which is independent of the HREM imaging parameters (for example, astigmatism correction) and, in high-angle patterns, to provide information on the crystal structure in the electron beam direction from a single crystal setting. The limited tilting facilities available on modern HREM machines makes this information difficult to obtain otherwise. Thus, for example, the component of a dislocation Burgers vector in the beam direction can be determined in favourable cases from CBED patterns (Carpenter and Spence 1982). The tetragonal distortion associated with phase transitions in metals has similarly been studied by Porter, Ecob, and Ricks (1983) using CBED patterns, and a study of nickel-based superalloys can be found in Fraser (1983).

Methods for combining the CBED and HREM techniques in older instruments have been discussed in the literature (Goodman and Olsen 1981) and a general review of the electron microdiffraction technique has recently appeared (Spence and Carpenter, 1986). Recent examples which demonstrate the power of this combination of techniques can be found in Yamamoto and Ishizuka (1983), Moodie and Whitfield (1984) (for the JEOL 200CX), and Fung and Yang (1984) (for the Philips EM400T). Figure 12.3 shows, as an example, the lattice image formed from the 'forbidden' $(4\bar{2}\bar{2})/3$ type reflections due to an intrinsic stacking fault lying normal to the incident beam in silicon. The microdiffraction pattern from this stacking fault is shown in Fig. 11.7, clearly revealing the forbidden reflections (Alexander, Spence, Shindo, Gottschalk, and Long 1986). The probe size used (about 4 nm in diameter) is smaller than the width of the stacking fault.

We consider now the microdiffraction patterns formed using field-emission STEM instruments, since the information these contain is closely related to that of an HREM image if an electron probe of near-atomic dimensions is used. The following material should be read in conjunction with that of Section 5.18. The size of the microdiffraction probe depends on the demagnification of the probe-forming lenses, on the electron source size, and on the focus setting, aperture size, and spherical aberration constant of the probe-forming lens. For field emission sources these last three factors are the most important, with the probe-broadening effect of spherical aberration becoming most important for large aperture sizes. In fact, it is a useful approximation to consider that the electron source in field-emission instruments is an idealized point emitter, in which case the focused probe is 'diffraction-limited' (i.e., it is the image of a point source, as formed by an imperfect lens). Thus, in the absence of spherical aberration, the probe has a size given approximately by $2r_s = d_p$ where

$$d_p = 1.22\lambda/\theta_c \tag{11.8}$$

where θ_c is the semi-angle subtended by the STEM objective aperture and shown in Fig. 11.8. In the presence of spherical aberration, there is no simple expression for probe size. However, computer calculations have shown (Spence 1978; Mory 1985) that, except at the optimum focus setting, the probe intensity distribution contains rather extended 'tails' and oscillations. In practice, the size of the STEM probe can be roughly estimated from the resolution of a STEM lattice image, or by forming a probe image

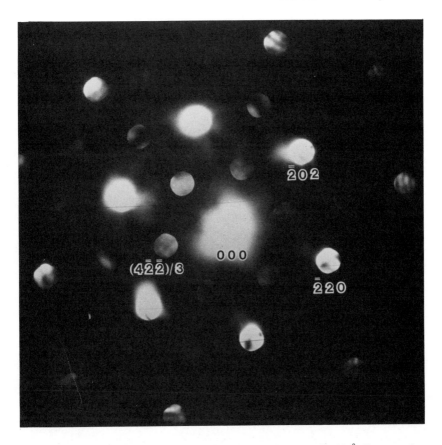

Fig. 11.7 Coherent electron microdiffraction pattern obtained with a 100 Å diameter electron probe from the stacking fault in silicon shown in Fig. 12.3. The probe size is smaller than the stacking fault width. The bulk (220) reflections and 'forbidden' $(4\bar{2}\bar{2})/3$ reflections due to the stacking fault are seen. (From Alexander *et al.* (1986).)

directly, on certain machines. On modern machines this probe size may be as small as 4 Å. Recent calculations by Colliex and coworkers (Mory 1985) have shown that the optimum focus needed to form the most compact probe (for example, for microanalysis) is approximately $\Delta f = -0.75 C_s^{1/2} \lambda^{1/2}$. The principle of reciprocity nevertheless indicates that the Scherzer focus setting gives structure images under STEM lattice-imaging conditions.

The assumption that the probe is diffraction-limited corresponds to the requirement that the objective aperture (C2 in Fig. 11.8) be coherently filled. This is so if the coherence width

$$X_a = \lambda/2\pi\theta_s \tag{11.9}$$

in the plane of the aperture is larger than the aperture. Here θ_s is the semi-angle subtended by the geometrical electron-source image (formed on

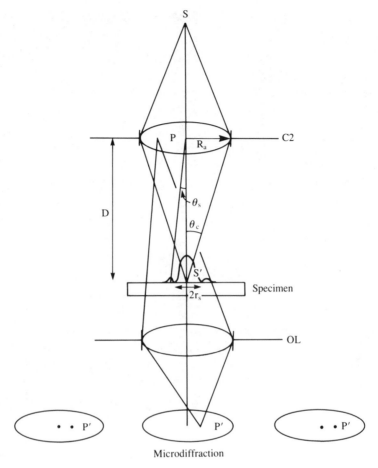

Fig. 11.8 Ray diagram for microdiffraction. Note that P′ is conjugate to P, and S is conjugate to the probe at S′. Partial coherence effects are incorporated by integrating over the source at S. (From Spence and Carpenter (1986).)

the specimen) at the aperture. Thus the requirement for a coherently filled aperture depends on the electron source size, which is proportional to θ_s. Figure 11.8 shows a ray diagram for coherent CBED. Since P in this diagram is conjugate to the set of points P′, we do not expect that CBED patterns from perfect crystals should differ in the two extreme cases where C2 is either coherently or incoherently filled. However, these points are conjugate only at exact focus, so that the through-focus behaviour of coherent and incoherent CBED patterns differ greatly. For defects, the coherent and incoherent CBED pattern may be entirely different, since in the absence of the Bragg condition each source point P gives rise to displaced pattern of diffuse elastic scattering. For the coherent (incoherent) case the separate complex amplitudes (intensities) of all these patterns must be added together. Partial coherence effects can be accounted for by integrating over an extended source. Each different such source point produces a probe at a slightly different position on the sample. However we have seen that for perfectly crystalline samples the diffraction pattern is independent of probe position (if the orders do not overlap), and therefore of source size. For microdiffraction from defects (or the case of overlapping orders), the diffraction pattern will depend strongly on source size and coherence conditions.

For a discussion of the close relationship between coherent microdiffraction and lattice-imaging in STEM, the reader is now referred to Section 5.18 and Fig. 5.34. In the coherent case, a method for the computer-simulation of microdiffraction patterns has been described (Spence 1978) in which the probe wavefunction is used as the boundary condition within an artificial superlattice at the entrance surface of the specimen, and the multislice multiple-Bragg-scattering algorithm used. In a series of papers, experimental microdiffraction patterns from various materials have been compared with the results of these calculations (see, for example, Zhu and Cowley (1982)). For the incoherent case, a method of microdiffraction pattern simulation for defects has also been described, based on the use of the popular diffraction-contrast software under the column approximation, by taking advantage of the reciprocity theorem (Carpenter and Spence 1982). A fuller discussion can be found in Spence and Carpenter (1986), which provides a review of electron microdiffraction and CBED.

The similarities between the problems involved in the interpretation of HREM images and coherent CBED patterns become clear when we consider the case of CBED patterns formed using a probe which is 'smaller' than the crystal unit cell. Then, as shown in Section 5.18, the CBED orders will overlap, and interference effects will be seen within the region of overlap which depend on the focus setting and spherical aberration constant of the probe-forming lens (Cowley and Spence 1979). Thus, the problem of determining atom positions by a comparison of computed and experimental coherent microdiffraction patterns involves a similar set of adjustable parameters (atom coordinates and instrumental parameters) to those involved in HREM image-matching. This applies both for defects or, in the microdiffraction case, for a particular identified group of atoms within a unit cell. However, it appears that the determination of instrumental parameters from coherent microdiffraction patterns with overlapping orders (the 'electron Ronchigram') is more straightforward and may be done in 'real time' (Cowley 1981). These 'Ronchigrams' may also be interpreted as point-projection shadow images of the crystal lattice (at larger defocus), or as electron holograms. Figure 11.9 (from Cowley 1981) shows such a pattern from $Ti_2Nb_{10}O_{29}$ in which, since the probe size is 'smaller' than the unit cell size, the symmetry of the pattern is not that of the crystal as a whole, but rather possesses the local symmetry about the probe (Cowley and Spence 1979). The patterns are seen to repeat as the probe is moved by a lattice

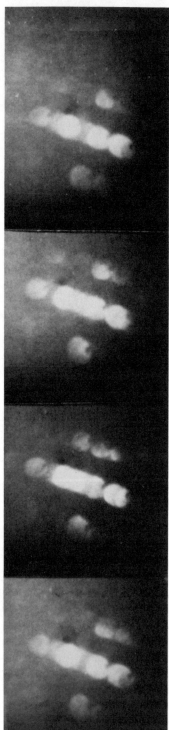

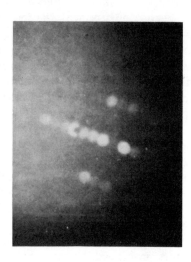

Fig. 11.9 Coherent convergent-beam microdiffraction patterns from a thin crystal of $Ti_2Nb_{10}O_{29}$, formed using a 0.3 nm diameter electron probe within the 2.8 nm periodicity of the crystal unit cell. The patterns are seen to repeat as the probe is moved from one side of the unit cell to the other (From Cowley (1981).)

translation vector (28 Å in length). An analysis of the propagation of energy through a crystal in a diffraction-limited probe under conditions of multiple electron scattering can be found in Marks (1985), and Humphreys and Spence (1979).

Since the intensity within the region of overlapping orders is sensitive to probe position, mechanical movement of the specimen or probe movement due to electronic instabilities is as important for coherent microdiffraction as for HREM. However, it can be shown that the loss of contrast in a coherent microdiffraction pattern due to vibration is a function of the scattering angle and focus setting. In general, for patterns with overlapping orders (or from defects) the finest detail which can be extracted is expected to be about the same as that which would appear in the corresponding STEM lattice image. High-angle scattering is most sensitive to instabilities.

For many problems in materials science, these interferometric effects and their dependence on instrumental parameters which result from the use of the smallest probes constitute an unnecessary complication. A more fruitful approach is then to observe characteristic disturbances in certain non-overlapping orders diffracted by interrupted or distorted planes of the crystal lattice. Then, in the spirit of 'g.b' analysis in TEM, by noting which of the microdiffraction spots are unaffected, it may be possible to determine fault vectors or to classify defect types using a probe only slightly 'larger' than the lattice period (Zhu and Cowley 1983). Several examples of this powerful approach have now appeared in the literature. Earlier dynamical calculations and experiments on coherent microdiffraction from metal catalyst particles (Cowley and Spence 1981), had shown that the normal CBED discs frequently appear as annular rings, possibly also broken up into segments of arc. These may appear simply as small blobs of intensity around the geometrical perimeter of the CBED disc. This effect is only observed when the probe lies near an edge or discontinuity in the lattice. This is a coherent interference effect, and is not due to dynamical scattering in the crystal. This intensity modulation within the CBED disc may make the identification of the geometrical outline of the disc difficult, and so complicates the indexing of coherent microdiffraction patterns from small particles. Further complications arise due to additional structure in the disks due to twinning, which is common in small particles. However, the usefulness of this effect lies in the fact that discs due to diffraction from planes unaffected by a defect will not show this 'splitting'. We now consider an example.

The patterns shown in Fig. 11.10 were obtained from samples of Cu_3Au containing antiphase domain boundaries of three possible types, depending on the near-neighbour coordination of atoms in the boundary. The electron probe size is about 5 Å, smaller than the average domain size. Since the domains represent a discontinuity in the Cu_3Au superlattice, the expected annular splitting is seen in the first-order superlattice reflections only (e.g., those nearest the central beam in Fig. 11.10(c)). These are the reflections for which $\mathbf{g} \cdot \mathbf{R} \neq 0$, where $\mathbf{R}$ is the translation vector between neighbouring

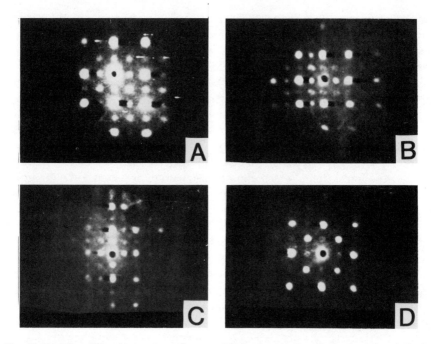

Fig. 11.10 Microdiffraction patterns from antiphase domain boundaries in ordered Cu_3Au, obtained using a sub-nanometer probe: (A) from the crystal, probe not on boundary; (B) probe on 'good' boundary' (C) probe on 'two good' boundary; (D) probe on 'bad' boundary. (From Zhu and Cowley (1982).)

domains. By analysing all the possible boundary arrangements, these authors were able to draw up a classification scheme relating spot splitting to boundary type (Zhu and Cowley 1983).

This work provides an example of a powerful general approach to a problem which would be insoluble using the HREM technique, since lattice images of Cu_3Au antiphase boundaries have proven extremely difficult to interpret. A similar approach has been applied to the case of stacking faults and twin boundaries in FCC metals, where the determination of fault vectors from spot splittings observed in coherent microdiffraction has also been demonstrated (Zhu and Cowley 1983). Other examples of this general approach include coherent microdiffraction studies of diamond platelets (Cowley and Osman 1984), of GP Zones in Al–Cu alloys (Zhu and Cowley 1985) and of metal–ceramic interfaces (Lodge and Cowley 1984). The method has also been applied to the study of twinning and defects in small metal particles (Roy, Messier, and Cowley 1981).

In all this work, the ability to record the weak microdiffraction fine structure efficiently using an axial parallel detection system and image intensifier has proved crucial. Many of these effects cannot be observed using the serial 'Grigson' detection scheme. An optical system has also proved useful for on-line dissection of the patterns and to provide a flexible method of varying the shape of the various STEM detectors used (Cowley and Spence 1979; Cowley 1985).

The use of computed Fourier transforms of small regions of HREM images as a microdiffraction technique is discussed in Tomita, Hashimoto, Ikuta, Endoh, and Yokota (1985). This method is subject to all the limitations outlined in Sections 3.5 and 6.9.

11.5 Cathodoluminescence in STEM

The analysis of the electronic and atomic structure of individual defects in crystals requires the development of a technique which is (1) sensitive to the very low concentrations of impurities which may be electrically important (usually present in concentrations too low to be detected by EDS, ELS or HREM); (2) capable of sufficient spatial resolution to isolate individual defects; and (3) able to provide sufficiently high spectral energy resolution to study the electronic states of interest. At present the cathodoluminescence (CL) technique in STEM provides the most favourable combination of these properties. Here the optical emission excited by the electron beam in passing through a thin sample is collected by a small mirror and passed to a conventional optical spectrometer for analysis, as shown in Fig. 11.11. By forming a small electron probe and plotting the CL intensity within a small spectral range as a function of electron probe position, a scanning monochromatic CL image may also be obtained, giving a spatial map of the impurity or electronic state of interest. For example, Roberts (1981) has used this method to relate structural inhomogeneities to luminescence in ZnS phosphors.

The usefulness of the STEM–CL technique for the study of semiconductor defects was first demonstrated by the pioneering work of Petroff and others (Petroff, Logan, and Savage 1980). Here a systematic study of dislocations in GaAs was undertaken, using the increased accelerating voltage of a high-voltage STEM instrument to allow the study of thicker specimens. A correlation between the electrical activity and type of dislocation was obtained in this work. More recently, systems have been fitted both to dedicated field-emission STEM instruments, and to conventional TEM–STEM machines.

The spatial resolution possible in STEM–CL appears to depend greatly on the specimen thickness for thin samples. In general it might be expected to be given by

$$d_r = (d_p^2 + d_g^2 + d_D^2)^{1/2} \tag{11.10}$$

where d_p is the electron probe diameter, d_g is the electron–hole pair generation volume, and d_D the carrier diffusion length (Pennycook 1981). For very thin transmission samples, d_g (the 'beam spreading' discussed in connection with EDS) and d_p are small compared with d_D, which is dominated by surface recombination. As a rough approximation one then

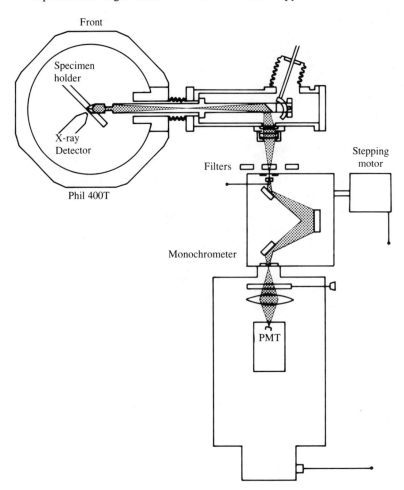

Fig. 11.11 Diagram of the experimental arrangement used to observe cathodoluminescence from samples in a TEM instrument at low temperature. The optical path from the ellipsoidal mirror is shown stippled. (From Yamamoto *et al.* (1984).)

obtains (Brown 1984)

$$d_r = t/\sqrt{6} \tag{11.11}$$

for a thin specimen of thickness *t*. For such very thin specimens, the emission intensity is very small, requiring efficient light-collection optics. The spatial resolution of a monochromatic scanning CL image is not influenced by the optical resolution limit of the light-collection mirror or lens; however this does affect the amount of stray light which is collected, and hence the signal-to-noise ratio in the image. From such a small volume (it may be as small as a few thousand cubic Ångstroms) a spectral resolution of perhaps 10 Å may be obtained from very low concentrations of impurities, and this spectral information may be collected together with the corresponding HREM image. In most applications, however, spectra are recorded from a region of about one micrometre in diameter in combination with diffraction-contrast images, and this ability to correlate CL spectra with high-resolution TEM images is the most important advantage of the STEM–CL technique over conventional scanning electron microscope (SEM)–CL work, and over the photoluminescence (PL) method. Much lower temperatures, however, can be obtained in PL, with a corresponding gain in emission intensity.

In addition to the semiconductor work referred to above, there have been several studies of MgO and diamond by this method. Using a tapered silver tube as a light-collection element fitted to the Vacuum Generators HB5 STEM, Pennycook, Brown, and Craven (1980) obtained CL images of individual dislocations in type IIb diamond in correlation with their transmitted electron images. In this material, almost all the optical luminescence arises from dislocations, but not all dislocations are luminescent. Both 60° and screw dislocations were found to be luminescent. By subtracting a spectrum recorded from the defect of interest from one recorded nearby, the effects of stray luminescence, filament light, and impurity luminescence may be eliminated. In a development of this work, Yamamoto, Spence, and Fathy (1984) have used a system based on that developed by Roberts (1981) to obtain weak-beam images and spectra from individual dislocations of known type in diamond. The apparatus (shown in Fig. 11.11) has been fitted to a Philips EM 400 TEM–STEM instrument. This apparatus allows spectra to be collected over a range of temperatures down to about 25 K. Example results from this work are shown in Fig. 11.12, where the TEM image of a single line defect and its associated optical emission spectrum are shown together with the scanning monocromatic CL image used to identity the defect. The spectrum was recorded at 89 K. In most materials, CL is found to increase in intensity at low temperatures, and the resulting fine structure observed may be more readily correlated with theoretical models. In this work, in addition, the polarization of the CL was measured from individual dislocations of known type and degree of dissociation. The dislocation emission (which occurs at a wavelength of 435 nm) is confirmed to be polarized along the dislocation line (rather than along its Burgers vector) direction, and to form a broad band whose width is 0.412 eV. No clear correlation between CL activity and dislocation type or degree of dissociation was found, though a one-dimensional 'donor–acceptor' model for dislocation core CL is suggested as consistent with all the experimental findings. A high kink density was found to reduce the CL intensity.

The limited aim of most of the STEM–CL work undertaken so far has been to correlate a spectral signature with a particular defect which can be imaged at high resolution by TEM. More detailed quantitative interpretation of CL spectra has yet to be attempted. This must follow closely along the lines of the theory of photoluminescence. Only at liquid-helium

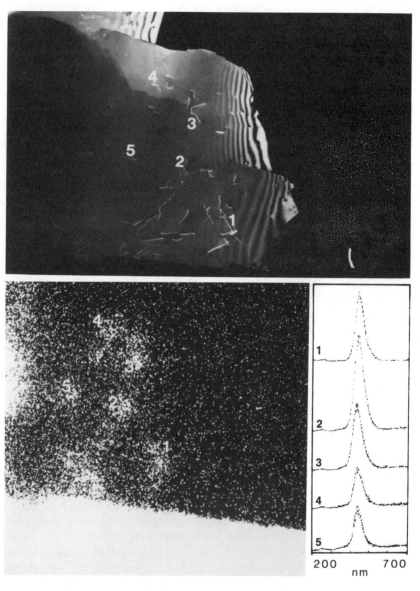

Fig. 11.12 Scanning monochromatic CL images (lower panel) and corresponding dark field
($\bar{2}20$) images (upper panel) of dislocations in type IIb diamond recorded at 89 K with the
apparatus of Fig. 11.11. Inset are the optical emission spectra taken from each of the individual
dislocations shown. The dislocation types are (1) screw, (2) edge, (3) screw, (4) edge, (5)
screw. (From Yamamoto *et al.* (1984).)

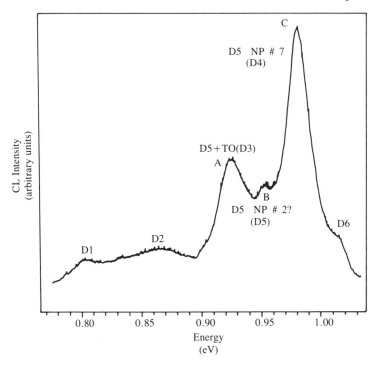

Fig. 11.13 Infrared CL spectrum recorded at 23 K from a group of straight dislocations in silicon using a similar apparatus to that shown in Fig. 11.11, but with an intrinsic germanium detector. The transverse optical (TO) and no-phonon (NP) lines are indicated. (From Graham *et al.* (1987).)

temperatures and for very simple, well-characterized systems would there appear to be any possibility of obtaining quantitative agreement between the intensities of experimental CL spectral and theoretical predictions. The identification of defect levels and phonon lines according to their energy is, however, more straightforward. However, since it gives the most sensitive spectral signature obtainable in correlation with HREM images, the technique shows considerable promise in many areas of microscopy. These include mineralogy, catalysis and semiconductor studies. For example, it is likely that measurements of CL polarization could be related to the crystallographic site of impurities, while the extension of CL detectors into the infrared region has opened up the possibility of work on silicon and other small-bandgap semiconductors. As final example, Fig. 11.13 shows the infrared CL emission spectrum obtained at 25 K from a group of straight dislocations in silicon, using an apparatus similar to that shown in Fig. 11.11. An intrinsic germanium detector was used, cooled to 77 K. The interpretation of these CL lines, and their relationship to those observed from similar samples by photoluminescence, are discussed in the literature (Graham, Spence, and Alexander 1987).

References

Alexander, H., Spence, J. C. H., Shindo, D., Gottschalk, P., and Long, N. (1986). Forbidden reflection lattice imaging for the determination of kink densities. *Phil. Mag.* **A53**, 627.

Ahn, C. C. and Krivanek, O. L. (1982). *EELS Atlas* [Available from Gatan Inc., 780 Commonwealth Drive, Warrendale, PA 15086 USA.]

Batson, P. E., Kavanagh, K. L., Woodall, J. M., and Mayer, J. W. (1986). Electron energy loss scattering near a single misfit dislocation at the GaAs/GaInAs interface. *Phys. Rev. Lett.* **57**, 2729.

Berger, S. D. and Pennycook, S. J. (1982). Detection of nitrogen at (100) platelets of diamond. *Nature (London)* **298**, 635.

Bethe, H. (1930). Zur Theorie des Durchgangs schneller Korpuskularstrahlen durch Materie. *Ann. Phys. (Leipzig)* **5**, 325.

Bianconi, A., Incoccia, L., and Stipcich, S. (1983). *EXAFS and near edge structure*, Springer Tracts in Chemical Physics, Vol. 27. Springer-Verlag, New York.

Blake, R. G., Jostsons, A., Kelly, P. M., and Napier, J. G. (1978). The determination of extinction distances by STEM. *Phil. Mag.* **A37**, 1.

Bourdillon, A. J., Self, P. G. and Stubbs, W. M. (1981) Crystallographic orientation effects in energy dispersive X-ray analysis. *Phil. Mag.* **44**, 1335. See also Bourdillon, A. J. (1984) Localisation and X-ray emission. *Phil. Mag.* **A50**, 839.

Bourret, A. and Colliex, C. (1982). Combined HREM and STEM microanalysis on decorated dislocation cores. *Ultramicroscopy* **9**, 183.

Brown, M. (1984). Personal communication. See also Schockley, W. (1953). *Electrons and holes in semiconductors.* Van Nostrand, New York.

Buseck, P., Cowley, J. M., and Eyring, L. (1988). *High resolution electron microscopy.* Oxford University Press, Oxford.

Cardona, M. and Ley, L. (eds) (1978). *Photoemission in solids I and II.* Springer Topics in Applied Physics, Vols 26 and 27. Springer Verlag, Berlin.

Carpenter, R. and Spence, J. C. H. (1982). Three dimensional strain information from CBED patterns. *Acta Crystallogr.* **A38**, 55.

Chan, H. (1984). Calcium site occupancy in $BaTiO_3$. *Mat. Res. Soc. Symp. Proc.* Vol. 13, p. 345. Materials Research Soc., Pittsburgh.

Cherns, D., Howie, A., and Jacobs, M. H. (1973). Characteristic X-ray production in thin crystals. *Z. Naturforsch.* **28a**, 565.

Christensen, K. K. and Eades, J. A. (1986). On 'parallel' illumination conditions for ALCHEMI. In *Proc. 44th Annual EMSA meeting,* (ed. G. Bailey), p. 622. San Francisco Press, San Francisco.

Cockayne, D., Hons, A., and Spence, J. C. H. (1981). Gliding dissociated dislocations in hexagonal CdS. *Phil. Mag.* **A42**, 773.

Colliex, C., Manoubi, T., Gasgnier, M., and Brown, L. M. (1985). Near-edge fine structures on EELS core-loss edges. In *Scanning electron microscopy—1985* Part 2, p. 289. SEM Inc. A. M. F. O'Hare, Chicago.

Cowley, J. M. (1981). Coherent interference effects in STEM and CBED. *Ultramicroscopy* **7**, 19.

Cowley, J. M. (1985). A new detector system for the HB5 STEM. *Proc. EMSA 1985* (ed. G. Bailey), p. 134. San Francisco Press, San Francisco.

Cowley, J. M. and Osman, A. (1984). Nanodiffraction from platelets in diamond. *Ultramicroscopy* **15**, 311.

Cowley, J. M. and Spence, J. C. H. (1979). Innovative imaging and microdiffraction in STEM. *Ultramicroscopy* **3**, 433.

Cowley, J. M. and Spence, J. C. H. (1981). Converent beam electron microdiffraction from small crystals. *Ultramicroscopy* **6**, 359.

Craven, A. J. and Colliex, C. (1977). The effect of energy-loss on phase contrast. In *Developments in electron microscopy and analysis,* Inst. Phys. Conf. Ser. No. 36, p. 271. Institute of Physics, Bristol.

Craven, A. J., Gibson, J. M., Howie, A., and Spalding, D. R. (1978). Study of single-electron excitations by electron-microscopy. *Phil. Mag.* **A38**, 519.

Crewe, A. V., Langmore, J. P., and Isaacson, M. S. (1975). Resolution and contrast in the scanning transmission electron microscope. In *Physical aspects of electron microscopy and microbeam analysis* (ed. B. M. Siegel and D. R. Beaman) p. 247. Wiley, New York.

Disko, M. M., Spence, J. C. H., Sankey, O., and Saldin, D. (1986). The electron energy loss structure of Be_2C. *Phys. Rev.* **B33**, 5642.

Durham, P., Pendry, J., and Hodges, C. H. (1982). Calculation of X-ray absorption near-edge structure, XANES. *Comput. Phys. Commun.*, **25**, 193.

Eades, J. A. (1984). Zone-axis diffraction patterns by the Tanaka method. *J. Electron Microsc. Tech.* **1**, 279.

Eades, J. A., Shannon, M. D., and Buxton, B. F. (1983). In *Scanning electron microscopy 1983* Vol. III (ed. O. Johari), p. 1051. A. M. F. O'Hare, Chicago.

Egerton, R. F. (1984). Parallel-recording systems for electron energy-loss spectroscopy (EELS). *J. Electron Microsc. Tech.* **1**, 37.

Egerton, R. F. (1986). *Electron energy-loss spectroscopy in the electron microscopy.* Plenum Press, New York.

Egerton, R. F. and Egerton, M. (1983). An electron energy-loss bibliography. In *Scanning electron microscopy 1983* (ed. O. Johari), p. 119. A. M. F. O'Hare, Chicago.

Fathers, D. and Rez, P. (1979). A transport equation theory of electron backscattering. In *SEM 1979* Vol. II (ed. O. Johari), p. 55. A. M. F. O'Hare, Chicago.

Feurbacher, B., Fitton, B., and Willis, R. F. (1978). *Photoemission and the electronic properties of surfaces,* p. 111. Wiley, New York.

Fraser, H. L. (1983). Applications of CBED to Ni-based superalloys, *J. Microsc. Spectrosc. Electron* **8**, 431.

Fung, K. K. and Yang, C. Y. (1984). Combined CBED and HREM in a Philips EM400T analytical electron microscopy. *Ultramicroscopy* **13**, 333.

Goodman, P. (1975). A practical method of 3-dimensional space-group analysis using convergent-beam electron-diffraction. *Acta Crystallogr.* **A31**, 804.

Goodman, P. and Olsen, A. (1981). Combining CBED with HREM. *Ultramicroscopy* **6**, 101.

Goodman, P. and Warble, C. (1987). The top-bottom N-beam phase contrast effect from finite crystals. Use of the effect in HREM studies of surfaces. *Phil. Mag.* (in press).

Graham, R., Spence, J. C. H., and Alexander, H. (1987). Infrared cathodoluminescence studies from dislocations in silicon in TEM. In *Mat. Res. Symp. Proc: Characterization of defects in materials.* (ed. B. Siegel, R. Sinclair, J. Wertman). North-Holland, New York.

Grovenor, C. R. M., Batson, P. E., Smith, D. A., and Wong, C. (1984). As segregation to grain boundaries in Si. *Phil. Mag.* **A50**, 409.

Herrmann, K.-H. and Krahl, D. (1982). The detection quantum efficiency of electron image recording systems. *J. Microsc.* **127**, 17.

Howie, A. (1963). Inelastic scattering of electrons by crystals: (I) The theory of small-angle inelastic scattering. *Proc. R. Soc. Lond.* **271A**, 268.

Howie, A. (1979). Image contrast and localised signal selection techniques. *J. Microsc.* **117**, 11.

Humphreys, C. J. and Spence, J. C. H. (1979). 'Wavons'. In *Proc. EMSA 1979* (ed. G. Bailey), p. 554. Claitor, Baton Rogue.

Iijima, I. and Ichihashi, T. (1985). Motion of surface atoms on small gold particles. *Japan J. Appl. Phys.* **24**, L125.

Johnson, D. W. and Spence, J. C. H. (1974). Determination of the single-scattering probability distribution from plural-scattering data. *J. Phys. D: Appl. Phys.* **7**, 771.

Joy, D. C., Romig, A. D., and Goldstein, J. I. (eds) (1986). *Principles of analytical electron microscopy.* Plenum Press, New York.

Kainuma, Y. (1955). The theory of Kikuchi patterns. *Acta. Crystallogr.* **8**, 247.

Korobochko, Y. S., Kosmach, V. F. and Mineev, V. I. (1965). Coherent Bremsstrahlung. *Sov. Phys.* JETP **21**, 834.

Koster, A. J., Van den Bos, A. and Van der Mast, K. D. (1987). *Ultramicros.* **21**, 209.

Krishnan, K. M. (1988). Atomic site and species determinations using channelling and related effects in analytical electron microscopy. *Ultramicroscopy* **24**, 125.

Krishnan, K. M., Rez, P., and Thomas, G. (1985). Crystallographic site-occupancy refinements in thin-film oxides by channelling-enhanced microanalysis. *Acta Crystallogr.* **B41**, 396.

Krishnan, K. M., Rez, P., Thomas, G., Kokata, Y., and Hashimoto, H. (1986). The combined effect of acceleration voltage and incident beam orientation on the characteristic X-ray production in thin crystals. *Phil. Mag.* **B53**, 339.

Krivanek, O. L., Stadelman, P., Higgs, A., Chen, C., and Disko, M. M. (1983). Is full computer automation of HREM feasible? In *Proc. 41st EMSA meeting* (ed. G. Bailey), p. 404. San Francisco Press, San Francisco.

Krivanek, O. L., Ahn, C. C., and Keeney, R. B. (1986). A parallel detection electron spectrometer using quadropole lenses. *Ultramicroscopy* **22**, 103.

Leapman, R. D., Grunes, L. A., and Fejes, P. L. (1982). Study of the L_{23} edges in the 3d transition metals and their oxides by electron-energy-loss spectroscopy with comparisons to theory. *Phys. Rev.*, **B26**, 614.

Leapman, R., Fejes, P. L., and Silcox, J. (1983). Orientation dependence of core edges from anisotropic materials determined by inelastic scattering of fast electrons. *Phys. Rev.* **B28**, 2361.

Lindner, H., Sauer, W., Engel, W., and Kambe, K. (1986). Near-edge structure in electron energy loss spectra of MgO. *Phys. Rev.* **B33**, 22.

Lodge, E. A. and Cowley, J. M. (1984). The surface diffusion of silver on MgO. *Ultramicroscopy* **13**, 215.

Marks, L. D. (1985). STEM probe spreading. *Mat. Res. Soc. Symp. Proc.* Vol. 41, p. 247. Elsevier, New York.

Maslen, V. W. and Rossouw, C. J. (1984). Implications of $(e, 2e)$ scattering for inelastic electron diffraction in crystals. I. Theoretical. *Phil. Mag.* **A49**, 735.

Mochel, M. E. and Mochel, J. M. (1986). A C.C.D. imaging and analysis system for the HB5 STEM. In *Proc. 44th Annual EMSA meeting*, (ed. G. Bailey) p. 616. San Francisco Press, San Francisco.

Moodie, A. F. and Whitfield, H. J. (1984). Combined CBED and HREM in the electron microscope. *Ultramicroscopy* **13**, 265.

Mory, J. (1985). PhD. Thesis, University of Paris.

Otten, M. and Buseck, P. R. (1987). Axial ALCHEMI: a method for assessing crystallographic site occupancies in garnet. *Ultramicroscopy* **23**, 151.

Petroff, P., Logan, R. A., and Savage, A. (1980). Nonradiative recombination at dislocation in III-V, *Phys. Rev. Lett.* **44**, 287.

Pennycook, S. J. (1981). Investigation of the electronic effects of dislocations by STEM. *Ultramicroscopy* **7**, 99.

Pennycook, S. J. (1987) Impurity lattice and sublattice location by electron channeling. In: *'Scanning Electron Microscopy 1987'.* (ed. O. Johari) p. 217. A. M. F. O'Hare, Chicago.

Pennycook, S. J., Brown, L. M., and Craven, A. J. (1980). Observation of cathodoluminescence at single dislocation by STEM. *Phil. Mag.* **A41**, 589.

Porter, A. J., Ecob, R., and Ricks, R. (1983). Coherency strain fields magnitude and symmetry. *J. Microsc.* **129**, 327.

Raether, H. (1980). *Excitation of plasmons and interband transitions by electrons.* Springer Tracts in Modern Physics, Vol. 88. Springer-Verlag, New York.

Reese, G. M., Spence, J. C. H., and Yamamoto, N. (1984). Coherent bremsstrahlung from kilovolt electrons in zone axis orientations. *Phil. Mag.* **A49**, 697.

Rez, P. and Williams, D. B. (1982). Electron microscope/computer interactions: A general introduction. *Ultramicroscopy* **8**, 247.

Roberts, S. H. (1981). In *Microscopy of semiconductor materials 1981* (ed. A. G. Cullis and D. C. Joy), Inst. Phys. Conf. Ser. No. 6, p. 377. Institute of Physics, Bristol.

Rossouw, C. J. and Maslen, V. W. (1987). Localisation and ALCHEMI for zone axis orientations. *Ultramicroscopy* **21**, 277.

Roy, R. A., Messier, R., and Cowley, J. M. (1981). Fine structure of gold particles in thin films prepared by metal-insulator co-sputtering. *Thin Solid Films* **79**, 207.

Saldin, D. K. and Rez, P. (1986). The theory of the excitation of atomic inner shells in crystals by fast electrons. *Phil. Mag.* **55**, 481.

Saxton, O., Smith, D., and Erasmus, S. J. (1983). Procedures for focusing, stigmating and alignment in HREM. *J. Microsc.* **130**, 187.

Schuman, H. (1981). Parallel recording of electron energy-loss spectra. *Ultramicroscopy* **6**, 163.

Self, P. G. and Buseck, P. R. (1983). Low-energy limit to channelling effects in the inelastic scattering of fast electrons. *Phil. Mag* **A48**, L21.

Shindo, D., Hiraga, K., Hirabayashi, M., Tokiwa, A., Kikuchi, M., Syono, Y., Nakatsu, O., Kobayashi, N., Muto, Y. and Aoyagi, E. (1987). Effect of Co substitution on Tc in $YBa_2Cu_3O_7$. *Jap. J. Appl. Phys.* **26**, L1667.

Sinclair, R., Yamashita, T., and Ponce, F. A. (1981). Atomic motion on the surface of a cadmium telluride single-crystal. *Nature (London)* **290**, 386.

Smith, D. and Marks, L. D. (1985). Direct atomic imaging of solid surfaces: III. Small particles and extended Au surfaces. *Ultramicroscopy* **16**, 101.

Spence, J. C. H. (1978). Approximations for the calculation of CBED patterns. *Acta Crystallogr.* **34**, 112.

Spence, J. C. H. (1985). The structural sensitivity of ELNES. *Ultramicroscopy* **18**, 165.

Spence, J. C. H. (1987). In *High resolution electron microscopy* (ed. P. Buseck, J. M. Cowley, L. Eyring). Oxford University Press, Oxford (in press).

Spence, J. C. H. and Carpenter, R. (1986). Electron microdiffraction. In *Elements of analytical electron microscopy* (ed. D. Joy, A. Romig, J. Hren, H. Goldstein) Chap. 3. Plenum Press, New York.

Spence, J. C. H. and Graham, R. (1986). Cold Alchemi. In *Mat. Res. Soc. Symp. Proc.* Vol. 62, p. 153. Materials Research Society, Pittsburgh.

Spence, J. C. H. and Lynch, J. (1982). STEM microanalysis by transmission electron energy-loss spectroscopy in crystals. *Ultramicroscopy* **9**, 267.

Spence, J. C. H. and Reese, G. (1986). Pendellösung radiation and coherent bremsstrahlung. *Acta Crystallogr.* **A42**, 577.

Spence, J. C. H. and Taftø, J. (1983). ALCHEMI: A new technique for locating atoms in small crystals. *J. Microsc.* **130**, 147.

Spence, J. C. H. and Zuo, J. M. (1988). A new parallel detector for quantitative electron diffraction with large dynamic range. *J. Sci. Intrum.* (in press).

Spence, J. C. H., Kuwabera, M. and Kim, Y. (1988). *Ultramicros.* (in press).

Spence, J. C. H., Krivanek, O. L., Taftø, J., and Disko, M. (1981). The crystallographic information in electron energy loss spectra. In Inst. Phys. Conf. Ser. No. 61, p. 253. Institute of Physics, Bristol.

Spence, J. C. H., Disko, M., Higgs, A., Wheatley, J., and Hashimoto, H. (1982). A digital on-line diffractometer and image processor for HREM. *Proc. Tenth. Int. Congr. Elec. Microsc.* (Hamburg), p. 520.

Spence, J. C. H., Titchmarsh, J., and Long, N. (1985). Coherent bremsstrahlung—new peaks in EDS: A new unavoidable artifact in thin-crystal X-ray microanalysis. In *Microbeam Analysis—1985* (ed. J. T. Armstrong), p. 349. San Francisco Press, San Francisco.

Steeds, J. and Vincent, R. (1983). Use of high-symmetry zone axis in electron diffraction in determining crystal point and space groups. *J. Appl. Crystallogr.* **16**, 317.

Taftø, J. (1982). The cation distribution in a $(Cr, Fe, Al, Mg)_3O_4$ spinel as revealed from the channelling effect in electron induced X-ray emission. *J. Appl. Crystallogr.* **15**, 378.

Taftø, J. (1984). Absorption edge fine structure study with subunit cell spatial resolution using electron channelling. *Nucl. Instrum. Methods* **B2**, 733.

Taftø, J. and Buseck, P. R. (1982). Quantitative study of Al–Si ordering in an orthoclase feldspar using an analytical transmission electron microscope. *Am. Mineral.* **68**, 944.

Taftø, J. and Buseck, P. R. (1983). Quantitative study of Al–Si ordering in an orthoclase feldspar using an analytical transmission electron microscope. *Am. Mineral.* **68**, 944.

Taftø, J. and Lehmfuhl, G. (1982). Direction dependence in electron energy loss spectroscopy from single crystals. *Ultramicroscopy* **2**, 287.

Taftø, J. and Liliental, S. (1982). Studies of the cation distribution in $ZnCr_xFe_{2-x}O_4$ spinels. *J. Appl. Crystallogr.* **15**, 260.

Taftø, J. and Spence, J. C. H. (1982). Crystal site location of iron and trace elements in an Mg–Fe-olivine using a new crystallographic technique. *Science* **218**, 49.

Taftø, J. and Spence, J. C. H. (1983). Crystal site location of dopants in semiconductors. *J. Appl. Phys.* **54**, 5014.

Taftø, J. and Zhu, J. (1982). Electron energy-loss near edge structure (ELNES), a potential technique in the studies of local atomic arrangements. *Ultramicrocsopy* **9**, 349.

Taftø, J., Krivanek, O. L., Spence, J. C. H., and Honig, J. M. (1982). Is your spinel normal or inverse? *Proc. Tenth Int. Congr. Elec. Microsc.* (Hamburg) Vol. 1, p. 615.

Taftø, J., Clarke, D. R., and Spence, J. C. H. (1983). Determination of crystal site occupancy in an improved cynroc-D ceramic using analytical electron microscopy. *Mat. Res. Soc. Symp. Proc.* Vol. 15, p. 9. Elsevier, New York.

Taftø, J., Suenaga, M., and Welch, D. O. (1984). Crystal site determination of dilute alloying elements in polycrystalline Nb_3Sn. *J. Appl. Phys.* **55**, 4330.

Tanaka, M., Sekii, H., and Nagasawa, T. (1983). Space-group determination by dynamic extinction in convergent-beam electron diffraction. *Acta Crystallogr.* **A39**, 825.

Teo, B. and Joy, D. (1981). *EXAFS spectroscopy*. Plenum, New York.

Tomita, M., Hashimoto, H., Ikuta, T., Endoh, H., and Yokota, T. (1985). Improvement and application of the fourier-transformed pattern from a small area of high resolution electron microscope images. *Ultramicroscopy* **16**, 9.

Vechio, K. S. and Williams, D. B. (1986). Experimental conditions affecting coherent bremsstrahlung in X-ray microanalysis. *J. Microsc.* **147**, 15.

Waddington, W. G., Rez, P., Grant, I., and Humphreys, C. J. (1986). White lines in the energy loss spectra and X-ray absorption spectra of 3d transition metals. *Phys. Rev.* **B34**, 17.

Wallenberg, L. R., Bovin, J., and Smith, D. (1985). Atom hopping on small gold particles imaged by high resolution electron microscopy. *Naturwissenschaften* **72**, S539.

Williams, D. B. and Joy, D. C. (1984). *Analytical electron microscopy 1984* San Francisco Press, San Francisco.

Yamamoto, N. and Ishizuka, K. (1983). Analysis of the incommensurate structure of $Sr_2Nb_2O_7$ by HREM and CBED. *Acta Crystallogr.* **B39**, 210.

Yamamoto, N., Spence, J. C. H., and Fathy, D. (1984). Cathodoluminescence and polarization studies from individual dislocations in diamond. *Phil. Mag.* **B49**, 609.

Zhu, J. and Cowley, J. M. (1982). Microdiffraction from antiphase domain boundaries in Cu_3Au. *Acta Crystallogr.* **A38**, 718.

Zhu, J. and Cowley, J. M. (1983). Microdiffraction from stacking faults and twin boundaries in f.c.c. crystals. *J. Appl. Crystallogr.* **16**, 171.

Zhu, J. and Cowley, J. M. (1985). Study of the early stage precipitation in Al-4%Cu by microdiffraction and STEM. *Ultramicroscopy* **18**, 419.

Zuo, J. M., Spence, J. C. H. and O'Keefe, M. (1988). Bonding in GaAs. *Phys. Rev. Letts.* (in press).

12

APPLICATIONS OF HIGH-RESOLUTION
ELECTRON MICROSCOPY

The past decade has seen enormous growth in the number and variety of applications of the HREM technique. Since this trend is likely to continue, the following summary of some recent applications is certain to become rapidly out of date. However, the first decade of a new field usually contains its most exciting and enduring discoveries, and I have therefore tried to select examples from many fields of materials science which either illustrate fundamental points of image interpretation, or have provided important new insights into condensed-matter physics. This chapter may therefore form the basis for a bibliography of HREM work in various areas covering the decade which has passed since the technique first became useful in materials physics. It cannot claim, however, to be an exhaustive review of the many hundreds of papers on HREM which have appeared since about 1974; however, much of this work may be traced through the references given here.

A number of excellent reviews of high-resolution work have appeared during this period. These include the 1979 Nobel Symposium on HREM (*Chemica Scripta*, **14**, No. 1–5) and special issues of *Ultramicroscopy* as follows. Volume **8**, No. 1/2 (1982) is devoted to the application of HREM to chemical problems; Volume **14**, No. 1/2 (1984) contains papers on the atomic-scale structure of interfaces; Volume **18**, No. 1–4 (1985) contains about 50 papers describing applications of HREM in various fields. The *Journal of Microscopy* Volumes **129** and **130** (1982) are also devoted to HREM work. General review articles on HREM have been written by Neumann, Pasemann, and Heydenreich (1982), Cowley (1978), Thomas (1982), and Fryer (1979) (for chemistry); Hutchinson, Jefferson, and Thomas (1977), and Buseck and Cowley (1983) (for minerals). The text by Buseck, Cowley, and Eyring (1988) is also devoted primarily to HREM studies. A review of applications of the Berkeley atomic-resolution micro-scope can be found in Thomas (1986).

The taxonomy which follows necessarily involves some overlap of areas (particularly 'ceramics' and 'oxides'). Studies of particular types of 'defects' (e.g., interfaces, point defects, line defects, planar defects) have been included under the material type in which they occur. They are also discussed in other portions of this book.

12.1 Layer compounds, intercalates, and incommensurate modulated structures

One of the most important applications of high-resolution electron micros-copy in the last decade has been to the study of incommensurate superlattice structures. These include both charge-density-wave (CDW) materials (such as some of the transition-metal dichalcogenides), and materials (such as oxides) in which the ordering of planar or point defects allows variations in stoichiometry to be accommodated. The superstructures and polytypes found in binary alloys described in Section 12.9 also form a closely related class of materials from a diffraction and HREM imaging point of view. Common to all three classes of materials is the idea of a 'modulated structure' (Cowley, Cohen, Salaman, and Wuensch 1979) in which a periodic or a partially periodic perturbation of a crystal structure exists, with a repetition distance appreciably greater than the basic unit cell (or 'subcell') dimension. This perturbation may be a periodic strain field (for charge-density-waves), or a composition modulation (for oxides and alloys), or both. For the oxides, (and some sulphides and carbides) the perturbation may take the form of regularly or irregularly spaced planar defects, which thereby allow families of structures of variable stoichiometry to be generated. (The average chemical composition will then not be that of the subcell). These systems are discussed in Section 12.7. The mean distance between planar faults (the superlattice 'period', as deduced from diffraction evidence) may not be an integral multiple of the corresponding subcell dimensions. The analysis of these oxide structures therefore has a good deal in common with that of the first class of materials, the charge-density wave materials. These materials have been analysed historically from a different point of view, in terms of Fermi surface instabilities and the Peierls transition mechanism. Here, a long-period modulation of the conduction-electron gas (with period h, say), known as a charge-density wave (CDW) causes a periodic lattice displacement of the ion cores with a similar period. Since the CDW period depends on the structure of the Fermi surface and its wavevector Q (which is largely independent of the sublattice period L), the superlattice may be incommensurate (i.e. $L/h \neq n/m$, where n and m are integers). Superlattice reflections will then also be seen in the diffraction pattern which do not form an integral subdivision of the regular sublattice reflections. In HREM images, however, the superlattice may be imaged directly (Gibson, Chen and McDonald (1983a), Kuwabara, Tomita, Hashi-moto and Endoh (1986a)), and it then becomes possible to determine both the phase relationship between the CDW and the subcell (Yamamoto and Ishizuka 1983), and to determine whether the modulation consists of a discommensurate or incommensurate superlattice. Some of the problems in imaging these structures were discussed in Section 5.4. This distinction cannot easily be made from diffraction evidence alone, since it can be shown that an incommensurate superlattice with abrupt discontinuities ('discom-mensurations') also gives rise to sharp diffraction spots. When the full three-dimensional structure and symmetry of a crystal containing, say, a two-dimensional CDW is considered, it will be appreciated that the resulting structures may become very complex. In some materials, phase transitions between commensurate and incommensurate structures occur at certain temperatures. A review of the properties of the transition-metal dichalcogenide layer structures in particular can be found in Wilson and Yoffe (1969). Since the lattice modulation which is imaged by HREM is rather weak, the important question arises of how the imaging conditions may be optimized for the weak superlattice reflections (Shindo 1982). These become stronger as the specimen thickness is increased; however, at larger thicknesses it becomes extremely difficult to align samples with sufficient accuracy for HREM work, and image interpretation is complicated.

A comprehensive review of this area of condensed-matter physics, with applications from many fields, including alloys, minerals, artificial multilayer films, oxides, and charge-density-wave materials can be found in Cowley, *et al.* (1979). Structure images of the commensurate phase of the charge-density-wave lattice modulation in $2H\text{-}TaSe_2$ recorded at 60 K have been used to provide support for a particular structural model and to locate the centre of symmetry in this material (Gibson, Chen, and McDonald 1983a). The local symmetry of the charge-density wave has been determined by convergent-beam electron diffraction to be orthorhombic—for a review of related CBED work, see Steeds, Bird, Eaglesham, McKernan, Vincent, and Withers (1985). Charge-density-wave effects in $1T\text{-}TaS_2$ are analysed by HREM in Kuwabara, Hashimoto, and Endoh (1986b), who find two types of commensurate superstructures. The ordering of metal vacancies in $TiS_{1.62}$ has similarly been shown to produce a superstructure of the $(2H)_2\text{-}2C$ type from a study of HREM images by Bando, Saeki, Onoda, Kawada, and Nakahira (1980b). A systematic study of charge-density-wave effects in $1T\text{-}TaS_2$ can be found in Van Tenderloo, Landuyt, and Amelinckx (1981) and earlier papers referenced therein. A study of $SbCl_5$-intercalated graphite can be found in Salamanca-Riba, Roth, Gibson, Kortan, Dressel-haus, and Birgenau (1986). The very large number of HREM studies of long-period modulated structures (or 'infinitely adaptive' structures (Kittel 1978)) which occur in other materials, such as the aurivillius layer structures, the ferrites with the magnetoplombite structure and many other non-stoichiometric oxides and minerals, are discussed in Section 12.7, and in the conference proceedings edited by Cowley *et al.* (1979). Amongst many others, the detailed studies of $Sr_2Nb_2O_7$ by Yamamoto and Ishizuka (1983) and of mixed oxides by Yagi and Roth (1978) have demonstrated the power of HREM work for the analysis of ordering on an atomic scale, and these papers can be particularly recommended for study. The HREM method appears to be the only technique at present capable of determining the positional phase of a charge-density wave. The value of images can also be appreciated by considering the finding (Fujiwara 1957) that a series of parallel domain boundaries occurring at irregular intervals of, for example, either four or five unit cells would give rise to *sharp* diffraction spots (together with some very weak diffuse scattering) with a spacing corresponding to, say 4.5 Å in real space if the distribution of four and five cell spacings occurred with relative frequencies such that the average spacing was 4.5 Å. A conclusion that the structure contained a periodic superlattice based on diffraction evidence would then be incorrect. Thus by 'solving' the phase problem of X-ray diffraction, imaging provides information which is extremely difficult to obtain by other methods.

12.2   Ceramics, diamond and oxide superconductors

High-resolution work on the atomic structure of nitrogen platelets in diamond has progressed steadily over the past decade, and may be traced through the most recent work at a point resolution of 1.6 Å by Barry (1986). Nitrogen has been identified in a particular platelet by electron energy-loss spectroscopy (Berger and Pennycook 1982), and microdiffraction work has been reported using a sub-nanometer probe (Cowley *et al.* 1984). The stacking fault energy of type IIa diamond has been measured by the weak-beam technique to be 279 ± 41 J m^{-2} (Pirouz, Cockayne, Sumida, Hirsch, and Lang 1983). Work on cathodoluminescence from individual dislocations in diamond is described in Section 11.5.

The continuing trend toward the replacement of steel components by those made from plastics, ceramics, and concrete has stimulated research into the microstructure of these materials. In particular, the growing importance of ceramics in materials science has led to a resurgence of interest in the atomic structure of their defects. The observation of intergranular amorphous phases has aroused considerable interest (Clarke and Thomas 1977). These phases, which may be just a few Ångstroms in width, are difficult to detect by other methods but exert a controlling influence on mechanical properties at high temperatures. Atomic-resolution images of crack tips in Si_3N_4 intergranular phases can be found in Kakibayashi, Shimotsu, and Nagata (1985), and images of crack tips in Sialon are shown at high resolution in Clarke, Lawn, and Roach (1986). HREM work on the ZrO_2/Al_2O_3 interface is reported by Kraus-Lanteri, Mitchell, and Heuer (1985) in their study of the phase-transformation toughening mechanism. A martensitic tetragonal-to-monoclinic phase transition occurs in the zirconia particles under the action of an applied external

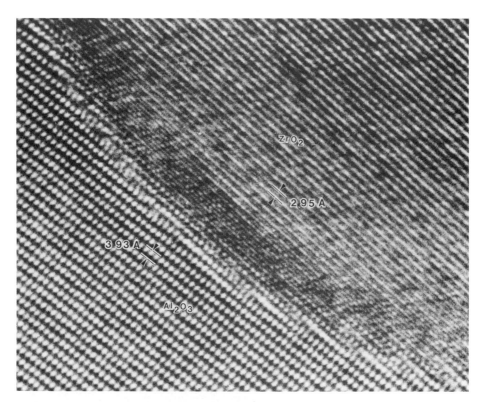

Fig. 12.1 The Al_2O_3–ZrO_2 interface. The micrograph was recorded at 200 kV. The ZrO_2 consists of finely dispersed particles which toughen the brittle Al_2O_3 host. (From Heuer, Kraus–Lanteri, Labun, Lanteri, and Mitchell (1985).)

stress, absorbing energy and so shielding crack tips from the stress. Intragranular interfaces are found to be incoherent while the intergranular interfaces contain a glassy phase. Figure 12.1 shows the ZrO_2/Al_2O_3 interface. The ordering of vacancies in vanadium carbide is studied by HREM in Yamamoto and Kumashiro (1982). The relationship between bulk structure and bounding surfaces in ZrO_2 is also analysed by HREM in Warble (1984).

With a point-resolution of less than 2 Å, modern HREM machines are capable of resolving individual atomic columns in many ceramics, including MgO. This capability is demonstrated in Hashimoto (1985). MgO is a valuable test sample for HREM work, since small smoke cubes in the [110] orientation provide an exact thickness calibration, allowing HREM images to be compared with calculations as a function of thickness (O'Keefe, Spence, Hutchinson, and Waddington 1985). Polytypes in SiC have been analysed in detail by Coene, Bender, Lovey, Van Dyck, and Amelinckx (1985b), and Van Tendeloo, Van Landuyt, and Amelinckx (1982c) and the sensitivity of these images to small atom displacements in thicker crystal have been analysed by Kuwabara, Endo, and Hashimoto (1986c). The study of polytypes in general appears to be one of the most powerful applications of the HREM technique. A combination of techniques (CBED, ELS, EDS, HREM) has been applied to the 15R and 12H polytypes in silicon aluminium oxynitride (sialon) by Bando, Mitomo, Kitami and Izumi (1987). These techniques have enabled the chemical composition, space-group and defect structure to be determined. A review of the application of HREM methods to ceramics has been given by Horiuchi (1985), including work at a resolution of about 0.17 nm on the $Si_3N_4 \cdot Y_2O_3$ crystalline–amorphous interface, on β-alumina, and on $2Nb_2O_5 \cdot 7WO_3$. A review of microanalytic studies of the component phases of toughened ceramics has been given by Bischoff and Ruhle (1985). The ZrO_2–ZrN system has been analysed by Van Tendeloo and Thomas (1985), who find monoclinic precipitates embedded in a rhombohedral matrix and an incommensurate modulated structure. A study of the ordering of oxygen vacancies and antiphase domains in mullite and sillimanite can be found in Nakajima, Morimoto, and Watanabe (1975). Crack tips have been found to be elliptical in amorphous SiO_2, and to be blunted (on an atomic scale) after wetting and relaxation, by Bando, Ito, and Tomozawa (1984). Metal–ceramic interfaces provide an important class of problems for HREM—the MgO–Ag interface is studied by HREM and microdiffraction in Lodge and Cowley (1984), who identify a new interfacial phase. Problems of de-bonding at fibre–ceramic interfaces in composite materials and at metal–ceramic interfaces in metal dispersoid systems are likely to be extensively studied by HREM in the future. As an example, the Al_2O_3–Nb interface has been analysed by Florjancic, Mader, Ruhle, and Turwitt (1985).

The vast literature on the new oxide superconductors contains many HREM studies (see, for example, Hewat, Dupuy, Bourret, Capponi and Marezio (1987), Hiraga, Shindo, Hirabayashi, Kikuchi and Syono (1987) and Beyers, Lim, Engler, Savoy, Shaw, Dinger, Gallagher and Sandstrom (1987)). HREM imaging has elucidated the role of twinning in $YBa_2Cu_3O_7$ in the tetragonal to orthorhombic phase transition (Barry (1988)). A comparison of computed and experimental images as a function of thickness in this material allows qualitative conclusions to be drawn about the oxygen ordering (Ourmazd and Spence (1987), see Frontispiece), and theoretical models to be distinguished. An entire issue of *J. Micros. Technique* has been devoted to these materials (*J. Micros. Tech.*, volume 8 No. 3 (1988)), which should be consulted for more information. Ion-beam damage may be minimized by thinning at low temperatures.

12.3 Icosahedral symmetry. Quasi-crystals

The discovery (Shechtman, Blech, Gratias, and Cahn 1984) of a class of materials with point-group symmetry $m\bar{3}\bar{5}$ which gives rise to apparently sharp diffraction patterns has aroused great interest amongst scientists in many countries. This point group is inconsistent with any three-dimensional translational group (but not with the six-dimensional groups), and it was first suspected that these alloys (of which rapidly solidified Al_6Mn is an example) were heavily twinned structures. However, HREM work (Portier, Schechtman, Gratias and Cahn 1985) has shown this to be unlikely, and the structure is not periodic in three dimensions, although it may be obtained as a projection of a periodic lattice in six dimensions. The planar spacings in particular directions follow the (non-periodic) Fibonacci series, and projection onto two dimensions is consistent with the Penrose tilting. An infinite Fibonacci series can be shown to produce sharp diffraction maxima. Three-dimensional lattices have also been proposed, based on the stacking of two kinds of rhombohedra (Mackay 1982). Structures of this class have

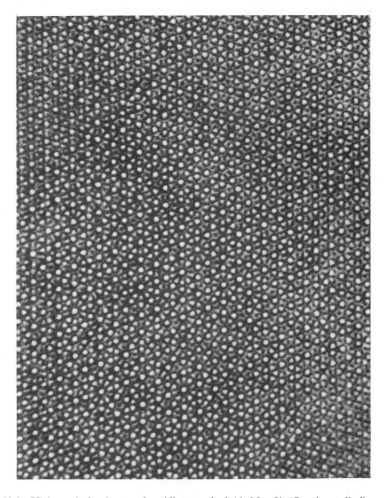

Fig. 12.2 High-resolution image of rapidly quenched $Al_{74}Mn_{20}Si_6$ 'Quasicrystal' alloy, viewed along the fivefold axis. The diffraction pattern shows tenfold symmetry. (From Hiraga, K. Hirabayashi, M., Inoue, A. and Masumoto, T. (1985). *J. Phys. Soc. Japan*. **54**, 4077.)

been called 'quasi crystals'. Unlike glasses they give sharp diffraction patterns and are not elastically isotropic.

High-resolution electron microscopy is an ideal technique for the study of these non-periodic structures. Local fivefold symmetry axes have been identified in HREM images, and speculative models have been proposed for the atomic structure (Bursill and Peng 1985; Hiraga, Hirabayashi, Inoue, and Masumoto 1985). The relationship between these structures and the Frank–Casper phases has also been analysed (Yang and Kuo 1986). Current interest centres on the existence of domains (which should be identifiable from HREM images), on the identification of slip systems and Burger's vectors (in the six dimensions in which the structures are periodic!) and on the determination of atomic coordinates. The determination of the atomic structure is, however, extremely difficult given the current resolution of HREM machines (about 0.16 nm), the closed-packed nature of the structure, and the difficulty of obtaining a simply interpretable projection in a non-periodic structure. Figure 12.2 shows a recent image of $Al_{74}Mn_{20}Si_6$ viewed down the fivefold axis. A new metastable icosahedral phase with $m\bar{3}5$ symmetry has been found in rapidly quenched TiVNi alloys (Zhang, Ye, and Kuo 1985b). The low defect energy in quasicrystals makes reproducible work on these systems difficult.

An alternative structure, based on twinned cubic crystallites of a structure containing 820 atoms per cell has also been proposed (Pauling 1987). More work needs to be done to determine why the twins expected in HREM images of Pauling's model cannot be seen (they must be extremely small). The main evidence for this model comes from measurements of interplanar angles from HREM micrographs. Accuracy in these measurements depends on very precise setting of the projector lens stigmators, which ensures that a square object is not imaged as a rectangle. On the JEOL 4000 this can be reduced to less than 0.5% by careful adjustment. At the time of writing, no sufficiently accurate electron imaging or diffraction measurements have been made to distinguish the Pauling and quasicrystal models.

12.4   Semiconductors

Ourmazd, Ahlborn, Ibeh, and Honda (1986a) have pointed out that the newest generation of HREM instruments allow the individual atomic columns in many semiconductors to be resolved in several projections, so that a true structure image can be formed for the first time (see Appendix 4). The series of volumes reporting the proceedings of the Microscopy of Semiconductor Materials Conferences (e.g., Fraser, Maher, Humphreys, Hetherington, Knoell, and Bean (1985)) provide extensive coverage of much semiconductor work in recent years. Hutchinson (1984) has also reviewed HREM work on semiconductors.

In an extensive series of papers, Bourret, Desseaux, and co-workers have analysed the core structures of partial dislocations and Lomer dislocations in several semiconductors and metals. Work on Lomer dislocations in silicon and germanium is reported in Bourret, Desseaux, and Renault (1982), who provide evidence for a new atomic model of the core structure, stabilized by impurity atom segregation to the core. Images of partial dislocation cores viewed along the dislocation line have also been reported by Sato, Hiraga, and Sumino (1980), Olsen and Spence (1981), and Bourret, Desseaux, and D'Anterroches (1981), who used these to distinguish the shuffle and glide models for the core of the 30° partial. Electronic band structure calculations have also been based on the atomic models supported by HREM evidence (Northrup, Cohen, Chelikowsky, Spence, and Olsen 1981). Images of dislocations in silicon viewed normal to the line have also been reported (Alexander, Spence, Shindo, Gottschalk, and Long 1986) using lattice fringes formed from the 'forbidden' ($\bar{4}22$)/3 type reflections due to the stacking fault in an attempt to identify kinks on the partial dislocation cores (see Fig. 12.3). Similar studies of grain boundaries in specially grown bicrystals of silicon and germanium have been reported (see Bourret and

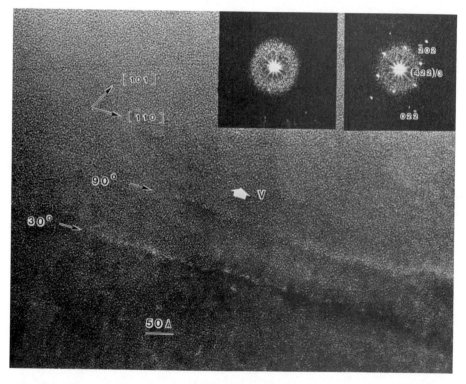

Fig. 12.3 Structure image of a stacking fault ribbon between 90° and 30° partial dislocation in silicon recorded at 200 kV. This Scherzer focus (see inset) lattice image was formed from 'forbidden' ($\bar{4}22$)/3 type reflections (shown inset and in Fig. 11.7) due to the stacking fault. Thickness is about 60 Å. The arrow indicates the direction of motion of the dislocation before it came to rest. Individual kinks can almost be seen, obscured by surface roughness. (From Alexander et al. (1986).)

Bacmann (1985) for a review). The atomic structural units at the boundaries have been elucidated, and reconstruction is observed. Tilt boundaries have been studied for a range of tilts between 0.5° and 5°, with a common [110] axis. Incommensurate interfaces are found to degenerate into commensurate portions separated by defects such as steps and dislocations, which have been the subject of separate analysis (see, for example Bourret and Desseaux (1979)).

Metal–semiconductor interfaces have been the subject of many HREM investigations, often with the aim of identifying the defects responsible for Fermi-level pinning which may occur at Schottky barriers, according to the theories of Bardeen and others. As examples, the following work may be cited: Cherns, Hetherington, and Humphreys (1984) on the $NiSi_2$–(001)Si interface; Cherns, Anstis, Hutchison, and Spence (1982) on the $NiSi_2$–(111)Si (see Fig. 12.4) in which an atomic model for a metal–semiconductor interface was proposed from HREM images; Gibson, Tung, and Poate (1983b), who investigated the important effects of vacuum conditions during preparation, and also studied the $CoSi_2$–(111)Si system. A review of all this work can be found in Gibson, Tung, Pimentel, and Joy (1985a), and readers are also referred to the special issue of *Ultramicroscopy* (**14**, No. 1/2 (1984)) devoted to the atomic structure of interfaces. Zhang, Kuo, and Wu (1986) have studied the Pd_2Si–(111)Si interface by HREM. For $NiSi_2$–(111)Si it has been found that the metal atoms nearest the interface are sevenfold coordinated, whereas for $CoSi_2$–(111)Si they appear to be in fivefold coordination. Work on the more difficult problem of metal contacts on GaAs has more recently commenced (Liliental-Weber, Gronsky, Washburn, Neuman, Spicer, and Weber 1986). Here both ohmic and Schottky barrier (rectifying) contacts may be formed, depending on the defect structure of the interface. The $Si-SiO_2$ interface has been imaged by Liliental, Krivanek, Goodnick, and Wilmsen (1985), and the degree of surface roughness correlated with electrical measurements on carrier mobility. The interface between amorphous germanium and *in situ* laser-annealed crystalline germanium has been analysed in HREM work by Cesari, Nihoul, Marfaing, Marine, and Mutaftschiev (1985). The buried Si(111) 5×5 reconstructed surface has also been observed in profile imaging work by Ourmazd, Taylor, Beuk, Davidson, Feldman, and Mannaerts (1986b). The degree of roughness at GaAs–(100)Si interfaces prepared by molecular beam epitaxy is studied at the atomic level by Hull, Rosner, Koch, and Harris (1986).

The study of oxygen precipitation in silicon has a long history (for a review, see Oehrlein and Corbett (1983). In a series of papers, Bourret and co-workers (see Bourret, Thibault-Desseaux and Siedman (1983), also Ponce, Yamashita, and Hahn (1983) and Bergholz, Hutchison, and Pirouz 1986) have elucidated the structure of the main defects formed under various heat-treatment conditions, and analysed the thermodynamic stability and transformation mechanisms for the phases observed (see Bourret (1984) for a review). Figure 12.5 shows an HREM image of one of the

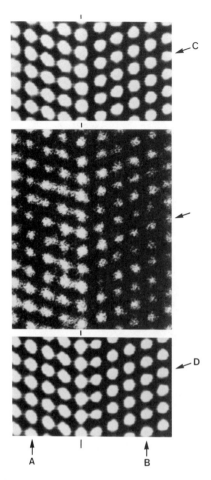

Fig. 12.4 Experimental (central panel) and computed images of the $NiSi_2$–Si (111) interface viewed along [110]. Thickness about 10 nm, 200 kV accelerating voltage. A better match is obtained with the upper panel, in which the interfacial plane of nickel atoms is in sevenfold coordination. (From Cherns *et al.* (1982).)

ribbon-like defects formed during low temperature heat treatment ($T <$ 700°C). This is believed to be hexagonal silicon, with the wurtzite structure (Bourret 1987). Similar structures are seen around indents in silicon at around 500°C (Pirouz, Chain and Samuels (1987)). Energy-loss spectroscopy has been used to identify oxygen at dislocations in germanium (see Section 11.2). Electron microdiffraction work on these radiation-sensitive precipitates is discussed in Section 11.4 (see also Kim, Spence, Long, Bergholz, and O'Keefe 1987).

HREM imagees of faulted dislocation dipoles in germanium (Chiang, Carter, and Kohlstedt 1980) and silicon (Spence and Kolar 1979) have been used to measure the stacking fault energy in these materials, and images of stacking fault tetrahedra in ion-implanted silicon can be found in Coene,

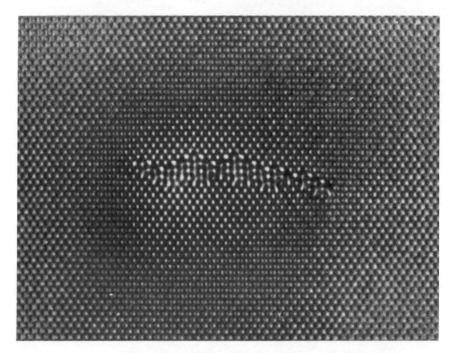

Fig. 12.5 Small precipitate viewed along [110] in a silicon crystal annealed at 650°C for 5 days. It is believed to be a hexagonal phase of silicon (wurzite structure). The identification of phases from microcrystal images such as these is extremely difficult, but has dramatically improved our understanding of oxygen precipitation phenomena in silicon. (From Bourret (1987).) A study of many samples with different controlled heat treatments is required.

Bender, and Amelinckx (1985a). Detailed calculations and comparisons with experiment for the asymmetric $(g, -g)$ contrast of weak-beam images of intrinsic stacking faults in silicon can be found in Wilson and Cockayne (1985).

Attempts to image point defects in silicon by HREM are described in Zakharov, Paseman, and Rozhanski (1982). The formidable problem of distinguishing the effects of small changes in specimen thickness on HREM images from those of bulk defects has yet to be satisfactorily solved.

An extensive literature exists on the HREM imaging of semiconductor multilayer superlattices. The possibilities for spatial mapping of composition variations in $Ga_xAl_{1-x}As$ multilayers using (200) lattice images was first analysed by Olsen, Spence, and Petroff (1980). Greatly improved images have been obtained by Hetherington, Barry, Bi, Humphreys, Grange, and Wood (1985) with the beam in the plane of the interface. These workers used two-dimensional fringes to study the degree of interfacial roughness. Specimens containing (001) interfaces, prepared by molecular-beam epitaxy were cross-sectioned and tilted through about 45° to allow lattice images to be obtained from the four composition-sensitive {200} reflections. An orientation was used in which the more thickness-sensitive {111} reflections are weak. This selective tilting technique would appear at present to offer the best hope for distinguishing thickness changes from composition changes in HREM images. Kakibayashi and Nagata (1985) have described a similar method based on Pendellösung fringes from 90°-cleaved wedge crystals, in which the thickness can be determined geometrically (see the discussion of imaging in MgO cubes in Section 5.13). Methods of distinguishing the two sublattices in GaAs, InP, and GaP by using a choice of focus or thickness which enhances the contribution of the composition-sensitive (200) reflections are discussed in Olsen et al. (1980) and Ourmazd, Rentschler, and Taylor (1986c). Work on strained layer Ge_xSi_{1-x} films has been reported by Fraser et al. (1985) and Gibson, Hull, and Bean (1985b). The abruptness of the $Ge_xSi_{1-x}(100)Si$ multilayer interface is investigated on an atomic scale by Hull, Gibson, and Bean (1985).

Several researchers have studied the defect structure of III–V and II–VI semiconductors by HREM methods. Dislocations in GaAs are reported by Tanaka and Jouffrey (1984) and in CdS by Echigoya, Pirouz, and Edington (1982), who also observe grain boundaries. It has not yet proved possible to distinguish the α, β type dislocation cores in these materials by HREM alone, but a CBED method has been developed which allows the polarity of the non-centrosymmetric zinc-blende structure to be determined (Taftø and Spence 1982). This has been combined with HREM work on faulted dislocation dipoles in GaAs (thereby allowing the α and β types to be distinguished) by DeCoonan and Carter (1987). Work on the electron-beam-induced phase transformations of CdS to Cd and crystalline CdS (from amorphous CdS) are reported in Ehrlich and Smith (1986). An extensive atomic-resolution study of the fascinating sequence of radiation-induced recrystallization effects in CdTe has been published by Ponce, Yamashita, and Sinclair (1980). The CdTe–InSb interface is analysed in Chew, Williams, and Cullis (1983) and Hutchison, Waddington, Cullis, and Chew (1986). The CdTe–GaAs interface has been studied by Cullis, Chew, Hutchison, Irvine, and Giess (1985). Measurements of stacking-fault energies in CdTe, CdS, and ZnTe are reported in Lu and Cockayne (1986a) who find similar separations for α and β type dislocations and agreement with a partially ionic model of the charge on the dislocation lines. The beam-induced motion of dislocations in CdS is analysed in Lu and Cockayne (1986b), who explain the enhanced mobility of α dislocations in terms of point-defect pinning and dislocation vibration in terms of the interaction of charged double kinks. HREM studies of the line and planar defects present in these crystals and ZnSe are also described in Sinclair, Ponce, Yamashita, and Smith (1983). Defects in ZnS are described in HREM work by Qin, Li, and Kuo (1986). The structure of two polytypes of the tetrahedrally coordinated Cu_2ZnGeS_4 structure have been determined using the powerful combination of HREM and convergent-beam electron diffraction (Moodie and Whitfield 1986) (see Section 11.4 for a fuller discussion of this technique). A similar family of structures which can be derived from the zinc blende structure have been analysed by Frangis, Van Tendeloo,

Manolikas, Spyridelis, Van Landuyt, and Amelinckx (1985) (for $AgIn_5Se_8$) and De Graaf, Bakker, Van Hemert, Van Landuyt, and Amelinckx (1984) (for $\beta MnGa_2S_4$). Metastable forms of germanium and silicon with the 'wurtzite' structure have been observed following various heating and pressure treatments (for a summary of this work, see Pirouz, Chain, and Samuels 1987).

12.5 Organic crystals

Section 6.9 discusses the application of minimum-exposure and image-processing techniques to radiation-sensitive materials, and should be read in conjuction with this section. Here we do not consider work on biological samples.

Metal phthalocyanines have provided an excellent HREM test sample for organic crystal studies for many years. In addition, these thin molecular films possess useful photoconductive properties and other interesting electronic properties. Modern work is summarized in Kobayashi, Fujiyoshi, and Uyeda (1982), who observed grain boundaries, twinning, and edge dislocations in these thin films. Defects and crystal-growth mechanisms have been studied in halogenated phthalocyanines by Smith, Fryer, and Camps (1986). The sensitivity of these images to variations in the degree of ionicity assumed in calculations has been analysed by Fujiyoshi, Ishizuka, Tsuji, Kobayashi, and Uyeda (1983). The observation of domains in Langmuir–Blodgett films is described in Fryer, Hann, and Eyres (1985) and Hann, Gupta, Fryer, and Eyres (1985). Dorset and Zemlin (1985) have studied radiation damage at low temperatures in paraffin crystals used for lattice imaging. They find no detectable damage at 0.25 nm resolution and doses of 10 electrons per squre Ångstrom. Edge dislocations have been observed in monolamellar paraffin crystals imaged at about 15 K and 0.25 nm resolution by Zemlin, Rueber, Beckmann, Zeitler, and Dorset (1985). The structure of lamellar paraffin eutectics is reviewed in Dorset (1987). Images have also been obtained from Ag·TCNQ (Uyeda, Kobayashi, Ishizuka, and Fujiyoshi 1980).

12.6 Mineralogy

Our ideas about the thermodynamic history and solid-state reactions which can occur in minerals have been radically transformed over the last decade as a result of the large amount of HREM work which has been done. Almost invariably, crystals whose idealized structure had been determined by X-ray diffraction were found on examination by HREM to contain a rich variety of point, line, and planar defects (see reviews mentioned in the introduction to this chapter). These defects frequently facilitate phase transformations, and influence diffusion, stoichiometry, and other pro-

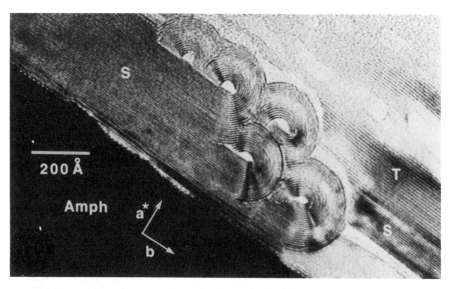

Fig. 12.6 Intergrowth of the serpentine structures (S), lizardite (planar layers), and chrysotile (curved layers). Talc (T) can also be identified. (From Veblen (1985).)

perties. Figure 12.6 shows intergrown phases of planar serpentine (lizardite) (S), chrysotile (asbestos) and talc (T), all magnesium silicates (Veblen and Buseck 1979a). The rolled layers are chrysotile.

A review of HREM work on silicates can be found in Veblen (1985). This review in particular should be consulted by new research workers interested in the study of silicates by HREM. Buseck and Cowley (1983) have reviewed HREM work on modulated and intergrowth structures. Modulated structures arise from cation site ordering or from slight atomic displacements and may provide intermediate phases during phase transformations. The following is a partial listing of recent HREM work which characterizes the defect structure, exsolution, and phase-transformation mechanisms of particular minerals: on the transformations of various non-crystalline carbonaceous materials to graphite (Buseck and Bo-Jun 1985); on exsolution in sodium mica wonesite (Veblen 1983); on the superstructures in plagioclase feldspars (Nakajima *et al.* 1977); on the defect structures of pinakiolite, ludwigite, orthopinakiolite, and takeuchiite (Bovin, O'Keefe, and O'Keefe 1981a); on the humite minerals, relating these structures and revealing superstructure (White and Hyde 1982); on manganese oxide tunnel structures (romanechite, hollandite, and nsutite (used in dry-cell batteries)) (Turner and Buseck (1979, 1983)); on leucophoenicite (White and Hyde 1983); on the oxyborates and their intergrowths, twins, and new phases (Bovin and O'Keefe 1981b); on chain width disorder in the bipyribole minerals (Veblen and Buseck 1979b); on fibrous amphiboles (Cressey, Whittaker, and Hutchison 1982); on stacking disorder in clinochlore chlorite (Spinnler, Self, Iijima, and Buseck 1984). Disorder in mica is analysed in Iijima and Buseck (1978).

In much of this work the challenge is to relate the observed defects to a transformation mechanism at the atomic level, and to relate this to the thermodynamic history of the mineral and its environment. Few new mineral crystal structures have been determined by HREM, in view of the difficulty of determining the experimental parameters (specimen thickness, focus setting) with sufficient accuracy, and, until very recently, owing to resolution limitations. In many cases, however, the fundamental 'building block' for the structure (e.g., SiO_4 tetrahedron or MO_6 octahedron) may be known and interatomic bond length information may be available to assist the analysis. An example of such a determination of a new crystal structure (takeuchiite) can be found in Bovin, O'Keefe, and O'Keefe (1981b). Further work on the structure determination of minerals by HREM can be found in Li and Hashimoto (1984), who propose using dynamical extinction effects in cebaite, and in Van Dyck, Tambuyser, Van Landuyt, and Amelinckx (1976), where imaging in dumortierite is analysed in detail.

In passing, two promising new techniques for the study of minerals which are compatible with HREM work should be mentioned. These are the use of measured 'chemical shifts' in electron energy-loss inner-shell edge spectra for the determination of oxidation state (Otten, Miner, Rask, and Buseck 1985) and the use of the ALCHEMI technique for locating the crystal site-occupancy of impurities. Both of these methods are discussed in Chapter 11.

12.7 Solid-state chemistry. Oxides

The observation by HREM of crystallographic shear planes (CS) and Wadsley defects in the Magneli phases in the late 1960s and early 1970s has stimulated a large literature on phase-transformation mechanisms and the role of planar defects in accommodating non-stoichiometry in oxides. This homologous series of transition-metal oxides M_nO_{3n-2} is based on the MO_6 octahedron and forms a family of structures which can be described in terms of ordered CS operations. These result in the systematic elimination of oxygen from the basic structure without introducing oxygen vacancies. Non-stoichiometry results from disorder amongst the CS planes, and the introduction of oxygen vacancies is not essential, as had been thought previously. Thus the earlier principle of site-conservation following the introduction of vacancies has been shown not to hold, with important consequences for entropy calculations and thermodynamic considerations. Ordering of the CS planes results in the formation of new phases. Early work on this topic has been reviewed by Allpress and Sanders (1973), Anderson (1973), and Iijima (1975b), and a theoretical approach is outlined in Stoneham and Durham (1973). More recent work is summarized in Sundberg (1980). Figure 5.28 shows a typical image of a niobium oxide. Perhaps the most extensive study has been that of reduced rutile (Blanchin, Bursill, and Smith (1985) and references therein). Here specimens prepared under controlled thermal and deformation conditions have been used to observe pairs of CS planes, platelet defects, and the atomic structure of the termination of the CS planes. The displacement vectors of CS planes are also deduced from HREM images in this work, which also provides structural models for the cation interstitial and anion vacancy.

In a similar way, reduction in the many oxides based on the tungsten bronze structure has been extensively studied, and atomic mechanisms have been proposed for the loss of oxygen. A recent review is given in Kihlborg and Sharma (1982). As examples we cite the following: Horuichi (1982) on $2Nb_2O_5 \cdot 7WO_3$; Kihlborg (1978) on disorder and superstructures in Cs_xWO_3; Yagi and Roth (1978) on the microphases present in the $Rb_2O-Ta_2O_5$, $Rb_2O-Nb_2O_5$ and $K_2O-Ta_2O_5$ mixed oxides; and Horiuchi, Muramatsu, and Matsui (1980) on $3Nb_2O_5 \cdot 8WO$.

Phase transitions in the rare-earth sesquioxides have been studied by Ben Salem, Dorbez, and Yangui (1984) and others. The higher oxides of cerium, praeseodymium, and terbium are complex structures incorporating oxygen vacancies in a fluorite substructure in which the metal atom positions are preserved. Intermediate phases with the generic formula R_nO_{2n-2} incorporate slabs of fluorite-related-oxygen deficient material in a regular way to form the ordered intermediate phases. Reviews of work on these materials can be found in Boulesteix (1983), Eyring (1988), and Eyring, Dufner, Goral, and Holladay (1985).

General HREM work on reduction and phase-transition mechanisms, non-stoichiometry, and the identification of new microphases has been reported as follows: on $Bi_2W_2O_9$, Bando, Watanabe, Sekikawa, Goto, and Horiuchi (1979b); on $Bi_7Ti_4NbO_{21}$, Horiuchi, Kikuchi, and Goto (1977), on $GeNb_9O_{25}$, Skarnulis, Iijima, and Cowley (1976); on $Na_2Ti_9O_{19}$, Bando, Watanabe, and Sekikawa (1979a); on the barium ferrite layer structures, McConnel, Hutchison, and Anderson (1974); on hollandite, Bursill and Wilson (1977); on the structure of V_2O_4 fibres, Fujiyoshi, Ishizuka, and Uyeda (1977); on MgF_2-NbO_5, Hutchison, Lincoln, and Anderson (1974); on the hexagonal ferrites, Hirotsu, Nakamura, Mizutani, Nagakura, and Nakamura (1983); on the ordering of potassium ions in cubic $KsBO_3$, Yagi and Cowley (1978); on $Na_2Ti_9O_{19}$, Bando, Watanabe, and Sekikawa (1980a); on uranium oxides, Zakharov, Gribeluk, and Vainshtein (1983); on the reduction of geothite to haematite, Watari, Delavignette, Van Landuyt, and Amelinckx (1983); on V_2O_3, Tanaka and Jouffrey (1984); on β'' aluminas, Bovin (1979); on Yb_3S_4, Ddiaz and Hyde (1983); on the $KNbO_3-Nb_2O_5$ system, Lundberg and Sundberg (1986); and on the structure of $NaNb_7O_{18}$, Marinder and Sunberg (1984). Guan, Hashimoto, and Kuo (1985) have reported and analysed images of copper oxide, and determined the optimum thickness for HREM imaging of the oxygen sublattice using dynamical effects. (This procedure depends on an absence of defects in the crystals.)

Perhaps the most promising and exciting possibility for the application of the HREM method has been to studies of the observation of individual point defects. A review of the some approaches to this problem is given in Iijima and Cowley (1978), and examples can be found in Iijima (1975b) (for point defects in $Nb_{12}O_{29}$) and in Section 12.9.

The structure of the Al/Al_2O_3 interface has been analysed in HREM work by Timsit, Waddington, Humphreys, and Hutchison (1985), while the disorder in human tooth enamel (hydroxyapatite crystals) which may act as dissolution sites is described in Bres, Barry, and Hutchison (1984). The effects of electron-beam irradiation at 400 kV has been found by Smith and Bursill (1985) to metallize the surface of tungsten oxide in recent work at a point-resolution of 0.17 nm.

12.8 Surfaces

With the steady improvement in the quality of electron microscope vacuum systems, meaningful studies of crystal surfaces have recently become possible by HREM techniques for the first time. This follows the pioneering ultra-high-vacuum TEM work on surfaces and small particles at lower resolution which has continued for many years (Poppa 1983; Barna, Radnoczi, and Safran; Klaua and Bethge 1985). The use of the electron microscope for surface imaging in the reflection mode was first attempted many years ago (Ruska 1933). Sporadic development of that technique has continued since then. Diffraction-contrast reflection images of bulk crystals can now be obtained using TEM instruments at a spatial resolution of about 0.9 nm (Hsu 1983; Uchida, Lehmpfuhl, and Jager 1984), or lattice images at higher resolution (Tanishiro, Takayanagi, and Yagi 1986). This powerful technique therefore permits *in situ* studies of surface phase transitions, such as that shown in Osakabe, Tanishiro, Yagi, and Honjo 1981 for the silicon 7×7 (111) to 1×1 surface reconstruction. General reviews of the application of electron microscopy or microdiffraction to surface imaging can be found in Yagi, Takanayagi, and Honjo (1982), Cowley (1987), Smith (1986), Venables, Spiller, Fathers, Harland, and Hanbucken (1983), Howie (1983), Klaua and Bethge (1985), and in two issues of *Ultramicroscopy* dedicated to the topic (**17** (1985) and **11**, No. 2/3 (1983)).

In addition to the application to surface science of reflection electron microscopy (REM) using TEM instruments, five other important new techniques have been developed for surface imaging in electron microscopy (see Fig. 12.7). These include (1) the scanning reflection electron microscopy (SREM) method (Cowley 1983); (2) the HREM profile imaging method (Sinclair *et al.* 1982; Marks and Smith 1983); and (3) transmission lattice-imaging of reconstructed surface layers using reflections which are forbidden in the bulk (Gibson *et al.* 1985c). (4) High resolution scanning electron microscopy in which secondary electrons are detected, and the LEED electron microscope (Bauer 1985). The REM and SREM techniques are extremely powerful, since they may be readily applied to heated or cooled bulk samples, and, in SREM, provide microdiffraction information

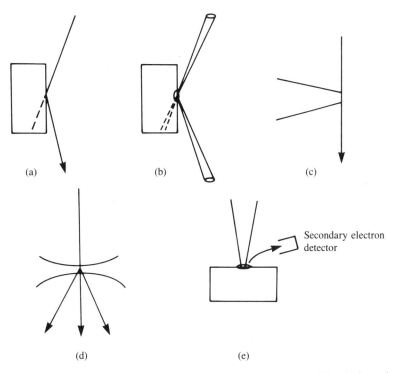

Fig. 12.7 Five electron-imaging techniques which have been developed for high-resolution imaging of surfaces: (a) reflection electron microscopy (REM); (b) scanning reflection electron microscopy (SREM); (c) HREM profile imaging; (d) conventional HREM using clean samples with reconstructed surfaces; (e) high-resolution SEM imaging. A LEED electron microscope (Bauer (1985)) has also been constructed. Techniques (c) and (d) require thinned samples, the remainder use bulk material.

from sub-nanometer regions on local surface crystallography. In addition, the X-ray microanalysis technique may be combined with SREM. Computational techniques for the calculation of dynamical REM images of defects are described in Peng and Cowley (1986), Cowley and Peng (1985), and Shuman (1977). The extensive literature on RHEED theory, although generally restricted to diffraction (rather than imaging or microdiffraction) from perfect crystal surfaces (rather than defects) is also relevant. Lattice images of the Si(111) 7×7 reconstruction were first demonstrated in the REM mode by Tanishiro *et al.* (1986), and this is illustrated in Fig. 12.8. Here the 7×7 reconstructed cell can be seen crossing a surface step. Antiphase domains have also been observed. This work has greatly expanded our understanding of surface structure and of phase-transformation mechanisms on surfaces, and, in particular, on the role which defects such as surface steps play in them. A novel application of the REM technique has been to semiconductor multilayer films (Yamamoto and Muto 1984).

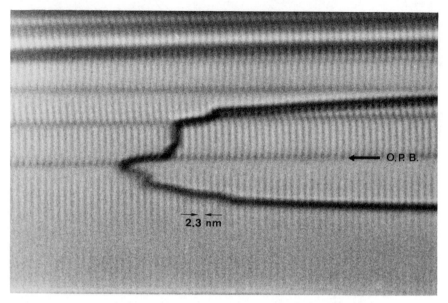

Fig. 12.8 Lattice image of the Si (111) 7 × 7 structure recorded in the reflection mode (REM) in a UHV/TEM instrument. The incident beam direction is close to [1̄10]. Over-focus increases up the page. Note the foreshortening and differing orthogonal magnifications. Surface phase transitions are readily studied by this method. (From Tanishiro *et al.* (1986).)

Quantitative transmission electron diffraction from thin samples with clean reconstructed surfaces appears also to be a powerful new technique, in which the weak dynamical coupling of surface superlattice beams to bulk beams may allow a kinematic interpretation (Spence 1983). Using this technique, Takayanagi, Tamishiro, Takahashi, and Takahashi (1985) have derived a model for the atomic struture of the Si(111) 7 × 7 surface which is now widely accepted, and which is consistent with evidence from scanning tunnelling microscopy and other techniques. The origin of the 'forbidden reflections' used is closely related to the origin of those which arise due to termination effects (Cherns 1974; Nihoul, Abdelmoula, and Metois 1984). An example of this important effect is shown in Fig. 12.3, and the history of its use is outlined in Alexander *et al.* (1986). Weak-beam dark-field techniques (Kambe and Lehmpfuhl 1975) and axial bright-field methods in transmission (Iijima 1981; Moodie and Warble 1967) have also proved powerful.

The HREM profile imaging method has also proved useful for the direct observation of reconstructed surfaces. Figure 12.9 (inset) shows the gold (110) surface viewed in profile, showing surface reconstruction (Marks and Smith 1983).

The field of ultra-high-vacuum (UHV) electron microscopy appears to be growing rapidly, and at least one company now offers a commercial UHV TEM. A UHV modification to the Philips EM430, including facilities for

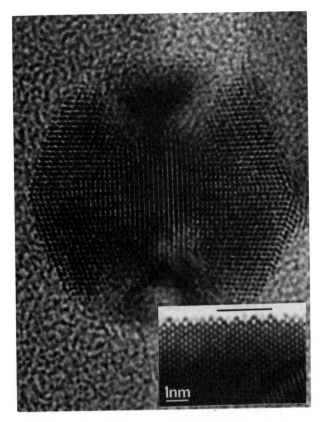

Fig. 12.9 Icosahedral particle of silver imaged at 0.18 nm point-resolution (500 kV). Atomic columns are black. A twin is indicated. Inset shows the partially reconstructed (110) surface of a similar gold particle, showing the 2×1 structure in which the period is doubled along [100]. Surface relaxations have been measured from these images, which are apparently consistent with STM and X-ray work. (From Marks and Smith (1983).)

in-situ specimen preparation, is described in Swan, Jones, Krivanek, Smith, Venebles, and Cowley (1987). The technical problems of heating, tilting, and *in situ* specimen preparation, evaporation, or exchange in a UHV instrument are considerable. However, only the scanning tunnelling microscopy (STM) method appears to offer similar possibilities for the study of individual surface defects and their role in surface phase transformations. By comparison with the STM method, the electron microscopy techniques offer slightly higher lateral resolution, a wider field of view, more straightforward image interpretation, the ability to search rapidly and 'zoom-in' on significant features, and the ability to study sub-surface features such as 'surface' stacking faults (in the profile HREM mode). The STM, on the other hand, induces little radiation damage, possesses superior resolution normal to the surface, provides spectroscopic information on an atomic scale, and operates over a wide range of pressures (or in liquids). A scanning tunnelling microscope in the form of a side-entry holder for use in the Philips EM400 transmission electron microscope has been described (Spence 1988). The electron microscope is operated in the reflection mode, and the STM tip axis is horizontal.

All this work has provided some of the most important developments in HREM work over the last decade, with exciting possibilities for the future. Just as electron microscope studies of bulk crystals have demonstrated the crucial role of defects in controlling bulk properties, so we may anticipate that future developments in UHV HREM work will elucidate the fundamental role which surface defects play in the processes of crystal growth, sublimation, epitaxy, surface reconstruction, and surface phase transformations. Some of these defects include surface steps, emerging bulk dislocations, vacancies and physisorbed or chemisorbed adatoms, surface stacking faults, surface antiphase domains, surface kinks, emerging bulk grain boundaries, and all the possible combinations of these. The additional exciting possibilities for the study of metal–semiconductor, metal–ceramic, and metal–metal interfaces prepared *in situ* are likely to prove equally significant. As an example, Takayanagi, Kolb, Kambe, and Lehmpfuhl (1980), have studied *in situ* evaporated lead monolayers on silver (111).

12.9 Metals

HREM work on metals may be broadly divided into two categories; studies of individual defects and work on ordered alloys and stacking disorder. We consider first work on small defects.

Stacking fault tetrahedra have been imaged at atomic resolution by Ajika, Hashimoto, and Takai (1985a). Variations in atomic column contrast are also attributed to bulk vacancies. Guinier–Preston zones have been imaged in Al–Cu alloys by Casanove-Lahana, Dorignac, and Jouffrey (1981) and Ajika, Endoh, Hashimoto, and Tomita (1985b), both supporting the model of a copper monolayer. Extensive dynamical calculation indicating the optimum conditions for the imaging of split interstitials in gold and aluminium have been reported by Fields and Cowley (1978). The core structure of edge dislocations in titanium have been studied by De Crecy, Bourret, Naka, and Lasalmonie (1983) and the stacking fault energy (150 erg cm^{-2}) was determined. The dissociation model of Tyson is supported. Low-angle tilt boundaries in aluminium containing edge and 60° dislocations are studied in Penisson and Bourret (1979). The measurement of small rigid-body displacements across boundaries in metals is discussed for the case of twins in copper by Wood, Stobbs, and Smith (1984), who find that an accuracy of about 2 per cent can be obtained from lattice images. Grain boundaries in molybdenum are studied by HREM in Penisson, Gronsky, and Brosse (1982). Work on the structural changes which accompany tempering in carbon steel is reported in Nagakura, Hirotsu, Kusunoki, Suzuki, and Nakamura (1983). Self, O'Keefe, and Stobbs (1983) report an

HREM study of the σ phase in duplex stainless steel. A review of applications of the Berkeley atomic resolution microscope to problems such as HfN precipitates in Mo and N precipitates in iron is given in Gronsky, Thomas, and Westmacott (1985). Work on Au–Sn and Au–In alloys is reviewed in Eyring *et al.* (1985).

The study of ordering in binary alloys has a long history extending well beyond the field of this book. For a review, see Cowley *et al.* (1979); theoretical treatments may be traced through the work of Cowley and Wilkins (1972). Here the shape of the Fermi surface is related to the pair-potential which controls ordering in these systems. An extensive literature exists on the study of antiphase boundaries and superstructures in ordered alloys by HREM. For example, Van Tendeloo and Amelinckx (1982b) have reported a new superstructure in $Au_{35}Mn_{13}$; work on stacking disorder in Cu–Al alloys is reported in Lovey, Van Tendeloo, Van Landuyt, Delaey, and Amelinckx (1984). Ordering mechanisms and domain structure in various binary alloys have also been studied in other systems as follows: Ni_3Mo (Schryvers, Van Tendeloo, and Amelinckx 1985b); Cu_4Ti (Zhang, Ye, and Kuo 1985b); Nb_3Sn (Takeda, Yoshida, and Hashimoto 1985b); AuMg alloys (Van Tendeloo and Amelinckx 1982a; Shindo, Hiraga, and Hirabayashi 1983); AuCd alloys (Hiraga *et al.* 1977); Pd_3Mn (Schryvers, Van Landuyt, Van Tendeloo, and Amelinckx 1982); Pt_3V (Schryvers and Amelinckx (1985); and for long-period superstructures in CuAu(II) (Takeda and Hashimoto 1985). An extensive study of the long-period superstructures in AuCd alloys is reported in Hiraga, Shindo, and Hirabayashi (1981). A review of a large amount of HREM work which has been done on the tetrahedrally close-packed Frank–Casper phases in superalloys can be found in Kuo, Ye, and Li (1986). They identify the σ, Laue, and μ phases (which cause embrittlement in these alloys, used for gas-turbine engine blades), and discuss their relationship to quasi-crystals (Yang and Kuo 1986). The important question of how to optimize imaging conditions for a weak superlattice modulation and corresponding computing approximations is discussed in Shindo (1982). The power of electron microdiffraction techniques (using sub-nanometer sized probes) for the study of antiphase domain boundaries in alloys should also not be overlooked (see Section 11.4).

12.10   Non-crystalline solids

The study of the atomic structure of 'amorphous' materials by electron microscopy has continued for many years, and has recently been reviewed by Howie (1988) and Saito (1984). Many papers have addressed, for example, the question of the degree of order in amorphous silicon (Gibson and Howie 1978; Krakow, Ast, Goldfarb, and Siegel (1976); Smith, Stobbs, and Saxton 1981; Ourmazd, Bean and Phillips 1985). Much of this work has sought to distinguish the two major competing models for non-crystalline solids—the microcrystalline model and the continuous random network model.

There can be no doubt that microcrystals have been seen in some images. However, patches of fringes may appear by chance in the projected structure, and the probability of this increases with thickness (Cochran 1973). Full consideration must be given to all electron optical artifacts (Fan, Cowley, and Spence 1987). The size of these patches of fringes, which may or may not reflect the existence of microcrystals, is found to depend sensitively on the preparation conditions of the films, including the hydrogen pressure during evaporation. In one case at least, in samples prepared by plasma chemical vapour deposition (CVD) in the presence of hydrogen (Furakawa, Seki, and Maeyama 1986), the evidence for microcrystals is overwhelming. It would appear that (1) the newest atomic-resolution HREM machines provide the best evidence available on the degree of crystallinity; though the projection problem remains intractable; and (2) microcrystallite size depends sensitively on preparation conditions. HREM work in this field is unfortunately frequently dependent on subjective judgements about the size and existence of fringe patches. An attempt to quantify the degree of order, based on the autocorrelation function of HREM images, can be found in Fan and Cowley (1985). Similar approaches based on electron microdiffraction patterns have been attempted. An ingenious but subjective technique due to Krivanek, Gaskell, and Howie (1976) uses a random phase-plate in the back-focal plane of an optical bench to test for 'perfect disorder' in the reconstructed image.

It is known that crystalline structures in nature generally possess a lower total energy than do random networks of similar atoms. The phenomenon of spontaneous crystallization has also been reported as the thickness of amorphous films is increased. An extensive literature exists on the theory of amorphous and glassy materials (see, for example, Elliott (1983)). In summary, it would appear that preparation conditions, which have only very recently become reproducible, exert a controlling influence on the structure of thin amorphous films, and that because of projection difficulties, HREM provides a useful characterization technique only for the thinnest films (e.g. less than 3 nm thick for 100 kV instruments) and at the very highest resolution obtainable. HREM work on amorphous SiO_2 can be found in Gibson and Dong (1980), in which pore size is related to water content.

12.11   Small particles, atomic clusters, catalysis

The atomic and electronic structures of the surfaces of catalysts are of crucial importance in determining their catalytic activity. For example, it is now believed that an alkaline environment may induce the 2×1 reconstruction of the Ag (110) surface, which favours the catalytic process. These structural changes on surfaces are driven by changes in electronic structure, which in turn are sensitive to the effects of promoters or poisoning agents.

Special high-energy sites, such as steps, ledges, surface vacancies, and corners may be particularly important. A considerable literature (e.g., *The Journal of Catalysis*) exists on the problem of explaining catalytic activity, and many spectroscopic techniques have been developed to analyse the electronic structure of catalysts. These methods generally provide information averaged over large areas and provide little information on surface crystallography or defects.

It has now become possible to determine the atomic arrangements on the surfaces of model catalyst particles directly by the HREM profile imaging technique in favourable cases (see also Section 12.8). Surface reconstructions and defects can therefore be observed directly. A review of this work and its relationship to other techniques in catalysis research can be found in Smith and Marks (1985), and in the issue of *Ultramicroscopy* devoted to this topic (**20**, Nos. 1 and 2, (1986)). Figure 12.9 shows a typical multiply-twinned small silver particle. Dislocations have been observed in icosahedral multiply-twinned particles, and twins, stacking faults, faceting, and surface reconstruction have all been observed (Marks and Smith 1983). More recently, fascinating video recordings have been obtained which show the movement of atoms near the surfaces of small particles in real time (Bovin, Wallenberg, and Smith 1985; Iijima and Ichihashi 1985). Details of the typical video recording apparatus used are given in Section 11.3. These spectacular atomic resolution 'movies' give direct evidence for the random-walk behaviour of surface atoms, and also show the formation of 'atomic clouds' outside surfaces. Individual atoms (or columns of atoms) are seen to 'hop' from one surface site to the next in the profile geometry. Microcrystals, containing twins and other defects, are seen to form from these 'atom' clouds before later evaporating. This work demonstrates the practicality of achieving atomic resolution from images recorded in about 1/30 second. Many of the effects observed, however, are directly induced by the electron beam, and may depend on the vacuum conditions. The provision of real-time image-recording facilities and image-intensifiers on the new generation of ultra-high-vacuum HREM machines will allow these beam-induced effects (and those due to the vacuum environment) to be isolated by studies in which the beam intensity is varied.

Many other catalysts have been studied by the HREM method. Work on an industrial tellurium–molybdenum oxide catalyst system is described in Gai, Boyes, and Bart (1982), who used an environmental cell to simulate reducing and oxidizing environments. No evidence for the idea that crystallographic shear planes may play a role in the catalytic process was obtained; however, this work demonstrates the power of *in situ* experiments in high-voltage machines. HREM profile imaging in catalytically active spinel $ZnCrFeO_4$ is described in Hutchison and Briscoe (1985), who observed steps, ledges, and reconstructions, some of which were beam-induced. Work on zeolites is described in Rai, O'Keefe, and Thomas (1985) and Millward, Thomas, and Glaeser (1985). Here, low-dose HREM imaging is proposed as a method for obtaining the projected structure of these crystallites, which cannot be studied by conventional X-ray methods. Evidence for the importance of a surface amorphous phase in vanadium oxide catalysts is given in HREM work by Andersson and Bovin (1985).

Since HREM techniques provide little chemical information, the complementary techniques of X-ray microanalysis (Lyman 1984) have proved useful in catalysis research. In addition, the powerful 'Z-contrast' technique, first proposed by Crew, Langmore, and Isaacson (1975) has been developed and applied to platinum and palladium catalysts by Treacy, Howie, and Wilson (1978). By using certain combinations of the high-angle dark-field STEM detector and energy-loss signals, these workers have developed techniques for eliminating unwanted diffraction-contrast effects and deriving an image signal which is strongly dependent on atomic number but little affected by thickness variations in the supporting substrate. The STEM instrument also becomes extremely powerful in small-particle research when used for microdiffraction, which allows the crystallography and twinning of individual particles to be determined (Monosmith and Cowley 1984; Cowley and Spence 1981). The cathodoluminescence technique in STEM (see Section 11.5) would also appear to hold considerable promise for catalysis research in view of its sensitivity to electronic structure, high energy resolution, and high spatial resolution.

In summary, it would appear that, with the provision of ultra-high-vacuum facilities and further developments in environmental stages, HREM work on catalysts and small particle will continue to develop as one of the most exciting areas of research in electron microscopy. Nevertheless one of the most important outstanding problems in research into very small clusters, the determination of their three-dimensional shape, cannot easily be solved by HREM techniques. For clusters consisting of a few dozen atoms, there appears to be no simple method of three-dimensional structure imaging, other than by relying on the similarity between particles found lying in different orientations.

12.12   Summary of applications of HREM

While the HREM technique has frequently been criticized for not being quantitative, there can be no doubt that the HREM images of defects and microphases that have been published in recent years have profoundly altered the way scientists think about the atomistic modelling of atomic processes in crystals. Put simply, HREM has shown over the last fifteen years that most crystals contain a greater variety and density of structural defects than was previously believed. These high-energy defect sites (in addition to crystal surfaces and their defects) frequently exert a controlling influence on the physical properties and phase-transformations mechanisms of crystals, so that an understanding of their structure provides an essential basis for theoretical modelling.

When the first edition of this book was written, the point-resolution of

HREM instruments was only slightly less than the dimensions of a typical unit cell or structural unit. For many minerals and oxides this resolution was sufficient to identify intergrowths and polytypes, and to allow much informed speculation on phase-transition mechanisms at the atomic level. True structure imaging at atomic resolution is now possible (see Appendix 4), and we are therefore starting to see the elucidation of the atomic structure of those special sites which control chemical, physical, and electronic properties. Some of the most dramatic results have been obtained in the following areas: the atomic structure of small particles and atomic clusters; the atomic mechanisms by which oxides accommodate changes in stoichiometry; the imaging of charge-density waves; the atomic structure of interfaces; the mechanisms of phase transitions on surfaces and in the bulk; and real-time recording of atomic motion on surfaces. The identification of polytypes, the structural analysis of long-period modulated structures, and the determination of the atomic structure of individual defects such as dislocation cores have also been important areas.

During this period, the limitations of the HREM method have also become more clearly apparent. These include the unwanted effects of radiation damage; the intractability of the projection problem; the lack of chemical information; and the need to study thin films which may be unrepresentative of bulk material. The availability of high-tilt goniometers for HREM machines may, however, address the projection problem by allowing images of the same area to be obtained in two or more crystal orientations. The supplementary techniques described in Chapter 11 go some way to overcoming these limitations, while the development of ultra-high-vacuum microscopes opens up the possibility for the first time of undertaking reproducible experiments in surface science using an electron microscope.

References

Ajika, N., Hashimoto, H., and Takai, Y. (1985a). Electron microscope observation of the structure and behaviour of stacking fault tetrahedra and single vacancies in gold crystals irradiated by 2 MeV electrons. *Phys. Stat. Sol.* **87**(a), 235.

Ajika, N., Endoh, H., Hashimoto, H., and Tomita, M. (1985b). Interpretation of atomic-resolution electron microscope images of Guiner–Preston zones in aluminium–copper alloys. *Phil. Mag.* **A51**, 729.

Anderson, J. S. (1973). On infinitely adaptive structures. *J. Chem. Soc. Dalton* **10**, 1107.

Andersson, A. and Bovin, J.-O. (1985). Amorphous vanadium oxide catalyst surface selective in the ammoxidation of 3-Picoline. *Naturwissenschaften* **72**, 209.

Alexander, H., Spence, J. C. H., Shindo, H., Gottschalk, H., and Long, N. (1986). Forbidden-reflection lattice imaging for the determination of kink densities on partial dislocations. *Phil. Mag* **A53**, 627.

Allpress, J. G. and Sanders, J. V. (1973). The direct observation of the structure of real crystals by lattice imaging. *J. Appl. Crystallogr.* **6**, 165.

Bando, Y., Watanabe, M., and Sekikawa, Y. (1979a). Structure analysis of $Na_2Ti_9O_{19}$ by 1 MV high-resolution electron microscopy. *Acta Crystallogr.* **B35**, 1541.

Bando, Y., Watanabe, A., Sekikawa, Y., Goto, M., and Horiuchi, S. (1979b). New layered structure of $Bi_2W_2O_9$ determined by 1 MV high-resolution electron microscopy. *Acta Crystallogr.* **A35**, 142.

Bando, Y., Watanabe, M., and Sekikawa, Y. (1980a). The structure of orthorhombic $Na_2Ti_9O_{19}$, a unit-cell twinning of monoclinic $Na_2Ti_9O_{19}$, determined by 1-MV high-resolution electron microscopy. *J. Solid-State Chem.* **33**, 413.

Bando, Y., Saeki, M., Onoda, M., Kawada, I., and Nakahira, M. (1980b). $(2H)_2$-2C type superstructure of $TiS_{1.62}$ determined by high-resolution electron microscopy. *J. Solid-State Chem.* **34**, 381.

Bando, Y., Ito, E., and Tomozawa, M. (1984). Direct observation of crack tip geometry of SiO_2 glass by high-resolution electron microscopy. *J. Am. Ceramic Soc.* **67**, C-36.

Bando, Y., Mitomo, M., Kitami, Y., and Izumi, F. (1987). Structure and composition analysis of silicon aluminium oxynitride polytypes by combined use of structure imaging and microanalysis. *J. Microscopy* (in press).

Barna, A., Barna, P. B., Radnoczi, G., and Safran, G. (1984). *In situ* UHV TEM study of the two-dimensional growth of Al_2Au phase on Al(111) surface. *Ultramicroscopy* **15**, 101.

Barry, J. (1986). Direct structure images of (100) platelets in diamond. *Proc. Eleventh Int. Cong. on Electron Microscopy,* Kyoto, p. 799.

Barry, J. (1988). Oxygen ordering and twinning in $YBa_2Cu_3O_7$. *J. Electron Micros. Technique* **8**, 325.

Batson, P. E. (1985). Inelastic scattering of fast electrons in clusters of small spheres. *Surf. Sci.* **156**, 720.

Bauer, E. (1985). The resolution of the low energy reflection electron microscope. *Ultramicroscopy* **17**, 51.

Ben Salem, M., Dorbez, R., and Yangui, B. (1984). Ferroelastic character and study by HREM of the mechanism of the hexagonal-monoclinic phase transition of rare earth sesquioxides. *Phil. Mag.* **A50**, 621.

Berger, S. and Pennycook, S. J. (1982). Detection of nitrogen at (100) platelets of diamond. *Nature (London)* **298**, 635.

Bergholz, W., Hutchison, J. L., and Pirouz, P. (1986). Enhanced oxygen diffusion and precipitation in silicon. *J. Microsc.* **141**, 143.

Beyers, R., Lim, G., Engler, E. M., Savoy, R. J., Shaw, T. M., Dinger, T. R., Gallagher, W. J. and Sandstrom, R. L. (1987). Crystallography and microstructure of $YBa_2Cu_3O_7$ perovskite based superconducting oxide. *Appl. Phys. Letts.* **50**, 1918.

Bischoff, E. and Ruhle, M. (1985). Microanalytical TEM studies of ceramic materials. *Mat. Res. Soc. Symp. Proc.* Vol. 41, p. 227. Elsevier, New York.

Blanchin, M. G., Bursill, L. A., and Smith, D. J. (1985). Extended defects in deformed rutile. *Phys. Stat. Sol.* **89(a)**, 559.

Boulesteix, C. (1983). High resolution and conventional electron microscopy studies of repeated wedge microtwins in monoclinic rare earth sesquioxides. *J. Microsc.* **129**, 315.

Bourret, A. (1984). Oxygen aggregation in silicon. *Thirteenth International Conference on Defects in Semiconductors* (ed. L. C. Kimberling and J. M. Parsey, Jr.) p. 129.

Bourret, A. (1987). Defects induced by oxygen precipitation in silicon: a new hypothesis involving hexagonal silicon. In *Electron microscopy of semiconductors 1987.* (I.O.P., Bristol). In press.

Bourret, A. and Bacmann, J. J. (1985). Atomic structure of grain boundaries in semiconductors studied by electron microscopy (Analogy and difference with surfaces). *Surf. Sci.* **162**, 495.

Bourret, A. and Desseaux, J. (1979). The low-angle [011] tilt boundary in germanium I. High-resolution structure determination. *Phil. Mag.* **A39**, 405.

Bourret, A., Desseaux, J., and D'Anterroches, C. (1981). Defect structure in CZ silicon and germanium studied by HREM. *Inst. Phys. Conf. Ser.* No. 60, p. 9. Institute of Physics, Bristol. [See also article by Anstis, Hirsch, Humphreys, Hutchison, and Ourmazd in the same proceedings, p. 15.]

Bourret, A., Desseaux, J., and Renault, A. (1982). Core structure of the Lomer dislocation in germanium and silicon. *Phil. Mag.* **A45**, 1.

Bourret, A., Thibault-Desseaux, J., and Seidman, D. N. (1983). Early stages of oxygen segregation and precipitation in silicon. *J. Appl. Phys.* **55**, 825.

Bovin, J.-O. (1979). High-resolution electron microscopy images of defects in Mg- and Li-Stabilized β''-aluminas. *Acta Crystallogr.* **A35**, 572.

Bovin, J.-O. and O'Keefe, M. (1981). Electron microscopy of oxyborates. II. Intergrowth and structural defects in synthetic crystals. *Acta Crystallogr.* **A37**, 35.

Bovin, J.-O., O'Keefe, M., and O'Keefe, M. A. (1981a). Electron microscopy of oxyborates. I. Defect structures in the minerals pinakiolite, ludwigite, orthopinakiolite and takeuchiite. *Acta Crystallogr.* **A37**, 28.

Bovin, J.-O., O'Keefe, M., and O'Keefe, M. A. (1981b). Electron microscopy of oxyborates. III. On the structure of takeuchiite. *Acta Crystallogr.* **A37**, 42.

Bovin, J.-O., Wallenberg, R., and Smith, D. J. (1985). Imaging of atomic clouds outside the surfaces of gold crystals by electron microscopy. *Nature (London)* **317**, 47.

Bres, E. F., Barry, J. C., and Hutchison, J. L. (1984). A structural basis for the carious dissolution of the apatite crystals of human tooth enamel. *Ultramicroscopy* **12**, 367.

Bursill, L. A. and Peng Ju Lin (1985). Penrose tiling observed in a quasi-crystal. *Nature (London)* **316**, 50.

Bursill, L. A. and Wilson, A. R. (1977). Electron-opical imaging of the hollandite structure at 3 Å resolution. *Acta Crystallogr.* **A33**, 672.

Buseck, P. R. and Bo-Jun, H. (1985). Conversion of carbonaceous material to graphite during metamorphism. *Geochim. Cosmochim. Acta* **49**, 2003.

Buseck, P. R. and Cowley, J. M. (1983). Modulated and intergrowth structures in minerals and electron microscope methods for their study. *Am. Mineral.* **68**, 18.

Buseck, P. R., Cowley, J. M., and Eyring, L. (1988). *High-resolution electron microscopy. Imaging and Analysis.* Oxford University Press, Oxford.

Casanove-Lahana, M. J., Dorignac, D., and Jouffrey, B. (1981). Computed high-resolution Guinier–Preston zone images in an Al–Cu alloy. *Inst. Phys. Conf. Ser.* No. 61, p. 377. Institute of Physics, Bristol.

Cesari, C., Nihoul, G., Marfaing, J., Marine, W., and Mutaftschiev, B. (1985). High-resolution electron-microscopy studies on laser-annealed unsupported amorphous germanium films. *J. Appl. Phys.* **57**, 5199.

Chiang, S.-W., Carter, C. B., and Kohlstedt, D. L. (1980). Faulted dipoles in germanium. A high-resolution transmission electron microscopy study. *Phil. Mag.* **A42**, 103.

Chew, N. G., Williams, G. M., and Cullis, A. G. (1983). Transmission electron microscope studies of heteroepitaxial CdTe on (001) InSb substrates. *Inst. Phys. Conf. Ser.* No. 68, 437.

Cherns, D. (1974). Direct resolution of surface atomic steps by transmission electron microscopy. *Phil. Mag.* **30**, 549.

Cherns, D., Anstis, G. R., Hutchison, J. L., and Spence, J. C. H. (1982). Atomic structure of the $NiSi_2$/(111)Si interface. *Phil. Mag.* **A46**, 849.

Cherns, D., Hetherington, C. J. D., and Humphreys, C. J. (1984). The atomic structure of the $NiSi_2$–(001)Si interface. *Phil. Mag.* **A49**, 165.

Clarke, D. R. and Thomas, G. (1977). Grain boundary phases in a hot-pressed MgO fluxed silicon nitride. *J. Am. Ceramic Soc.* **60**, 491.

Clarke, D. R., Lawn, B. R., and Roach, D. H. (1986). The role of surface forces in fracture. In *Fracture mechanics of ceramics* Vol. 8 (ed. R. C. Bradt, A. G. Evans, D. P. H. Hasselman, and F. F. Lange). Plenum Press, New York.

Cochran, W. (1973). Theory of electron micrographs of amorphous materials. *Phys. Rev.* **B8**, 623.

Coene, W., Bender, H., and Amelinckx, A. (1985a). High-resolution structure imaging and image simulation of stacking fault tetrahedra in ion-implanted silicon. *Phil. Mag.* **A52**, 369.

Coene, W., Bender, H., Lovey, F. C., Van Dyck, D., and Amelinckx, S. (1985b). On the influence of crystal orientation on the high-resolution image contrast of polytypes. *Phys. Stat. Sol.* **87**, 483.

Cowley, J. M. (1978). High-resolution electron microscopy of crystal defects and surfaces. *Annu. Rev. Phys. Chem.* **29**, 251.

Cowley, J. M. (1983). The STEM approach to the imaging of surfaces and small particles. *J. Microsc.* **129**, 253.

Cowley, J. M. (1987). Electron microscopy of surface structure. *Progress in surface science* Vol. 21(3), pp. 209–250. Pergamon, New York.

Cowley, J. M. and Peng, L.-M. (1985). The image contrast of surface steps in reflection electron microscopy. *Ultramicroscopy* **16**, 59.

Cowley, J. M. and Spence, J. C. H. (1981). Convergent beam electron microdiffraction from small crystals. *Ultramicroscopy* **6**, 359.

Cowley, J. M. and Wilkins, S. W. (1972). In *Interatomic potentials and simulation of lattice defects* (ed. P. C. Gehlen, J. R. Beeler, and R. I. Jaffee), p. 265. Plenum Press, New York.

Cowley, J. M., Cohen, J. B., Salamon, M. B., and Wuensch, B. J. (1979). *Modulated Structures-1979*. AIP Conference Proceedings (Series ed. Hugh C. Wolfe) No. 53.

Cowley, J. M., Osman, M. A., and Humble, P. (1984). Nanodiffraction from platelet defects in diamond. *Ultramicroscopy* **15**, 311.

Cressy, B. A., Whittaker, E. J. W., and Hutchison, J. L. (1982). Morphology and alteration of asbestiform grunerite and anthophyllite. *Mineral. Mag.* **46**, 77.

Crewe, A. V., Langmore, J. P., and Isaacson, M. S. (1975). In *Physical aspects of electron microscopy and microbeam analysis* (ed. B. M. Siegel and D. R. Beaman), p. 47. Wiley, New York.

Cullis, A. G., Chew, N. G., Hutchison, J. L., Irvine, S. H. C., and Giess, J. (1985). HREM studies of II–VI heteroepitaxial layers. *Inst. Phys. Conf. Ser.* No. 76, p. 29. Institute of Physics, Bristol.

DeCoonan, B. C. and Carter, C. B. (1987). Faulted dipoles in GaAs. *Appl. Phys. Lett.* **50**, 40.

De Crecy, A., Bourret, A., Naka, S., and Lasalmonie, A. (1983). High-resolution determination of the core structure of $1/3\langle11\bar{2}0\rangle\{10\bar{1}0\}$ edge dislocation in titanium. *Phil. Mag.* **A47**, 245.

De Graaf, M., Bakker, M., Van Hemert, M., Van Landuyt, J., and Amelinckx, S. (1984). Structure Models for β-MnGa$_2$S$_4$ as derived from electron diffraction and high-resolution electron microscopy. *J. Solid-State Chem.* **55**, 133.

Diaz, C. O. and Hyde, B. G. (1983). On the non-stoichiometric ytterbium sulphide phase 'Yb$_3$S$_4$'. *Acta Crystallogr.* **B39**, 569.

Dorset, D. L. (1987). Crystal structure of lamellar paraffin eutectics. *Macromolecules* (in press).

Dorset, D. L. and Zemlin, F. (1985). Structural changes in electron irradiated paraffin. *Ultramicroscopy* **17**, 229.

Echigoya, J., Pirouz, P., and Edington, J. W. (1982). Preliminary studies of crystal defects in cadmium sulphide by high-resolution transmission electron microscopy. *Phil. Mag.* **A45**, 455.

Ehrlich, D. J. and Smith, D. J. (1986). Electron beam stimulated nonthermal crystallization of CdS surface layers: Observations by real-time atomic-resolution electron microscopy. *Appl. Phys. Lett.* **48**, 1751.

Elliot, S. R. (1983). *Physics of Amorphous materials*. Longmans, Harlow.

Eyring, L. (1988). In *High-resolution electron microscopy* (ed. P. Buseck, J. M. Cowley, and L. Eyring) Oxford University Press, Oxford.

Eyring, L., Dufner, C., Goral, J. P., and Holladay, A. (1985). HREM studies of chemical reactions in thin films. *Ultramicroscopy* **18**, 253.

Fan, G. Y. and Cowley, J. M. (1985). Auto-correlation analysis of high-resolution electron micrographs of near-amorphous thin films. *Ultramicroscopy* **17**, 345.

Fan, G. Y., Cowley, J. M., and Spence, J. C. H. (1987). Comment on submicrocrystallites and the orientation proximity effect. *Phys. Rev. Lett.* **58**, 282.

Fields, P. M. and Cowley, J. M. (1978). Computed electron microscope images of atomic defects in f.c.c. metals. *Acta Crystallogr.* **A34**, 103.

Florjancic, M., Mader, W., Ruhle, M., and Turwitt, M. (1985). HREM and diffraction studies of an Al_2O_3/Nb interface. *J. de Phys.*, C4-129.

Frangis, N., Van Tendeloo, G., Manolikas, C., Spyridelis, J., Van Landuyt, J., and Amelinckx, S. (1985). A study of polytypism in $AgIn_5Se_8$ by combined electron microscopy techniques. *J. Solid-State Chem.* **58**, 301.

Fraser, H. L., Maher, D. M., Humphreys, C. J., Hetherington, C. J. D., Knoell, R. V., and Bean, J. C. (1985). The detection of local strains in strained layer superlattices. *Inst. Phys. Conf. Ser.* No. 76, p. 307. Institute of Physics, Bristol.

Fryer, J. (1979). *The chemical applications of transmission electron microscopy*. Academic Press, London.

Fryer, J., Hann, R. A., and Eyres, B. L. (1985). Single organic monolayer imaging by electron microscopy. *Nature (London)* **313**, 382.

Fujiwara, K. (1957). On the period of out-of-step of ordered alloys with anti-phase domain structure. *J. Phys. Soc. Japan* **12**, 7.

Fujiyoshi, Y., Ishizuka, K., and Uyeda, N. (1977). Structure analysis of a new type of vanadium oxide by high-resolution electron microscopy. *J. Electron Microsc.* **26**, 47.

Fujiyoshi, Y., Ishizuka, K., Tsuji, M., Kobayashi, T., and Uyeda, N. (1983). Charge density distribution from high-resolution molecular images. *Proceedings of the 7th International Conference on High Voltage Electron Microscopy*, Berkeley, CA.

Furukawa, S., Seki, M., and Maeyama, S. (1986). Crystallization of polysilane in binary Si:H alloys. *Phys. Rev. Lett.* **57**, 2029.

Gai, P. L., Boyes, E. D., and Bart, J. C. J. (1982). Electron microscopy of industrial oxidation catalysts. *Phil. Mag.* **A45**, 531.

Gibson, J. M. (1984). High-resolution electron microscopy of interfaces between epitaxial thin films and semiconductors. *Ultramicroscopy* **14**, 1.

Gibson, J. M. and Dong, D. W. (1980). Direct evidence of 1 nm pores in 'dry' thermal SiO_2 from high-resolution transmission electron microscopy. *J. Electrochem. Soc.* **127**, 2722.

Gibson, J. M. and Howie, A. (1978–79). Investigation of local structure and composition in amorphous solids by high-resolution electron microscopy. *Chem. Scr.* **14**, 109.

Gibson, J. M., Chen, C. H., and McDonald, M. L. (1983a). Ultrahigh-resolution electron microscopy of charge-density waves in $2H$-$TaSe_2$ below 100K. *Phys. Rev. Lett.* **50**, 1403.

Gibson, J. M., Tung, R. T., and Poate, J. M. (1983b). Structural studies of metal-semiconductor interfaces with high-resolution electron microscopy. *Mat. Res. Soc. Symp. Proc.* Vol. 14, p. 395. Elsevier, New York.

Gibson, J. M., Tung, R. T., Pimentel, C. A., and Joy, D. C. (1985a). Interfacial atomic structure and Schottky barrier height. *Inst. Phys. Conf. Ser.* No. 76, p. 173. Institute of Physics, Bristol.

Gibson, J. M., Hull, R., and Bean, J. C. (1985b). Elastic relaxation in transmission electron microscopy of strained-layer superlattices. *Appl. Phys. Lett.* **46**, 649.

Gibson, J. M., McDonald, M. L., and Unterwald, F. C. (1985c). Direct imaging of a novel silicon surface reconstruction. *Phys. Rev. Lett.* **55**, 1765.

Gronsky, R., Thomas, G., and Westmacott, K. H. (1985). High-resolution, high-voltage and analytical electron microscopy. *J. Metals* **37**, 36.

Guan, R., Hashimoto, H., and Kuo, K. H. (1985). Electron-microscopic study of the structure of metastable oxides formed in the initial stage of copper oxidation. III. $Cu_{64}O$. *Acta Crystallogr.* **B41**, 219.

Hann, R. A., Gupta, S. K., Fryer, J. R., and Eyres, B. L. (1985). Electrical and structural studies on Cu–TBP films. *Thin Solid Films* **134**, 35.

Hashimoto, H. (1985). Achievement of ultra-high resolution by 400 kV analytical atomic-resolution electron microscopy. *Ultramicroscopy* **18**, 19.

Hetherington, C. J. D., Barry, J. C., Bi, J. M., Humphreys, 'C. J., Grange, J., and Wood, C. (1985). High-resolution electron microscopy of semiconductor quantum well structures. *Mat. Res. Soc. Symp. Proc.* Vol. 27, p. 41. Materials Research Society, Pittsburgh.

Heuer, A. H., Kraus-Lanteri, S., Labun, P. A., Lanteri, V., and Mitchell, T. E. (1985). HREM studies of coherent and incoherent interfaces in ZrO_2-containing ceramics: A preliminary account. *Ultramicroscopy* **18**, 335.

Hewat, E. A., Dupuy, M., Bourret, A., Capponi, J. J., and Marezio, M. (1987). High resolution electron microscopy of the high temperature superconductor $YBa_2Cu_3O_7$. *Nature* **327**, 400.

Hiraga, K., Hirabayashi, M., and Shindo, D. (1977). High-resolution electron microscopy of long period superstructures of hexagonal Au–Cd alloys. *Proc. Fifth Int. Conf. on High Voltage Electron Microscopy*, Kyoto, p. 309.

Hiraga, K., Shindo, D., and Hirabayashi, M. (1981). High voltage HREM images of Au–Cd alloys. *J. Appl. Crystallogr.* **14**, 185.

Hiraga, K., Hirabayashi, M., Inoue, A., and Masumoto, T. (1985). Structure of Al–Mn quasicrystal studied by high-resolution electron microscopy. *J. Phys. Soc. Japan* **54**, 4077. [For $Al_{74}Mn_{20}Si_6$ (shown in Fig. 12.2) see *Proc. XIth Int. Congr. Electron Microsc.* (Kyoto) (1986) p. 163.]

Hiraga, K., Shindo, D., Hirabayashi, M., Kikuchi, M., and Syono, Y. (1987). High resolution electron microscopy of high-T_c superconductor Y-Ba-Cu-O. *J. Electron Microsc.* **36**, 261.

Hirotsu, Y., Nakamura, Y., Mizutani, J., Nagakura, S., and Nakamura, T. (1983). Multiple beam lattice imaging of hexagonal ferrites. *Trans. Japan Inst. Metals* **24**, 461.

Horiuchi, S. (1982). Reduction in a niobium tungsten bronze, $2Nb_2O_5 \cdot 7WO$, studied by 1MV HRTEM. *J. Appl. Crystallogr.* **15**, 323.

Horiuchi, S. (1985). Application of atomic-resolution electron microscopy to ceramic materials. *Am. Ceramic Soc. Bull.* **64**, 1590.

Horiuchi, S., Kikuchi, T., and Goto, M. (1977). Structure determination of a mixed-layer bismuth titanate, $Bi_7Ti_4NbI_{21}$, by super-high-resolution electron microscopy. *Acta Crystallogr.* **A33**, 701.

Horiuchi, S., Muramatsu, K., and Matsui, Y. (1980). Circular diffuse scattering from a niobium tungsten bronze, $3Nb_2O_5 \cdot 8WO_3$, studied by 1 MV high-resolution electron microscopy. *J. Appl. Crystallogr.* **13**, 141.

Howie, A. (1983). Surface reactions and excitations. *Ultramicroscopy* **11**, 141.

Howie, A. (1988). In *High-resolution electron microscopy* (ed. P. Buseck, J. M. Cowley, and L. Eyring). Oxford University Press, Oxford.

Hsu, T. (1983). Reflection electron microscopy of vicinal surfaces of f.c.c. metals. *Ultramicroscopy* **11**, 167.

Hull, R., Gibson, J. M., and Bean, J. C. (1985). Structure imaging of commensurate $Ge_xSi_{1-x}/Si(100)$ interfaces and superstructures. *Appl. Phys. Lett.* **46**, 179.

Hull, R., Rosner, S. J., Koch, S. M., and Harris, Jr., J. S. (1986). Atomic structure of the GaAs/Si interface. *Appl. Phys. Lett.* **49**, 1714.

Hutchison, J. L. (1984). High-resolution electron microscopy in the study of semiconductor materials. *Ultramicroscopy* **15**, 51.

Hutchison, J. L. and Briscoe, N. A. (1985). Surface profile imaging of spinel catalyst particles. *Ultramicroscopy* **18**, 435.

Hutchison, J. L., Lincoln, F. J., and Anderson, J. S. (1974). Multiple phases in the system $MgF_2-Nb_2O_5$ an electron microscope study of intergrowths, defects, and disordered crystals. *J. Solid-State Chem.* **10**, 312.

Hutchison, J. L., Jefferson, D. A., and Thomas, J. M. (1977). The ultrastructure of minerals as revealed by HREM. In *Surface and defect properties of solids*, Vol. 6 (ed. M. W. Roberts and J. M. Thomas), pp. 320–358. The Chemistry Society, London.

Hutchison, J. L., Waddington, W. G., Cullis, A. G., and Chew, N. G. (1986). High resolution imaging of the CdTe/(100)InSb interface: a lattice-matched hetero-epitaxial structure. *J. Microsc.* **142**, 153.

Iijima, S. (1975a). Ordering of the point defects in nonstoichiometric crysals of $Nb_{12}O_{29}$. *Acta Crystallogr.* **A31**, 784.

Iijima, S. (1975b). High-resolution electron microscopy of crystallographic shear structures in tungsten oxides. *J. Solid-State Chem.* **14**, 52.

Iijima, S. (1981). Observation of atomic steps of (111) surface of a silicon crystal using bright field electron microscopy. *Ultramicroscopy* **6**, 41.

Iijima, S. and Buseck, P. R. (1978). Experimental study of disordered mica structures by high-resolution electron microscopy. *Acta Crystallogr.* **A34**, 709.

Iijima, S. and Cowley, J. M. (1978). Studies of ordering using high-resolution electron microscopy. *J. de Phys. Colloq.* **C7**, 135.

Iijima, S. and Ichihashi, T. (1985). Motion of surface atoms on small gold particles revealed by HREM with real-time VTR system. *Japan. J. Appl. Phys.* **24**, L125.

Kakibayashi, H. and Nagata, F. (1985). Composition dependence of equal thickness fringes in an electron microscope image of $GaAs/Al_xGa_{1-x}As$ multilayer structure. *Japan. J. Appl. Phys.* **24**, L905.

Kakibayashi, H., Shimotsu, T., and Nagata, F. (1985). Fine structure of crack in yttrium oxide and aluminum oxide fluxed silicon nitride. *J. Electron Microsc.* **34**, 78.

Kambe, K. and Lehmpfuhl, G. (1975). Weak-beam technique for electron microscopic observation of atomic steps on thin single-crystal surfaces. *Optik* **42**, 187.

Kihlborg, L. (1978). Order, disorder, and defects in intergrowth tungsten bronzes. *Chem. Scr.* **14**, 187.

Kihlborg, L. and Sharma, R. (1982). Order and disorder in compounds with tungsten bronze structures. *J. Microsc. Spectrosc. Electron* **7**, 387.

Kittel, C. (1978). On infinitely adaptive crystal structures. *Solid-State Commun.* **25**, 519.

Kim, Y., Spence, J. C. H., Long, N., Bergholz, W., and O'Keefe, M. (1987). Oxygen precipitation in silicon—new phase. *J. Appl. Phys.* **62**, 419.

Klaua, M. and Bethge, H. (1985). Step contrast on ultrathin Au films investigated by TEM. *Ultramicroscopy* **17**, 73.

Kobayashi, T., Fujiyoshi, Y., and Uyeda, N. (1982). The observation of molecular orientations in crystal defects and the growth mechanism of thin phthalocyanine films. *Acta Crystallogr.* **A38**, 356.

Krakow, W., Ast, D. G., Goldfarb, W., and Siegel, G. M. (1976). Origin of the fringe structure observed in high-resolution bright-field electron micrographs of amorphous materials. *Phil. Mag.* **33**, 985.

Kraus-Lanteri, S. P., Mitchell, T. E., and Heuer, A. H. (1985). The structure of incoherent ZrO_2/Al_2O_3 interfaces. *J. Am. Ceramic Soc.*

Krivanek, O. L., Gaskell, P. H., and Howie, A. (1976). Seeing order in 'amorphous' materials. *Nature (London)* **262**, 454.

Kuo, K. H., Ye, H. Q., and Li, D. X. (1986). Tetrahedrally close-packed phases in superalloys: new phases and domain structures observed by high-resolution electron microscopy. *J. Mater. Sci.* **21**, 2597.

Kuwabara, M., Tomita, M., Hashimoto, H., and Endoh, H. (1986a). Study of the commensurate superstructure in $4Hb-TaS_2$ by high-resolution electron microscopy. *Japan. J. Appl. Phy.* **25**, L1.

Kuwabara, M., Hashimoto, H., and Endoh, H. (1986b). Direct observation of the superstructure of the nearly commensurate phase in $1T-TaS_2$ by high-resolution electron microscopy, *Phys. Stat. Sol.* **A96**, 39.

Kuwabara, M., Endoh, H., and Hashimoto, H. (1986c). Contrast anomaly in high-resolution electron microscope images of SiC. *Proc. XIth Int. Congr. on Electron Microscopy*, Kyoto, p. 809.

Li, F. H. and Hashimoto, H. (1984). Use of dynamical scattering in the structure determination of a minute fluorocarbonate mineral cebaite $Ba_3Ce_2(CO_3)_5F_2$ by high-resolution electron microscopy. *Acta Crystallogr.* **B40**, 454.

Lilental, Z., Krivanek, O. L., Goodnick, S. M., and Wilmsen, C. W. (1985). Correlation of Si–SiO$_2$ interface roughness with mosfet carrier mobility. *Mat. Res. Soc. Symp. Proc.* Vol. 37, p. 193. Materials Research Society, Pittsburgh.

Lilental-Weber, Z., Gronsky, R., Washburn, J., Newman, N., Spicer, W. E., and Weber, E. R. (1986). Schottky and ohmic gold contacts on GaAs. *Appl. Phys. Lett.* **49**, 1514.

Lodge, E. A. and Cowley, J. M. (1984). The surface diffusion of silver under high-resolution imaging conditions. *Ultramicroscopy* **13**, 215.

Lovery, F. C., Van Tendeloo, G., Van Landuyt, J., Delaey, L., and Amelinckx, S. (1984). On the nature of various stacking defects in 18R martinsite in Cu–Al alloys. *Phys. Stat. Sol.* **86**, 553.

Lu, G. and Cockayne, D. J. H. (1986a). Partial separations of extended α and β dislocations in II–VI semiconductors. *Phil. Mag.* **A53**, 307.

Lu, G. and Cockayne, D. J. H. (1986b). Dislocation structure and motion in CdS. *Phil. Mag.* **A53**, 297.

Lundberg, M. and Sundberg, M. (1986). Studies of phases in the KNbO$_3$–Nb$_2$O$_5$ system high-resolution electron microscopy and X-ray powder diffraction. *J. Solid-State Chem.* **63**, 216.

Lyman, C. E. (1984). Analytical electron microscopy of heterogeneous catalyst particles. *Am. Chem. Soc.*, 311.

Mackay, A. L. (1982). Crystallography and the penrose pattern. *Physica* **114A**, 609.

McConnell, J. D. M., Hutchison, J. L., and Anderson, J. S. (1974). Electron microscopy of the barium ferrite layer structures. *Proc. R. Soc. Lond.* A **339**, 1.

Marinder, B.-O. and Sundberg, M. (1984). The structure of NaNb$_7$O$_{18}$ as deduced from HREM images and X-ray powder diffraction data. *Acta Crystallogr.* **B40**, 82.

Marks, L. D. and Smith, D. J. (1983). Direct surface imaging in small metal particles. *Nature (London)* **303**, 316.

Millward, G. R., Thomas, J. M., and Glaeser, R. M. (1985). Probing the structure of zeolites by Fourier transform electron microscopy: Zeolite-L as a test case. *J. Chem. Soc., Chem. Commun.*, 962.

Monosmith, W. B. and Cowley, J. M. (1984). Electron microdiffraction from very small gold particles. *Ultramicroscopy* **12**, 177.

Moodie, A. F. and Warble, C. E. (1967). The observation of primary step growth in magnesium oxide by direct transmission electron microscopy. *Phil. Mag.* **16**, 891.

Moodie, A. F. and Whitfield, H. J. (1986). Determination of the structure of Cu$_2$ZnGeS$_4$ polytypes by lattice imaging and convergent beam electron diffraction. *Acta Crystallogr.* **B42**, 236.

Nagakura, S., Hirotsu, Y., Kusunoki, M., Suzuki, T.,and Nakamura, Y. (1983). Crystallographic study of the tempering of martensitic carbon steel by electron microscopy and diffraction. *Metall. Trans.* **14A**, 1025.

Nakajima, Y., Morimoti, N., and Watanabe, E. (1975). Direct observation of oxygen vacancy in mullite, 1.86Al$_2$O$_3$·SiO$_2$ by high-resolution electron microscopy. *Proc. Japan Acad.* **51**, 173.

Nakajima, Y., Morimoto, N., and Kitamura, M. (1977). The superstructure of plagioclase feldspars. *Phys. Chem. Minerals* **1**, 213.

Neumann, W., Pasemann, M., and Heydenreich, J. (1982). High-resolution electron microscopy of crystals. *Crystals, growth, properties and applications 7,* Springer-Verlag, Berlin.

Nihoul, G., Abdelmoula, K., and Metois, J. J. (1984). High-resolution images of a reconstructed surface structure on (111) gold platelets: interpretation and comparison with theoretical models. *Ultramicroscopy* **12**, 353.

Northrup, J. E., Cohen, M. L., Chelikowsky, J. R., Spence, J., and Olsen, A. (1981). Electronic structure of the unreconstructed 30° partial dislocation in silicon. *Phys. Rev.* **B24**, 4623.

Oehrlein, G. S. and Corbett, J. W. (1983). Defects in semiconductors II. (ed. S. Mahajan and J. W. Corbett), p. 107. North-Holland, New York.

O'Keefe, M. A., Spence, J. C. H., Hutchison, J. L., and Waddington, W. G. (1985). HREM profie image interpretation in MgO cubes. *Proceedings of the 43rd Ann. Proc. EMSA*, p. 64.

Olsen, A. and Spence, J. C. H. (1981). Distinguishing dissociated glide and shuffle set dislocations by high-resolution electron microscopy. *Phil. Mag.* **A43**, 945.

Olsen, A., Spence, J. C. H., and Petroff, P. (1980). Compositional analysis of III–V interface lattice images. In *Proc. 28th Ann. Meet. Electron Microsc. Soc. Am.* (ed. G. W. Bailey), p. 318. Claitor, Baton Rouge.

Osakabe, N., Tanishiro, Y., Yagi, K., and Honjo, G. (1981). Direct observation of the phase transition between the (7×7) and (1×1) structures of clean (111) silicon surfaces. *Surf. Sci.* **109**, 353.

Otten, M. T., Miner, B., Rask, J. H., and Buseck, P. R. (1985). The determination of Ti, Mn, and Fe oxidation states in minerals by electron energy-loss spectroscopy. *Ultramicroscopy* **18**, 285.

Ourmazd, A. and Spence, J. C. H. (1987). The detection of oxygen ordering in super-conducting cuprates. *Nature* **329**, 425.

Ourmazd, A., Bean, J. C., and Phillips, J. C. (1985). Submicrocrystallites and the orientational proximity effect. *Phys. Rev. Lett.* **55**, 1599.

Ourmazd, A., Ahlborn, K., Ibeh, K., and Honda, T. (1986a). Lattice and atomic structure imaging of semiconductors by high-resolution transmission electron microscopy. *Appl. Phys. Lett.* **47**, 685.

Ourmazd, A., Taylor, D. W., Bevk, J., Davidson, B. A., Feldman, L. C., and Mannaerts, J. P. (1986b). Observation of 5×5 surface reconstruction on pure silicon and its stability against native oxide formation. *Phys. Rev. Lett.* **57**, 1332.

Ourmazd, A., Rentschler, J. R., and Taylor, D. W. (1986c). Direct resolution and identification of sublattices in semiconductors by HREM. *Phys. Rev. Lett.* **57**, 3073.

Pauling, L. (1987). So-called icosahedral and decagonal quasicrystals are twins of an 820 atom crystal. *Phys. Rev. Lett.* **58**, 365.

Peng, L. M. and Cowley, J. M. (1986). A multislice approach to RHEED and REM. *Acta Cryst.* **A42**, 545.

Penisson, J. M. and Bourret, A. (1979). High-resolution study of [001] low-angle tilt boundaries in aluminium. *Phil. Mag.* **A40**, 811.

Penisson, J. M., Gronsky, R., and Brosse, J. B. (1982). High-resolution study of a $\Sigma = 41$ grain boundary in molybdenum. *Scr. Metall.* **16**, 1239.

Pirouz, P., Cockayne, D. J. H., Sumida, N., Hirsch, P., and Lang, A. R. (1983). Dissociation of dislocations in diamond. *Proc. R. Soc. Lond.* **A386**, 241.

Pirouz, P., Chain, R., and Samuels, J. (1987). An HREM investigation of indentation induced phase transformations of silicon. In *Proc. Fifth Int. Conf. on Properties and Structure of Dislocations in Semiconductors.* (ed. U. O'sipiyan). Academy of Sciences of the USSR, Moscow.

Ponce, F. A., Yamashita, T., and Sinclair, R. (1980). Structure imaging of faults in cadmium telluride. *38th Ann. Proc. EMSA,* San Francisco, p. 320.

Ponce, F. A., Yamashita, T., and Hahn, S. (1983). Structure of thermally induced microdefects in Czochralski silicon. *Appl. Phys. Lett.* **43**, 1051.

Poppa, H. R. (1983). Surface studies with clean supported metal particles and clusters. *Ultramicroscopy* **11**, 105.

Portier, R., Shechtman, D., Gratias, D., and Cahn, J. W. (1985). High-resolution electron microscopy of the icosahedral quasiperiodic structure in Al–Mn system. *J. Microsc. Spectrosc. Electron* **10**, 107.

Quin, L. C., Li, D. X., and Kuo, K. H. (1986). An HREM study of the defects in ZnS. *Phil. Mag.* **A53**, 543.

Rai, R. S., O'Keefe, M. A., and Thomas, G. (1985). HREM studies on a new zeolite $AlPO_4\#5$. *Proc. 43rd Ann. Meet. of Electron Microsc. Soc. Am.* (ed. G. W. Bailey), p. 372.

Ruska, E. (1933). Die electronenmikroskopische Abbildung electronenbestrahlter Oberflächen, *Z. Phys.* **83**, 492.

Saito, Y. (1984). The structural studies of amorphous Ge films prepared by vacuum deposition. *J. Phys. Soc. Japan* **53**, 4230.

Salamanca-Riba, L., Roth, G., Gibson, J. M., Kortan, A. R., Dresselhaus, G., and Birgeneau, R. J. (1986). Electron-beam-induced damage and structure of SbCl$_5$-graphite intercalation compounds. *Phys. Rev.* **B33**, 2738.

Sato, M., Hiraga, K., and Sumino, K. (1980). HVEM structure images of extended 60° and screw dislocations in silicon. *Japan. J. Appl. Phys.* **19**, L155.

Schryvers, D. and Amelinckx, S. (1985). New long period superstructure in Pt$_3$V with M = 3/2. *Mater. Res. Bull.* **20**, 367.

Schryvers, D., Van Landuyt, J., Van Tendeloo, G., and Amelinckx, S. (1982). High-resolution electron microscopy of coincidence patterns in the ordered Pd$_3$Mn alloy. *Phys. Stat. Sol.* **71**, K9.

Schryvers, D., Van Tendeloo, G., and Amelinckx, S. (1985b). On the ordering mechanism of Ni$_3$Mo a high-resolution electron microscopy study. *Phys. Stat. Sol.* **87**, 401.

Shechtman, D., Blech, I., Gratias, D., and Cahn, J. W. (1984). Metallic phase with long-range orientational order and no translational symmetry. *Phys. Rev. Lett.* **53**, 1951.

Shindo, D. (1982). HREM images of ordered alloys. *Acta Crystallogr.* **A38**, 310.

Shindo, D., Hiraga, K., and Hirabayashi, M. (1983). Double hexagonal superstructure of Au–Mg alloy with 20 at%Mg studied by HREM. *J. Appl. Crystallogr.* **16**, 233.

Shuman, H. (1977). Bragg diffraction imaging of defects at crystal surfaces. *Ultramicroscopy* **2**, 361.

Self, P. G., O'Keefe, M. A., and Stobbs, W. M. (1983). A transmission electron microscopical study of the σ phase in an iron–chromium steel. *Acta Crystallogr.* **B39**, 197.

Sinclair, R., Ponce, F. A., Yamashita, T., Smith, D. J., Camps, R. A., Freeman, L. A., Erasmus, S. J., Nixon, W. C., Smith, K. C. A., and Catto, C. J. D. (1982). Dynamic observation of defect annealing in CdTe at lattice resolution. *Nature (London)* **298**, 127.

Sinclair, R., Ponce, F. A., Yamashita, T., and Smith, D. J. (1983). High-resolution electron microscopy of II–VI compound semiconductors. *Inst. Phys. Conf. Ser.* No. 67, p. 103.

Skarnulis, A. J., Iijima, S., and Cowley, J. M. (1976). Refinement of the defect structure of 'GeNb$_9$O$_{25}$' by high-resolution electron microscopy. *Acta Crystallogr.* **A32**, 799.

Smith, D. (1986). High-resolution electron microscopy in surface science. In *Chemistry and physics of solid surfaces VI* (ed. R. Vanselow and R. Howe) Chapter 15. Springer-Verlag, New York.

Smith, D. J. and Bursill, L. A. (1985). 'Metallisation' of oxide surfaces observed by *in situ* high-resolution electron microscopy. *Ultramicroscopy* **17**, 387.

Smith, D. J. and Marks, L. D. (1985). Direct atomic imaging of solid surfaces. III. Small particles and extended Au surfaces. *Ultramicroscopy* **16**, 101.

Smith, D. J., Stobbs, W. M., and Saxton, W. O. (1981). Ultra-high-resolution electron microscopy of amorphous materials at 120 kV. *Phil. Mag.* **B43**, 907.

Smith, D. J., Fryer, J. R., and Camps, R. A. (1986). Radiation damage and structural studies: Halogenated phthalocyanines. *Ultramicroscopy* **19**, 279.

Spinnler, G. E., Self, P. G., Iijima, S., and Buseck, P. R. (1984). Stacking disorder in clinochlore chlorite. *Am. Mineral.* **69**, 252.

Spence, J. C. H. (1983). High energy transmission electron diffraction and imaging studies of the silicon (111) 7 × 7 surface structure. *Ultramicroscopy* **11**, 117.

Spence, J. C. H. and Kolar, H. (1979). Lattice images of dislocation dipoles in silicon. *Phil. Mag.* **A39**, 59.

Spence, J. C. H. (1988). A scanning tunneling microscope in a side-entry holder for reflection electron microscopy in the Philips EM400. *Ultramicroscopy* (in press).

Steeds, J., Bird, D. M., Eaglesham, D. J., McKernan, S., Vincent, R., and Withers, R. L. (1985). Study of modulated structures by transmission electron microscopy. *Ultramicroscopy* **18**, 97.

Stoneham, A. M. and Durham, P. J. (1973). The ordering of crystallographic shear planes: Theory of regular arrays. *Phys. Chem. Solids* **34**, 2127.

Sundberg, M. (1980). Structure and 'oxidation behavior' of W$_{24}$O$_{70}$, a new member of the {103} CS series of tungsten oxides. *J. Solid-State Chem.* **35**, 120.

Swan, P. R., Jones, J. S., Krivanek, O. L., Smith, D. J., Venebles, J. A., and Cowley, J. M. (1987). *Proc. 45th Ann. Meeting of Electron Microscopy Society of America.* G. W. Bailey, Ed. (San Francisco Press, San Francisco), p. 136.

Taftø, J. and Spence, J. C. H. (1982). A simple method for determining crystal polarity by CBED. *J. Appl. Crystallogr.* **15**, 60.

Takeda, M. and Hashimoto, H. (1985). Atomic structure of CuAuII studied by high-resolution electron microscopy. *Phys. Stat. Sol.* **87**, 141.

Takeda, M., Yoshida, H., and Hashimoto, H. (1985). Local tetragonality and atomic structure in Nb_3Sn superconductor studied by high-resolution electron microscopy. *Phys. Stat. Sol.* **87**, 473.

Takayanagi, K., Kolb, D. M., Kambe, K., and Lehmpfuhl, G. (1980). Deposition of monolayer and bulk lead on Ag(111) studied in vacuum and in an electrochemical cell. *Surf. Sci.* **100**, 407.

Takayanagi, K., Tanishiro, Y., Takahashi, M., and Takahashi, S. (1985). Structural analysis of Si(111)–7 × 7 by UHV transmission electron diffraction and microscopy. *J. Vac. Sci. Tech.* **A3**, 1502.

Tanaka, M. and Jouffrey, B. (1984a). Dissociated dislocations in GaAs observed in high-resolution electron microscopy. *Phil. Mag.* **A50**, 733.

Tanaka, M. and Jouffrey, B. (1984b). Lattice-image interpretation of a relatively-small-unit-cell crystal. *Acta Crystallogr.* **A40**, 143.

Tanishiro, Y., Takayanagi, K., and Yagi, K. (1986). Observation of lattice fringe of the Si(111)–7 × 7 structure by reflection electron microscopy. *J. Microsc.* **142**, (pt. 2), 211.

Thomas, J. M. (1982). Placing the applications of high-resolution electron microscopy to chemical problems into wider perspective. *Ultramicroscopy* **8**, 13.

Thomas, G. (1986). Some applications of electron microscopy in materials science. *Ultramicroscopy* **20**, 239.

Timsit, R. S., Waddington, W. G., Humphreys, C. J., and Hutchison, J. L. (1985). Structure of the Al/Al_2O_3 interface. *Appl. Phys. Lett.* **46**, 830.

Treacy, M. M. J., Howie, A., and Wilson, C. J. (1978). Z contrast of platinum and palladium catalysts. *Phil. Mag.* **A38**, 569.

Turner, S. and Buseck, P. R. (1979). Manganese oxide tunnel structures and their intergrowths. *Science* **203**, 456.

Turner, S. and Buseck, P. R. (1983). Defects in nsutite (γ-MnO_2) and dry-cell battery efficiency. *Nature (London)* **304**, 143.

Uchida, Y., Lehmpfuhl, G., and Jager, J. (1984). Observation of surface treatments on single crystals by reflection electron microscopy. *Ultramicroscopy* **15**, 119.

Uyeda, N., Kobayashi, T., Ishizuka, K., and Fujiyoshi, Y. (1980). Crystal structure of Ag·TCNQ. *Nature (London)* **285**, 95.

Van Dyck, D., Tambuyser, P., Van Landuyt, J., and Amelinckx, S. (1976). High-resolution electron microscopy of dumortierite. *Am. Mineral.* **61**, 1016.

Van Tendeloo, G., Landuyt, J. V., and Amelinckx, S. (1981). Real space observation of charge density waves in the incommensurate room temperature phase of TaS_2. *Phys. Stat. Sol.* **64(a)**, K105.

Van Tendeloo, G. and Amelinckx, S. (1982a). High-resolution electron microscopic and electron diffraction study of the Au–Mg system. *Phys. Stat. Sol.* **69**, 103.

Van Tendeloo, G. and Amelinckx, S. (1982b). A new long period superstructure with ideal composition $Au_{35}Mn_{13}$ derived from the DO_{22} structure, studied by means of high-resolution electron microscopy and electron diffraction. *Phys. Stat. Sol.* **71**, 185.

Van Tendeloo, G. and Thomas, G. (1985). High-resolution microscopy investigation of the system ZrO_2–ZrN. *Adv. Ceramics* **12**, 164.

Van Tendeloo, G., Van Landuyt, J., and Amelinckx, S. (1982). High-resolution study of polytypism: Application to SiC and Au_3Mn. *40th Ann. Proc. EMSA,* Washington, p. 540.

Veblen, D. R. (1983). Exsolution and crystal chemistry of the sodium mica wonesite. *Am. Mineral.* **68**, 554.

Veblen, D. R. (1985). Direct TEM imaging of complex structures and defects in silicates. *Ann. Rev. Earth Planet. Sci.* **13,** 119.

Veblen, D. R. and Buseck, P. R. (1979a). Serpentine minerals: intergrowths and new combination structures. *Science* **206,** 1398.

Veblen, D. R. and Buseck, P. R. (1979b). Chain-width order and disorder in biopyriboles. *Am. Mineral.* **64,** 687.

Venables, J., Spiller, G. D. T., Fathers, D. J., Harland, C. J., and Hanbucken, M. (1983). U.H.V.-S.E.M. studies of surface process: recent progress. *Ultramicroscopy* **11,** 149.

Warble, C. E. (1984). The relationship between bulk structure and bounding surfaces in ZrO_2. *Ultramicroscopy* **15,** 301.

Watari, F., Delavignette, P., Van Landuyt, J., and Amelinckx, S. (1983). Electron microscopic study of dehydration transformations. Part III: High resolution observation of the reaction process $FeOOH \rightarrow Fe_2O_3$. *J. Solid-State Chem.* **48,** 49.

White, T. J. and Hyde, B. G. (1982). Electron microscope study of the humite minerals: I. Mg-rich specimens. *Phys. Chem. Minerals* **8,** 55.

White, T. J. and Hyde, B. G. (1983). An electron microscope study of leucophoenicite. *Am. Mineral.* **68,** 1009.

Wilson, A. R. and Cockayne, D. J. H. (1985). Calculated asymmetry for weak beam intrinsic stacking fault images. *Phil. Mag.* **A51,** 341.

Wilson, J. A. and Yoffe, A. D. (1969). The transition metal dichalcogenides. Discussion and interpretation of the observed optical, electrical and structural properties. In *Advances in Physics 18,* p. 193. Taylor and Francis Ltd., London.

Wood, G. J., Stobbs, W. M., and Smith, D. J. (1984). Methods for the measurement of rigid-body displacements at edge-on boundaries using high-resolution electron microscopy. *Phil. Mag.* **A50,** 375.

Yagi, K. and Cowley, J. M. (1978). Electron microscopy study of ordering of potassium ions in cubic $KSbO_3$. *Acta Crystallogr.* **A34,** 625.

Yagi, K. and Roth, R. S. (1978). Electron-microscope study of crystal structures of mixed oxides in the systems $Rb_2O-Ta_2O_5$, $Rb_2-O-Nb_2O_5$ and $K_2O-Ta_2O_5$ with composition ratios near 1:3. II. Various intergrowth phases and two-dimensional ordering of pentavalent ions. *Acta Crystallogr.* **A34,** 773.

Yagi, K., Takayanagi, K., and Honjo, G. (1982). In situ UHV electron microscopy of surfaces. *Crystals, growth, properties, and applications 7,* p. 48. Springer-Verlag, Berlin.

Yamamoto, N. and Ishizuka, K. (1983). Analysis of the incommensurate structure of $Sr_2Nb_2O_7$ by electron microscopy and CBED. *J. Appl. Crystallogr.* **B39,** 210.

Yamamoto, N. and Kumashiro, Y. (1982). Vacancy ordering in vanadium carbide. In *Proc. 40th Ann. Meet. Electron Microsc. Soc. Am.* (ed. G. W. Bailey) p. 726. Washington, D.C.

Yamamoto, N. and Muto, S. (1984). Direct observation of $Al_xGa_{1-x}As/GaAs$ superlattices by REM. *Japan. J. Applied. Phys.* **23,** L806.

Yang, Q. B. and Kuo, K. H. (1986). A structural model of the icosahedral phase derived from its close relationship to the pentagonal Frank–Kasper phases. *Phil. Mag.* **B53,** L115.

Zakharov, N. D., Pasemann, M., and Rozhanski, V. N. (1982). Observations of point defects in silicon by means of dark-field lattice plane imaging. *Phys. Stat. Sol.* **71,** 275.

Zakharov, N. D., Gribeluk, M. A., and Vainshtein, B. K. (1983). Structure of U–W oxides investigated by means of 1 MV high-resolution electron microscopy. *Acta Crystallogr.* **B39,** 575.

Zemlin, F., Reuber, E., Beckmann, B., Zeitler, E., and Dorset, D. L. (1985). Molecular resolution electron micrographs of monolamellar paraffin crystals. *Science* **229,** 461.

Zhang, J. P., Ye, H. Q., Kuo, K. H., and Amelinckx, S. (1985a). A HREM study of the crystal structure of Cu_4Ti. *Phys. Stat. Sol.* **88,** 475.

Zhang, Z., Ye, H. Q., and Kuo, K. H. (1985b). A new icosahedral phase with $m\bar{3}\bar{5}$ symmetry. *Phil. Mag.* **A52,** L49.

Zhang, J., Kuo, K. H., and Wu, Z. Q. (1986). An HREM study of defects at the $Pd_2Si/Si(111)$ interface. *Phil. Mag.* **A53,** 677.

APPENDIX 1

The following FORTRAN program calculates values of the defocus Δf, spherical aberration constant Cs and astigmatism constants from optical diffractogram ring radii S_n. It is based on the method of Krivanek (*Optik,* **45**, p. 97) and was written by M. A. O'Keefe. As written, *it is restricted to the case of large underfocus settings,* but is easily modified for overfocus (Δf positive) values, as described below.

The method depends on the fact that minima will occur in the diffractogram intensity (given by eqn (3.36)) if

$$\chi(u) = 2\pi[\Delta f \lambda u^2/2 + C_s \lambda^3 u^4/4] = N\pi \qquad (A1.1)$$

Here N is any integer (including zero). On dividing by u^2, we find that a plot of N/u^2 against u^2 forms a straight line with slope $(Cs\lambda^3/2)$ and intercept $(\Delta f \lambda)$. The program uses linear regression to find the slope and intercept (and hence Cs and Δf) from the values of N and S_n supplied. The ring radii S_n must be specified in reciprocal Angstroms, and are related to the measured ring radii by eqn (8.5) or (8.6). Input data includes a title and the microscope accelerating voltage in kilovolts. Then follows one line of input data for each maximum or minimum, containing, in column 2, the number of the ring, in column 4 a zero (for a minimum) or a one (for a maximum), and, in the next sixteen columns, the radius of this maximum or minimum (in reciprocal Angstroms). The method of numbering the rings is indicated by the numbers across the page in Fig. A1.1. These numbers correspond to $-N$ in eqn (A1.1). (The program actually works with a new $n = 2N$ for convenience, evaluated after line 10, so that the slope and intercept are twice the values given above.)

The correct numbering of the rings requires some care. All the worked examples in Krivanek (1976) follow the system indicated in Fig. A1.1, which

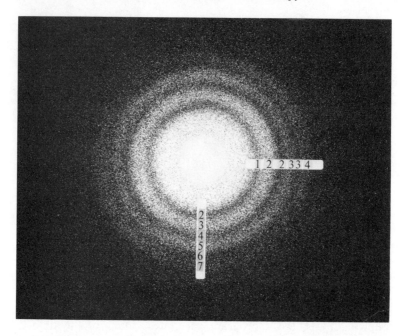

Fig. A1.1 Optical diffraction pattern of an electron image of a thin carbon foil suitable for use with the FORTRAN program given. The pattern shows negligible astigmatism and drift (see Section 8.7). Numbers running across the pattern are the numbers NRING needed as input to the program. The maximum of the first (inner) ring has not been used since its position is difficult to determine. The numbers running down the pattern are the quantities $n = 2$NRING-MAXMIN evaluated after line 10 of the program. The intensity in this pattern is given by eqn (3.36), together with the damping effects of partial coherence and chromatic aberration discussed in Chapter 4. Instrumental aberration constants are obtained by the program from patterns such as these.

is correct only for large underfocus (Δf negative). Figure A1.2 shows $\chi(u)/\pi$ plotted for three defocus values with $Cs = 1.8$ mm at 100 kV. The values of N from eqn. (A1.1) are indicated on the curves. It is seen that there are essentially three cases: 1. Positive focus, for which $N = 1, 2, 3, 4, \ldots$, 2. Near Scherzer focus, for which the first value of N must be 0 ($N = 0, 1, 2, 3, \ldots$), and 3. Larger negative focus, in which case (for the example shown) $N = -1, -2, -3, -3, -2, -1, 0, 1, 2, 3, 4, \ldots$. The turning point occurs at the stationary phase condition (see eqn (5.76))

$$U_0 = \sqrt{\frac{\Delta f}{C_s \lambda^2}} \tag{A1.2}$$

Here the declining values of N start to increase, and may or may not repeat at the minimum value (e.g. -3 above). The stationary phase condition may *sometimes* be identified on a diffractogram by a broad intense ring, however if this condition coincides with $\chi(u) = -m\pi$, a broad absence of intensity will result. In practice, as shown in Fig. A3.1, the turning point may be

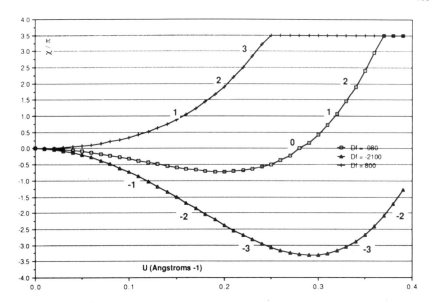

Fig. A1.2 The function $\chi(u)$ plotted in units of π for three focus settings $\Delta f = -980$ (Scherzer focus), -2100 and $+800$ Angstroms as indicated. Values of N in eqn (A1.1) are indicated. Here $Cs = 1.8$ mm and the accelerating voltage is 100 kV.

driven beyond the resolution of the microscope by choosing (from eqn A1.2)

$$|\Delta f| \geq C_s \lambda^2 u_0^2 \qquad (A1.3)$$

where u_0 is the highest spatial frequency in the diffractogram. Then the values of N will increase monotonically by unit increments with negative sign. These are the conditions for which the FORTRAN program is written. An additional internal consistency test also exists, since the program returns values of the standard deviation for slope and intercept. Because of the method of analysis, these values should *not* be taken as the errors in Cs and Δf. They may, however, be used to confirm that a good straight-line fit has been obtained, and that the values of N have therefore been correctly chosen.

The graphical method described in Krivanek's paper gives a more immediate indication of goodness of fit, so that in cases involving a turning point, suitable values of N can be found by trial and error. Figure A1.3 provides a copy of this nomogram, which may be duplicated using an enlarging photocopy machine. For the simplest case of large underfocus, choose $n = -2, -4, -6, -8, \ldots$ for minima (i.e. regions where the laser intensity is low) and $n = -1, -3, -5, -7, -9, \ldots$ for maxima. For overfocus, $n = 2, 4, 6, 8, \ldots$ for minima and $n = 1, 3, 5, 7, \ldots$ for maxima. Find corresponding values of n and u^2 on the nomogram, and plot the points. Join these by a straight line. The slope (ordinate increment divided by abscissa increment) when divided by the cube of the electron wavelength

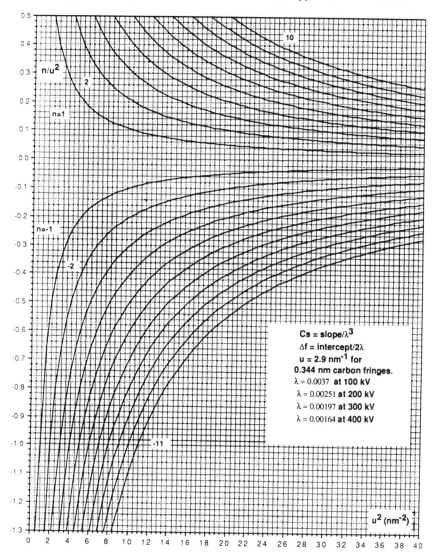

Fig. A1.3 Chart for the determination of C_s and Δf from optical diffractograms. See text for details.

(in nm) gives C_s in nanometers. The ordinate intercept divided by twice the wavelength (in nm) gives the defocus in nm. Astigmatism may be determined by taking measurements from the diffractogram in two orthogonal directions.

Cases involving a turning point can also be treated with the graph. For example, a straight line from $n/u^2 = -0.8$, $u^2 = 0$ to $u^2 = 18$, $n/u^2 = 0$ has n values -1, -2, -3, -3, -2, -1. The first three values of n suggest the direction of the straight line, the rest are guessed.

Neither the graph or the FORTRAN program below are useful near
Scherzer focus, since few minima are obtained. This is particularly true
modern medium-voltage (300–400 kV machines), since the information
olution limit approximately coincides with the Scherzer cutoff (see
pendix 3).

The best way to use the program for the calibration of a new microscope
herefore to record several through-focus series at focus settings satisfying
a (A1.3) (using manufacturer's data for C_s and U_0), from which the
allest focal step increment (and C_s) can be deduced. The Scherzer focus
uired for structure imaging can then be obtained by counting "clicks"
m the minimum contrast position, given by eqn (6.26) (see also Section
5).

RTRAN program to give defocus and spherical aberration constants from optical diffractograms

```
PROGRAM CSEPS
TO DETERMINE CS(MM) AND DEFOCUS (ANGSTROMS) GIVEN RING DIAMETERS FROM
DIFFRACTOGRAMS OF MICROGRAPHS OF THIN AMORPHOUS MATERIALS
    DIMENSION NAME(13),X(20),Y(20)
READ IN VOLTAGE IN KV
    READ (5,100) NAME,VOLTK
CALCULATE WAVELENGTH
    WAVL = 0.387818/SQRT(VOLTK * (1.0 + 0.978459E–3 * VOLTK))
    WAVCUB = (WAVL * *3) * 1.0E7
70 N = 0
20 CONTINUE
READ IN NUMBER OF RING (1 FOR FIRST, 2 FOR SECOND. . . .),
WHETHER RING IS MAXIMUM (MAXIMIN = 1) OR MINIMUM (MAXIMIN = 0)
RADIUS OF RING IN RECIPROCAL ANGSTROMS
    READ(5,110)NRING,MAXMIN,URAD
    IF(NRING,NE.0) GO TO 10
    IF(N.EQ.0) GO TO 80
    GO TO 30
10 CONTINUE
    IF(N.EQ.0) WRITE(6,200) NAME, VOLTK,WAVL
    N = N + 1
    USQ = URAD * URAD
    X(N) = USQ
    Y(N) = −FLOAT(2 * NRING-MAXMIN)/USQ
CHECK WHETHER MAXIMUM OR MINIMUM
    IF(MAXMIN.EQ.0) GO TO 25
    IF(MAXMIN.EQ.1) GO TO 35
    WRITE(6,260)
    GO TO 80
25 WRITE(6,220) NRING, URAD,X(N), Y(N)
    GO TO 20
35 WRITE(6,210) NRING,URAD,X(N), Y(N)
    GO TO 20
30 CONTINUE
```

```
C   IF N.GT.1 THEN FIT CURVE
        IF(N.FT.1) GO TO 40
        WRITE(6,230)
        GO TO 70
    40 CONTINUE
        SUMX = 0.0
        SUMY = 0.0
        SUMX2 = 0.0
        SUMXY = 0.0
        SUM2 = 0.0
        FN = FLOAT(N)
C   SUM OVER N POINTS
        DO 50 1 = 1,N
        XI = X(I)
        YI = Y(I)
        SUMX = SUMX + XI
        SUMY = SUMY + YI
        SUMX2 = SUMX2 + XI * XI
        SUMXY = SUMXY + XI * YI
        SUMY2 = SUMY2 + YI * YI
C   CALCULATE CURVE PARAMETERS
    50 CONTINUE
        DELTA = FN * SUMX2-SUMX * SUMX
        A = (SUMX2 * SUMY-SUMX * SUMXY)/DELTA
        B = (SUMXY * FN-SUMX * SUMY)/DELTA
        EPS = A/(2.0 * WAVL)
        CS = B/WAVCUB
        C = FN - 2.
C   IF NUMBER OF POINTS IS LESS THAN 3 THEN SKIP STD. DEV. CALCULATION
        IF(C.LT.1.E-3) GO TO 60
        VNCE = (SUMY2 + A * A * FN + B * B * SUMX2-2.0 * (A * SUMY + B * SUMXY-A * B * SUMX))/C
        SIGA = SQRT(VNCE * SUMX2/DELTA)
        SIGB = SQRT(VNCE * FN/DELTA)
        R = (FN * SUMXY-SUMX * SUMY)/(SQRT(DELTA * (FN * SUMY2-SUMY * SUMY)))
        SIGEPS = SIGA/(2.0 * WAVL)
        SIGCS = SIGB/WAVCUB
        WRITE(6,240) A,SIGA,B,SIGB,R,EPS,SIGEPS,CS,SIGCS
        GO TO 70
    60 CONTINUE
        WRITE(6,250)A,B,EPS,CS
        GO TO 70
    80 CONTINUE
        CALL EXIT
   100 FORMAT(13AG/F10.5)
   110 FORMAT(2I2,F12.6)
   200 FORMAT(1H1,13AG,/,15HOVOLTAGE (KV) = ,F8.2,10X,16HWAVELENGTH (A) = ,
      $ F9.6,//)
   210 FORMAT(1H0,10X,17HRADIUS OF MAXIMUM,12,2H = ,F7.4,17H (RECIP.ANGSTR
      $ OM),20X,3HX = ,F7.5,5X,3HY = ,F8.2)
   220 FORMAT(1HO,10X,17HRADIUS OF MINIMUM,12,2H = ,F7.4,17H (RECIP.ANGSTR
      $ OM),20X,3HX = ,F7.5,5X,3HY = ,F8.2)
   230 FORMAT(1HO,10X,41HONLY ONE POINT IN DATA SET-DATA IGNORED,/)
   240 FORMAT(1HO,10X,11HINTERCEPT = ,F8.2,2,H(,F5.2,11X,7HSLOPE = ,F8.
      $ 2,2H (,F6.2,1H),11X,12HCORR.COEFF. = ,F6.4,///,1HO,50X,20HDEFOCUS (S
      $ TD.DEV.) = ,F9.2,2H (,F6.2,6H) ANG.,5X,15HCS (STD,DEV.) = ,F6.3,2H (
      $ ,F4,3.5H) MM.,//)
```

```
 250 FORMAT(/,1H0,10X,
   $ 11HINTERCEPT = ,F8.2,5X,7HSLOPE = ,F8.2,39X,9HDEFOCUS = ,F9
   $ .2,5H ANG.,5X,4HCS = ,F6.3,4H MM.,//)
 260 FORMAT(1HO,20X,'***RING NOT LABELED AS MAXIMUM OR MINIMUM***')
```

Sample data and output from program (JEM100B, 100 KV)

```
TEST OF PROGRAM - O.L.'S DATA
VOLTAGE (KV) = 100.00    WAVELENGTH (A) = 0.037013
        RADIUS OF MAXIMUM 1 = 0.0770 (RECIP.ANGSTROM)    X = 0.00593    Y = -168.66
        RADIUS OF MINIMUM 1 = 0.1120 (RECIP.ANGSTROM)    X = 0.01254    Y = -159.44
        RADIUS OF MAXIMUM 2 = 0.1400 (RECIP.ANGSTROM)    X = 0.01960    Y = -153.06
        RADIUS OF MINIMUM 2 = 0.1660 (RECIP.ANGSTROM)    X = 0.02756    Y = -145.16
        RADIUS OF MAXIMUM 3 = 0.1890 (RECIP.ANGSTROM)    X = 0.03572    Y = -139.97
        RADIUS OF MINIMUM 3 = 0.2150 (RECIP.ANGSTROM)    X = 0.04622    Y = -129.80
        RADIUS OF MAXIMUM 4 = 0.2420 (RECIP.ANGSTROM)    X = 0.05856    Y = -119.53
        RADIUS OF MINIMUM 4 = 0.2890 (RECIP.ANGSTROM)    X = 0.08352    Y = -95.78
        INTERCEPT = -171.83 (0.85)    SLOPE = 908.90 ( 19.57)    CORR.COEFF. = 0.9986
DEFOCUS (STD.DEV.) = -2321.26 ( 11.50) ANG. CS (STD.DEV.) = 1.792 (0.039)MM
```

APPENDIX 2

Use of an absorption function to represent the objective aperture effect

The justification for the use of an absorption function $\phi_i(x, y) = \mu(x, y)t/2\sigma$ in Section 6.1 to account for the exclusion of scattering outside the objective aperture from the image can be seen as follows. The true image amplitude in the 'flat Ewald sphere' approximation (see Section 3.4) is

$$\psi_i(x, y) = [\exp(-i\sigma\phi_R(x, y))] * I(x, y)$$

where the $*$ denotes convolution and $I(x, y)$ is the microscope impulse response (equal to the transform of $A(u, v)$ in Section 3.2). Here $\phi_R(x, y)$ is the unsmoothed specimen electrostatic potential. The use of an absorption function predicts an image amplitude given by eqn (6.2) as

$$\psi_i'(x, y) = \exp(-i\sigma\phi_R(x, y) * I(x, y)) \exp(-\sigma\phi_i(x, y))$$

A comparison of the expansion to second order of these two expressions allows the absorption potential to be defined as

$$\sigma\phi_i(x, y) = \left(\frac{\sigma^2}{2}\right)\phi_R^2(x, y) * I(x, y) - \left(\frac{\sigma^2}{2}\right)[\phi_R(x, y) * I(x, y)]^2$$

which gives the correct limiting behaviour for small and large apertures. I am grateful to Professor J. M. Cowley for a discussion on this point. It is interesting to note that the use of an absorption function common in biological microscopy for the imaging of thick specimens at low resolution is based on a form of 'column approximation'. The image intensity deficit in the neighbourhood of a thin column of specimen is obtained by calculating the scattering from an equivalent specimen whose structure (mass thickness) is everywhere the same as that of the true specimen within this column.

410

APPENDIX 3

Resolution-limiting factors and their wavelength dependences

As mentioned in Section 4.2, it is necessary to distinguish two resolution limits in high-resolution electron microscopy. The first, the 'information-resolution limit', gives the ultimate band limit of the instrument and indicates the highest resolution detail which could in principle be extracted by image-processing techniques, or by comparisons with computed images. The second is the familiar 'point-resolution' limit at which images can be simply interpreted. The theory on which these two concepts are based is first reviewed, before discussing their differing wavelength dependences.

For specimen thicknesses at which the central beam is much stronger than any other diffracted beam, the combined effects of electronic instabilities and the use of an extended, incoherent effective disc source result in the imposition of a virtual aperture or 'damping envelope' on the objective-lens transfer function of the form (O'Keefe and Anstis, unpublished results; see also references for Section 4.2).

$$A(\mathbf{K}) = \exp\{-\pi^2\lambda^2\Delta^2\mathbf{K}^4/2\} \, \frac{2J_1 \, |2\pi\theta_c[\Delta f\mathbf{K} + \lambda(\lambda C_s - i\pi\Delta^2)\mathbf{K}^3]|}{|2\pi\theta_c[\Delta f \, \mathbf{K} + \lambda(\lambda C_s - i\pi\Delta^2)\mathbf{K}^3]|} \quad (A3.1)$$

A similar expression is given in Section 4.2 for a Gaussian source. Here λ is the electron wavelength, C_s is the spherical aberration coefficient, Δf is the defocus increment, $\mathbf{K}$ is the scattering vector ($|\mathbf{K}| = \sin\theta/\lambda = (u^2 + v^2)^{1/2}$ where θ is the scattering angle), and θ_c is the illumination semi-angle. In wedge-shaped specimens showing near-sinusoidal Pendellösung in the zone-axis orientation (few Bloch waves excited), eqn (A3.1) can be used at the many thicknesses for which the central beam has maximum intensity. It does not apply in general, however, to the computation of STEM bright-field high-resolution images unless these are formed under conditions reciprocally related to those for which it applies in TEM (see Section 5.21). The quantity Δ is defined by

$$\Delta = C_c Q = C_c \left[\frac{\sigma^2(V_0)}{V_0^2} + \frac{4\sigma^2(I_0)}{I_0^2} + \frac{\sigma^2(E_0)}{E_0^2} \right]^{1/2} \tag{A3.2}$$

where $\sigma^2(V_0)$ and $\sigma^2(I_0)$ are the variances in the statistically independent fluctuations of accelerating voltage V_0 and objective lens current I_0 respectively. Thus the root-mean-square value of the high-voltage fluctuation is equal to the standard deviation $\sigma(V_0) = [\sigma^2(V_0)]^{1/2}$. C_c is the lens chromatic aberration constant and the full width at half maximum height of the energy distribution of electrons leaving the filament is

$$\Delta E = 2.345 \sigma(E_0) = 2.345 [\sigma^2(E_0)]^{1/2}$$

For high-resolution imaging in which no objective aperture is used, eqn (A3.1) sets the limit to the highest resolution detail which can be extracted from an electron image. This detail, however, may not be simply related to the specimen structure, since it is obtained using an oscillating transfer function. An 'information-resolution limit' can thus be defined as the **K** value(s) $U_0(i)$ for which $A(\mathbf{K}) = 2J_1(2.6)/2.6 = \exp(-1)$. The solutions to the resulting cubic equation are, for $\Delta = 0$,

$$U_0(1) = S_1 + S_2$$
$$U_0(2) = -U_0(1)/2 + i\sqrt{3}(S_2 - S_2)/2 \tag{A3.3}$$
$$U_0(3) = -U_0(1)/2 - i\sqrt{3}(S_1 - S_2)/2$$

where

$$S_{1,2} = \left[\left\{ b/\theta_c \pm \left(\frac{a/\Delta f^3}{\lambda^2 C_s} + \frac{b}{\theta_c} \right)^{1/2} \right\}^{1/2} \middle/ (\lambda^2 C_s) \right]^{1/3}$$

with

$$a = 1/27 \quad \text{and} \quad b = (2.6/4\pi)^2$$

The outer cut-off $U_0(1)$ is preceded by one or two crossings of the line $f(\mathbf{K}) = \exp(-1)$ at $U_0(2)$ and $U_0(3)$ if $\mathrm{Im}(U_0(2)) = 0$ as indicated in Fig. A3.3(b). Physically, the effect of partial coherence is seen to attenuate contrast transfer in regions where the slope of $\chi(\mathbf{K})$ (see eqn (3.24)) is large and to preserve good contrast in regions where this slope is small, as shown in Fig. A3.1. The function $\chi(\mathbf{K})$ has a minimum within the contrast transfer intervals which occur between U_1 and U_2 where

$$U_{1,2} = \left[\left(\frac{8n+3}{2} \right)^{1/2} \pm 1 \right]^{1/2} C_s^{-1/4} \lambda^{-3/4} = k_n^{1/2} C_s^{-1/4} \lambda^{-3/4} \tag{A3.4}$$

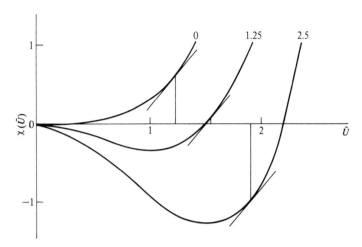

Fig. A3.1 The function $\chi(\bar{U})$ drawn for three focus values $\Delta f = d_i C_s^{1/2} \lambda^{1/2}$ where d_i are the numbers indicated on each curve. The ordinate is the generalized spatial frequency $\bar{U}$ commonly used in the literature, where $|\mathbf{K}| = \theta/\lambda = (C_s \lambda^3)^{-1/4} \bar{U}$. The attenuation of spatial frequencies due to limited spatial coherence is approximately proportional to the gradient of this curve (see eqn (4.8)). The curves explain how the band limit moves to higher spatial frequencies $\bar{U}$ with increasing defocus. If the attenuation is considered severe for all $\bar{U}$ beyond that for which $\chi(\bar{U})$ has the critical gradient shown by the three tangential line segments indicated, we see that this cut-off gradient occurs at higher spatial frequencies as the defocus is increased. (After Frank 1976.)

for a focus setting

$$\Delta f_n = \left(\frac{8n + 3}{2}\right)^{1/2} C_s^{1/2} \lambda^{1/2} \qquad (A3.5)$$

as shown in Fig. A3.3(a).

For $n = 0$, eqns (A3.4) and (A3.5) give the Scherzer conditions commonly used for the structure imaging of defects, and this defines the instrumental point-resolution (see eqn (6.17)). The cases $n = 0$ and $n = 4$ are shown for a typical modern instrument in Fig. A3.2. The value of n is equal to the number of minima which preceed the passband. We note that the width of these 'passbands' $\Delta U(n) = U_1 - U_2$ depends on C_s, λ, and n as discussed below. Here the slight effect of α and Δ on $\Delta U(n)$ has been neglected. Solutions similar to those given in eqn (A3.3) can also be obtained for $\Delta \neq 0$; however, the resulting expressions are cumbersome and of little practical use since these resolution-limiting effects can readily be treated separately.

For $\theta_c = 0$ and $A(\mathbf{K}) = \exp(-1)$, eqn (A3.1) gives (Fejes 1977),

$$U_0(\Delta) = \left(\frac{2}{\pi \lambda \Delta}\right)^{1/2} \qquad (A3.6)$$

as the information-resolution limit due to electronic instabilities alone. We

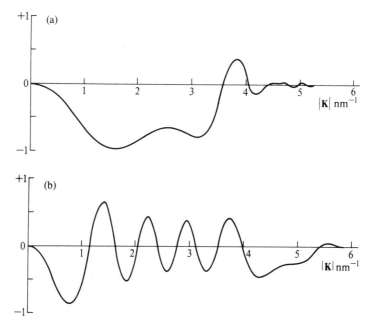

Fig. A3.2 Transfer functions for a modern high-resolution 100 kV instrument with $C_s = 0.7$ mm, beam divergence $\theta_c = 1$ mrad and $\Delta = 5.4$ nm. The functions shown here and in Fig. 4.3 are actually $A_i(\mathbf{K})\cos\chi(\mathbf{K}) + A_R(\mathbf{K})\sin\chi(\mathbf{K})$, where $A_i(\mathbf{K})$ and $A_R(\mathbf{K})$ are the imaginary and real parts of eqn (A3.1), since, for linear imaging, this can be shown to be the coherent transfer function following the method of Section 3.4. In fact the term involving $A_i(\mathbf{K})$ is very small, and the damping envelope can be taken to be the real part of eqn (A3.1) to a good approximation. Curve (a) is drawn for the Scherzer focus $\Delta f = -62.3$ nm while (b) is drawn for $\Delta f = -212.8$ nm ($n = 4$ in eqn (A3.5)). Since this second curve includes a passband extending beyond 5 nm^{-1}, these curves suggest that a reconstructed image could, in principle, be synthesized by image-processing techniques from a series of images recorded on such an instrument which would show a point-resolution of less than 0.2 nm. The zero-crossings of the transfer function and their effect on image noise greatly complicate this procedure.

note that it is the *product* $\lambda C_c Q$ which one wishes to minimize for highest 'information-limit' resolution. Assuming that images are recorded at the focus settings given by eqn (A3.5) so that partial coherence effects can be neglected in the neighbourhood of the passband $\Delta U(n)$, eqns (A3.4) and (A3.6) give

$$\Delta = \left(\frac{2}{k_n \pi}\right) C_s^{1/2} \lambda^{1/2} = C_c Q \qquad (A3.7)$$

as the electronic stability needed for inclusion of passband $\Delta U(n)$, where $k_n = (8n + 3/2)^{1/2} + 1$. For the commonly used Scherzer condition, this is

$$\Delta = 0.286 C_s^{1/2} \lambda^{1/2} = C_c Q \qquad (A3.8)$$

These results indicate that, in the absence of an objective aperture, differing factors will control the resolution of electron images at various accelerating voltages. For example, at 100 kV with $C_s = 0.7$ mm and $C_c = 1$ mm, eqn (A3.8) gives $Q = 1.5 \times 10^{-5}$ as the stability needed to obtain the Scherzer resolution limit ($n = 0$ in eqns (A3.4) and (A3.5)). This stability is comfortably exceeded by most modern 100 kV instruments, for which manufacturers typically claim $\sigma(V_0)/V_0 = 2 \times 10^{-6} = \sigma(I_0)/I_0$, and $\sigma(E_0)/E_0 = 5 \times 10^{-6}$ giving $Q = 6 \times 10^{-6}$. Electron energy-loss measurements give 0.7 eV $< \Delta E < 2.4$ eV as the electron gun-bias setting varies between maximum and minimum for a thermonic filament (see Fig. 7.5). Note that maximum gun bias here corresponds to minimum beam current. Thus, for such an instrument we have from eqn (A3.2)

$$\Delta = 10(4 + 16 + 9)^{1/2} = 5.4 \text{ nm} \tag{A3.9}$$

for maximum gun bias and

$$\Delta = 10(4 + 16 + 100)^{1/2} = 10.1 \text{ nm}$$

at minimum gun-bias setting. Since, for small illumination semiangles the effect of partial coherence is negligible in the region of the pass-bands given by eqn (A3.4), the resolution of 100 kV images recorded at the favourable focus settings of eqn (A3.5) is seen to be controlled by the gun-bias setting for small bias settings (high beam current) and by the objective-lens current stability for high bias settings (low beam current). This broad conclusion is independent of the objective-lens aberration coefficients. At maximum gun bias we have, from eqn (A3.6), for such an instrument, $U_0(\Delta) = 0.56$ Å^{-1}, which, by eqn (A3.4) would allow the fifth contrast transfer interval to be included. At minimum gun bias, $U_0(\Delta) = 0.395$ Å^{-1}. Thus, the use of very small illuminating apertures on 100 kV machines would appear to be a less important priority than the minimization of chromatic aberration. For structure images recorded at the Scherzer focus ($n = 0$ in eqn (A3.5)), however, the use of smaller illumination semi-angles will nevertheless introduce higher-resolution detail; however, the use of several images for subsequent analysis recorded at the focus settings of eqn (A3.5) allows the largest illumination aperture to be used (producing the most intense final image) for the smallest resolution penalty. We note also from eqn (A3.5) that for structure imaging under Scherzer conditions ($n = 0$ in eqns (A3.4) and (A3.5)) the rather large illumination semi-angle of $\theta_c = 2.3$ mrad can be used at 100 kV with $C_s = 0.7$ mm if the resolution limit due to partial coherence is to be equal to the Scherzer cut-off ($U_1 = 0.345$ Å^{-1}). Finally, as a matter of practical experience, it is generally found that conscientious monitoring of the gun-chamber vacuum is needed in order to maintain the manufacturer's quoted high-voltage stability.

At higher voltages the effects of chromatic aberration due to the electron source rapidly become unimportant. The best current values for eqn (A3.2) at an accelerating voltage of 1 MeV appear to be $\sigma(V_0)/V_0 = \sigma(I_0)/I_0 = 5 \times 10^{-6}$ with $\sigma(E_0)/E_0 = 1 \times 10^{-6}$ (Horiuchi, personal communication). With $C_c = 4.4$ mm such an instrument has

$$\Delta = 44(25 + 100 + 1)^{1/2} = 49.4 \text{ nm} \tag{A3.10}$$

giving, from eqn (A3.6), an information-resolution limit of 0.384 A^{-1}, set chiefly by objective-lens current instability. This $\lambda\Delta$ value of 4.3 A^2 is considerably poorer than the value of 2.0 A^2 for the 100 kV instrument cited in eqn (A3.9) at maximum gun bias. This suggests that, if comparable high-voltage stability can be obtained, a field-emission 100 kV instrument with, if possible, improved objective-lens current stability should give the highest information-resolution limit. Practical experience indicates that to obtain the required stability of emission, such a machine would require an ultra-high-vacuum system. The chief advantage of high-voltage instruments is thus seen to lie in the extended range of simply interpretable image detail which they can provide. This may be a crucial factor for the structure analysis of defects or small crystalline phases whose structure is completely

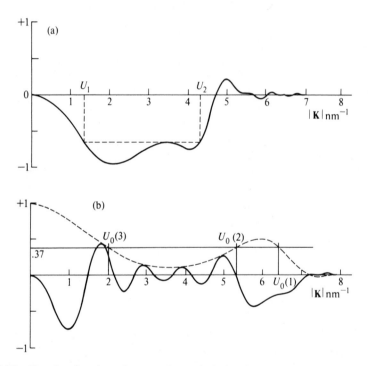

Fig. A3.3 Transfer functions for an advanced design for an electron microscope. The accelerating voltage chosen (300 kV) is a compromise between the various practical disadvantages of 1 MeV machines and the improvement in point-resolution accompanying a reduction in electron wavelength. Parameters used are; $V_0 = 300$ kV, $C_s = 1.5$ mm, $\Delta = 5.4$ nm, $\theta_c = 1$ mrad, defocus $\Delta f = -66.5$ nm (Scherzer focus) in (a). The electron wavelength is 0.001969 nm. Curve (b) is drawn for similar conditions with $\Delta f = -227.3$ nm ($n = 4$) and includes the damping envelope (shown dotted) given approximately by the real part of eqn (A3.1). This dotted curve clearly illustrates the dip in the damping envelope which occurs for large values of n where the gradient of $\chi(|\mathbf{K}|)$ can be large for $|\mathbf{K}|$ values smaller than the passband. The horizontal line in (b) is drawn for $f(|\mathbf{K}|) = \exp(-1)$ and the points $U_0(1)$, $U_0(2)$, $U_0(3)$, of eqns (A3.3) are indicated. In (a), the dotted lines indicate the method of defining the quantities U_1 and U_2 of eqn (A3.4) which determine the width and position of the passbands.

unknown so that comparisons of computed and experimental images would be prohibitively time-consuming.

In summary, the methods of image analysis and image simulation can most profitably be applied to images recorded on 100 kV machines at low gun bias and the focus settings of eqn (A3.5) at the current stage of instrumental development. For directly interpretable images one should use the highest accelerating voltage available for which electronic instabilities allow the detail of interest to be resolved (see eqn (A3.6)). Practical considerations such as image-viewing convenience, photographic emulsion and viewing-phosphor sensitivity, room height for the microscope and overall cost suggest that designs for a few hundred kilovolts achieve the best compromise. For example, a design with $V_0 = 300$ kV, $C_s = 1.5$ mm, and $\Delta = 5.4$ mm gives a Scherzer resolution limit ($n = 0$ in eqn (A3.4)) of 0.21 nm (point-to-point) or an information-resolution limit of $U_0(\Delta) = 0.79$ Å^{-1} (eqn A3.6). The allowed illumination semi-angle consistent with this Scherzer resolution is then 1.9 mrad, producing a conveniently intense final image for observation at high magnification. Transfer functions for this 300 kV design are shown in Fig. A3.3. The design stabilities used are $\sigma(V_0)/V_0 = \sigma(I_0)/I_0 = 2 \times 10^{-6}$, as for present-day 100 kV machines. Unlike 100 kV instruments, however, the quality of the images obtained on such an instrument would be rather insensitive to changes in the electron gun bias setting, since, at 300 kV, the last figure in the bracket of eqn (A3.9) changes insignificantly between the maximum and minimum gun bias settings. The highest resolution currently available on commercial machines is about 0.165 nm (point to point) at 400 kV.

References

Fejes, P. L. (1977). Approximations for the calculation of high-resolution microscope images of thin films. *Acta Crystallogr.* **A33,** 109.

Frank, J. (1976). Determination of source size and energy spread from electron micrographs using the method of Young's fringes. *Optik* **44,** 379.

APPENDIX 4

What is a structure image?

The term 'structure image' has been proposed by J. M. Cowley to describe that restricted set of electron lattice images which may be directly interpreted to some limited resolution in terms of the specimen's projected structure and *which were obtained using instrumental conditions which are independent of the structure,* and so require no *a priori* knowledge of the crystal structure. This last condition currently excludes most of the common lattice images of small-unit-cell crystals and all images obtained by 'matching' the microscope transfer function to the crystal reciprocal lattice. At present only the weak-phase object and the projected charge density approximations provide a prescription for structure-independent instrumental conditions which therefore give structure images.

The development of the newest generation of 'medium-voltage' HREM machines has now made structure imaging routinely possible in many materials (see, for example, Ourmazd, Ahlborn, Ibeh, and Honda. (1985) *Appl. Phys. Letts.* **47,** 685). The measured point-resolution of these instruments appears to be about 0.165 nm at 400 kV.

APPENDIX 5

The challenge of high resolution electron microscopy

This book is intended to teach students and research workers how to record and interpret atomic resolution transmission electron micrographs. However, a well-defined aim is the greatest stimulus to progress in human affairs. The following project therefore provides a thorough test of all the experimental skills taught in this book, together with a severe test of the underlying theory. Any student who completes it may reasonably claim to be an authority on high resolution electron microscopy!

Magnesium oxide crystals of sub-micron dimensions are easily made by burning magnesium ribbon in air (see section 8.1). A "holey carbon" grid passed through the smoke will collect particles for examination in the electron microscope. The particles form in perfect cubes, and can be found in the (110) orientation with the electron beam passing along the cube face diagonal. The thickness is therefore known exactly at every point in a lattice image. It is possible to record a structure image of this material on the newest electron microscopes, and to determine the focus setting from diffractogram analysis (see section 8.7). The image can then be matched as a function of thickness against computed images by the methods described in sections 5.6 or 5.7. All the thickness-dependent contrast reversals in the lattice fringes due to multiple scattering should be correctly reproduced in the computed images.

Challenging aspects of this project include (1) the variation of focus along the lower face of the crystal, (2) the possible need for "absorption" corrections in the calculations for the thick regions (see Goodman, P. and Lehmpfuhl, G. *Acta Cryst.* **22,** p. 14 (1967)). (3) Difficulties in matching the fresnel fringes along the crystal edge of varying thickness, where a "profile image" of the MgO surface will be seen. (4) Refinement of the atomic scattering factors for ionicity effects. (5) The need for very accurate alignment of the crystal. To the author's knowledge, a thorough analysis of this problem has yet to be published, however preliminary work is referenced in section 12.2. Ultimately, the credibility of the dynamical imaging theory given in this book depends on its ability to reproduce these experimental results. A comprehensive survey of metal smoke particles, including MgO, can be found in R. Uyeda, *Morphology of crystals,* pp. 367–508 (Ed. I. Sunagawa. Terra Scientific Publishing Co., Tokyo) 1987.

This project represents a summation of practically all the useful knowledge in this book

INDEX